Energy Implications of Thermal Comfort in Buildings Considering Climate Change

Energy Implications of Thermal Comfort in Buildings Considering Climate Change

Editors

Daniel Sánchez-García
David Bienvenido-Huertas

MDPI

Basel • Beijing • Wuhan • Barcelona • Belgrade • Novi Sad • Cluj • Manchester

Editors
Daniel Sánchez-García
University of Cádiz
Cádiz
Spain

David Bienvenido-Huertas
University of Granada
Granada
Spain

Editorial Office
MDPI AG
Grosspeteranlage 5
4052 Basel, Switzerland

This is a reprint of articles from the Special Issue published online in the open access journal *Applied Sciences* (ISSN 2076-3417) (available at: https://www.mdpi.com/journal/applsci/special_issues/7ANNCIYU5V).

For citation purposes, cite each article independently as indicated on the article page online and as indicated below:

Lastname, A.A.; Lastname, B.B. Article Title. *Journal Name* **Year**, *Volume Number*, Page Range.

ISBN 978-3-7258-1907-2 (Hbk)
ISBN 978-3-7258-1908-9 (PDF)
doi.org/10.3390/books978-3-7258-1908-9

Cover image courtesy of Daniel Sánchez-García

Contents

About the Editors

Daniel Sánchez-García

Daniel Sánchez-García is Assistant Professor in the Department of Mechanical Engineering and Industrial Design at the University of Cádiz. His area of expertise includes climate change in the building sector, adaptive thermal comfort, energy poverty, energy efficiency, and the design of nearly zero-energy buildings. He is the author of more than 25 research papers and is a recognized reviewer of various international indexed journals. He is the developer of the Python library "accim", which applies setpoint temperatures based on adaptive comfort models in EnergyPlus building energy models; with this, he won the 2022 CIBSE Building Simulation Award.

David Bienvenido-Huertas

David Bienvenido-Huertas is Associate Professor in the Department of Building Construction at the University of Granada. His area of expertise covers climate change in the building sector, adaptive thermal comfort, heat transfer, fuel poverty, energy conservation measures, and the design of nearly zero-energy buildings. He is the author of more than 80 research papers and is a recognized reviewer of various international indexed journals.

Preface

The landscape of thermal comfort in buildings is undergoing significant transformation due to the need for energy efficiency and climate resilience. This Special Issue, entitled "Energy Implications of Thermal Comfort in Buildings considering Climate Change", compiles crucial research addressing these challenges and opportunities, collecting studies and solutions aimed at enhancing thermal comfort while mitigating the environmental impacts associated with building energy consumption.

A key theme is recognizing that climate change significantly challenges the maintenance of thermal comfort in buildings. As global temperatures rise and extreme weather events become more frequent, the demand for cooling and heating in buildings is set to increase dramatically. This has substantial implications for energy consumption and greenhouse gas emissions, necessitating a re-evaluation of current building practices and technologies.

This Special Issue focuses on enhancing thermal comfort and energy efficiency in buildings through innovative technologies and strategies. Here, researchers explore advanced materials such as phase-change materials (PCMs), advanced insulation, and dynamic building envelopes, which enhance thermal performance and adapt to changing climates. For example, PCMs in building envelopes reduce peak cooling and heating loads, minimizing energy consumption while enhancing comfort. Passive design strategies that leverage natural ventilation, solar gain, and thermal mass offer sustainable alternatives to mechanical systems, promoting building resilience across different climates. Optimizing orientation, shading, and ventilation minimizes energy input, maintaining comfortable indoor temperatures effectively. Understanding the impact of occupants' behavior on thermal comfort and energy use is crucial; therefore, this Special Issue highlight the achievement of significant energy savings through tailored design and behavioral interventions. Retrofitting existing buildings with upgraded insulation, windows, and renewable energy systems is vital for reducing energy consumption and carbon emissions. These measures align with global sustainability goals, ensuring that buildings are more energy-efficient and resilient in the face of climate change challenges.

As the Guest Editor of this Special Issue, I am struck by the breadth and depth of the research contributions. These studies advance our understanding of thermal comfort and energy efficiency and provide practical solutions for real-world settings. From innovative materials and technologies to passive design strategies and policy recommendations, this Special Issue offers a comprehensive exploration of sustainable and resilient buildings.

In conclusion, this Special Issue serves as a vital resource for researchers, practitioners, policymakers, and stakeholders in the building sector. It underscores the urgency of addressing the implications of thermal comfort in the context of climate change and offers a wealth of knowledge and inspiration for driving progress in this critical field. I extend my gratitude to the contributors for their exceptional research and insights, with this Special Issue inspiring further advancements and collaborations in the pursuit of sustainable and resilient buildings.

Daniel Sánchez-García and David Bienvenido-Huertas
Editors

applied sciences

MDPI

litorial

:nergy Implications of Thermal Comfort in Buildings :onsidering Climate Change

aniel Sánchez-García [1,*] and David Bienvenido-Huertas [2]

[1] Department of Electrical Engineering, University Carlos III of Madrid, 28903 Leganés, Spain
[2] Department of Building Construction, University of Granada, 18071 Granada, Spain; dbienvenido@ugr.es
* Correspondence: dsgarcia@ing.uc3m.es

Extreme weather events and rising global temperatures are signs of the urgent threat that climate change poses to our planet [1]. Energy used for heating, cooling, and sustaining thermal comfort in buildings is a crucial factor in causing this issue [2]. It is crucial to look at the energy implications of attaining thermal comfort in buildings in the context of climate change as the world struggles with the urgency of lowering greenhouse gas emissions [3].

An essential component of human wellbeing, thermal comfort has a direct bearing on our level of well-being and productivity [4]. A comfortable interior atmosphere is provided by buildings, which normally maintain a temperature range between 20 and 25 °C. However, it is becoming harder to sustain thermal comfort in the face of climatic change. Temperature extremes become more frequent and intense due to climate change, requiring more energy to cool spaces [5]. In a vicious loop, increasing energy demand exacerbates greenhouse gas emissions, which worsen climate change. The traditional fixed-temperature approach to thermal comfort is challenged by the idea of adaptive comfort, which is especially relevant considering climate change, given the energy-saving opportunities it promises in present [6] and future scenarios [7]. It implies that people can adjust to a wider range of temperatures and their comfort temperature changes depending on the outdoor temperature fluctuations of the previous days [8,9]. Encouraging tenants to adopt new habits, such as using fans instead of air conditioning, can cut energy usage [10].

To reduce the energy implications of thermal comfort, building design is crucial. Innovative design techniques can lessen the need for mechanical heating and cooling, such as passive solar design [11] and green roofs [12]. The reduction in energy use also depends on improvements in building materials and technology, such as energy-efficient HVAC systems [13] and smart windows [14].

To reduce buildings' carbon impacts, we must switch to renewable energy sources. The use of fossil fuels may be decreased using sustainable energy sources for heating and cooling, such as solar panels [15], wind turbines [16], and geothermal systems [17]. Smart grid technology can optimise energy use and lower peak demand [18].

Promoting energy-efficient construction practises heavily depends on government regulations and incentives. The battle against climate change must include creating laws requiring better energy efficiency requirements, tax incentives for green construction practises, and financial support for renewable energy installations [19].

In the context of climate change, the energy implications of maintaining thermal comfort in buildings cannot be overstated. Cooling-related energy demand will keep rising as temperatures rise and severe weather events become more common, boosting greenhouse gas emissions. A multifaceted strategy incorporating creative building design, behavioural modifications, the integration of renewable energy sources, and helpful regulations is needed to overcome this situation. To create sustainable solutions prioritising both human satisfaction and environmental responsibility, researchers, architects, legislators, and the general public must work together to lessen the energy costs associated with maintaining

itation: Sánchez-García, D.; ienvenido-Huertas, D. Energy nplications of Thermal Comfort in uildings Considering Climate hange. *Appl. Sci.* **2023**, *13*, 10708. ttps://doi.org/10.3390/ pp131910708

eceived: 23 September 2023
.ccepted: 25 September 2023
ublished: 26 September 2023

thermal comfort in buildings, cut down on carbon emissions, and construct a more resilient and sustainable future in the face of climate change.

Author Contributions: Conceptualization, D.S.-G. and D.B.-H.; methodology, D.S.-G. and D.B.-H.; validation, D.S.-G. and D.B.-H.; formal analysis, D.S.-G. and D.B.-H.; investigation, D.S.-G. and D.B.-H.; resources, D.S.-G. and D.B.-H.; data curation, D.S.-G. and D.B.-H.; writing—original draft preparation, D.S.-G. and D.B.-H.; writing—review and editing, D.S.-G. and D.B.-H.; visualization, D.S.-G. and D.B.-H.; supervision, D.S.-G. and D.B.-H. All authors have read and agreed to the published version of the manuscript.

Funding: This research received no external funding.

Conflicts of Interest: The authors declare no conflict of interest.

References

1. Meinshausen, M.; Smith, S.J.; Calvin, K.; Daniel, J.S.; Kainuma, M.L.T.; Lamarque, J.; Matsumoto, K.; Montzka, S.A.; Raper, S.C.B.; Riahi, K.; et al. The RCP Greenhouse Gas Concentrations and Their Extensions from 1765 to 2300. *Clim. Change* **2011**, *109*, 213–24 [CrossRef]
2. Yang, L.; Yan, H.; Lam, J.C. Thermal Comfort and Building Energy Consumption Implications—A Review. *Appl. Energy* **2014**, *11* 164–173. [CrossRef]
3. Sánchez-García, D.; Rubio-Bellido, C.; del Río, J.J.M.; Pérez-Fargallo, A. Towards the Quantification of Energy Demand and Consumption through the Adaptive Comfort Approach in Mixed Mode Office Buildings Considering Climate Change. *Energ Build.* **2019**, *187*, 173–185. [CrossRef]
4. De Dear, R.J.; Akimoto, T.; Arens, E.A.; Brager, G.; Candido, C.; Cheong, K.W.D.; Li, B.; Nishihara, N.; Sekhar, S.C.; Tanabe, S. et al. Progress in Thermal Comfort Research over the Last Twenty Years. *Indoor Air* **2013**, *23*, 442–461. [CrossRef] [PubMed]
5. Holmes, M.J.; Hacker, J.N. Climate Change, Thermal Comfort and Energy: Meeting the Design Challenges of the 21st Centur *Energy Build.* **2007**, *39*, 802–814. [CrossRef]
6. Bienvenido-Huertas, D.; Rubio-Bellido, C.; Pérez-Fargallo, A.; Pulido-Arcas, J.A. Energy Saving Potential in Current and Future World Built Environments Based on the Adaptive Comfort Approach. *J. Clean. Prod.* **2020**, *249*, 119306. [CrossRef]
7. Bienvenido-Huertas, D.; Sánchez-García, D.; Rubio-Bellido, C. Influence of the RCP Scenarios on the Effectiveness of Adaptive Strategies in Buildings around the World. *Build. Environ.* **2022**, *208*, 108631. [CrossRef]
8. De Dear, R.J.; Brager, G.S. Developing an Adaptive Model of Thermal Comfort and Preference. *ASHRAE Trans.* **1998**, *104*, 145–16
9. McCartney, K.J.; Nicol, J.F. Developing an Adaptive Control Algorithm for Europe. *Energy Build.* **2002**, *34*, 623–635. [CrossRef]
10. Knudsen, M.; Rissetto, R.; Carbonare, N.; Wagner, A.; Schweiker, M. Comfort and Economic Viability of Personal Ceiling Fan Assisted by Night Ventilation in a Renovated Office Building. *Buildings* **2023**, *13*, 589. [CrossRef]
11. Askari, M.; Jahangir, M.H. Evaluation of Thermal Performance and Energy Efficiency of a Trombe Wall Improved with Dual Phase Change Materials. *Energy* **2023**, *284*, 128587. [CrossRef]
12. Mousavi, S.N.; Gheibi, M.; Wacławek, S.; Behzadian, K. A Novel Smart Framework for Optimal Design of Green Roofs in Buildings Conforming with Energy Conservation and Thermal Comfort. *Energy Build.* **2023**, *291*, 113111. [CrossRef]
13. Li, Z.; Meng, Q.; Wei, Y.; Zhang, L.; Sun, Z.; Lei, Y.; Yang, L.; Yan, X. Dynamic Room Temperature Setpoints of Air-Conditioning Demand Response Based on Heat Balance Equations with Thermal Comfort Model as Constraint: On-Site Experiment and Simulation. *J. Build. Eng.* **2023**, *65*, 105798. [CrossRef]
14. Hong, S.; Choi, S.; Jeong, S. Building Energy Savings by Developing Complex Smart Windows and Their Controllers. *Appl. Sci.* **2023**, *13*, 9647. [CrossRef]
15. Li, Y.; Jia, Z.; Zhang, X.; Liu, Y.; Xiao, F.; Gao, W.; Xu, Y. Energy Flexibility Analysis and Model Predictive Control Performances of Space Heating in Japanese Zero Energy House. *J. Build. Eng.* **2023**, *76*, 107365. [CrossRef]
16. Assareh, E.; Dejdar, A.; Ershadi, A.; Jafarian, M.; Mansouri, M.; Roshani, A.S.; Azish, E.; Saedpanah, E.; Lee, M. Techno-Economic Analysis of Combined Cooling, Heating, and Power (CCHP) System Integrated with Multiple Renewable Energy Sources and Energy Storage Units. *Energy Build.* **2023**, *278*, 112618. [CrossRef]
17. Mehmood, S.; Lizana, J.; Friedrich, D. Low-Energy Resilient Cooling through Geothermal Heat Dissipation and Latent Heat Storage. *J Energy Storage* **2023**, *72*, 108377. [CrossRef]
18. Charalambous, C.; Heracleous, C.; Michael, A.; Efthymiou, V. Hybrid AC-DC Distribution System for Building Integrated Photovoltaics and Energy Storage Solutions for Heating-Cooling Purposes. A Case Study of a Historic Building in Cyprus. *Renew. Energy* **2023**, *216*, 119032. [CrossRef]
19. Aleluia Reis, L.; Vrontisi, Z.; Verdolini, E.; Fragkiadakis, K.; Tavoni, M. A Research and Development Investment Strategy to Achieve the Paris Climate Agreement. *Nat. Commun.* **2023**, *14*, 3581. [CrossRef] [PubMed]

applied sciences

MDPI

Article

The Intersection of Architectural Conservation and Energy Efficiency: A Case Study of Romanian Heritage Buildings

Simona Elena Șerban [1], Tiberiu Catalina [1,2,*], Razvan Popescu [1] and Lelia Popescu [1]

1 Faculty of Building Services, Technical University of Civil Engineering, 021414 Bucharest, Romania; lelialetitia@yahoo.com (L.P.)
2 National Institute for Research-Development in Construction, Urbanism and Sustainable Territorial Development—INCD URBAN-INCERC, 021652 Bucharest, Romania
* Correspondence: tiberiu.catalina@gmail.com; Tel.: +40-763-915-461

Abstract: In Europe, it is estimated that 14% of existing buildings were built before 1919, whereas 26% were built before 1945. In Romania, about 31% of the buildings date from before 1961, contributing to the current stock of old buildings with historic and architectural value in the country. This paper illustrates the current state of buildings with historic and architectural value in Romania, alongside a case study of a representative administrative building in Câmpulung, Romania. The analysis of the Town Hall building in Câmpulung, Romania, demonstrates that potential energy savings of up to 47.53% can be achieved by implementing interventions such as upgrading windows, insulating the attic, and installing photovoltaic panels. The highest energy reduction is obtained by replacing the window glass with a value of 18.16% with attic insulation with a value of 16.1%. This paper also presents indoor measurements of temperature and humidity in different offices positioned in the north and the south. The study conducted on the south façade office revealed consistent temperatures ranging from 21.7 °C to 24.4 °C, with an average of 23.31 °C. However, the humidity levels fluctuated considerably, ranging from 17.1% to 39.1%, with an average of 26.89%. The sun-exposed section of the building saw relatively stable temperature conditions, but the varying humidity levels could have a detrimental impact on the quality of the indoor atmosphere and potentially decrease the effectiveness of the workforce. By contrast, the north façade office exhibited lower and more fluctuating temperatures, ranging from 19.8 °C to 23.6 °C, with an average of 21.74 °C. Additionally, it had higher and more stable humidity levels, ranging between 19.5% and 41.7%, with an average of 29.83%. A thermographic analysis was performed on the north façade of the Câmpulung Town Hall, utilizing thermal imaging technology to detect areas of heat loss, and thus identifying the energy inefficiency problems of the building's exterior. The investigation found notable variations in temperature, especially around the windows, where temperatures could be as high as 14.1 °C, highlighting the insufficiency of the building's antiquated timber-framed windows in preventing energy loss.

Keywords: energy efficiency; historic buildings; heritage conservation; thermography; energy audit

Citation: Șerban, S.E.; Catalina, T.; Popescu, R.; Popescu, L. The Intersection of Architectural Conservation and Energy Efficiency: A Case Study of Romanian Heritage Buildings. *Appl. Sci.* **2024**, *14*, 4835. https://doi.org/10.3390/app14114835

Academic Editors: David Bienvenido Huertas and Daniel Sánchez-García

Received: 18 April 2024
Revised: 6 May 2024
Accepted: 8 May 2024
Published: 3 June 2024

1. Introduction

Buildings account for approximately 40% of the European Union's total energy consumption and generate 36% of greenhouse gas emissions. In Europe, it is estimated that 14% of the existing building stock was built before 1919, and 26% was built before 1945 [1]. Heritage architecture is an element that emphasizes personality, culture, and history, being a symbol for European cultural heritage and the identity of European society, giving it an identifiable character [1]. The existing building stock thus needs to be preserved through continuous transformation and optimization while presenting the opportunity to modernize and optimize energy production in all sites of European cultural heritage, both on a large scale and locally. In this approach, the perspective is changing, and the existing building fund is beginning to be seen as a reusable resource, optimal for interventions th will provide space for contemporary activities in the future; this has also been demonstrate by the European Union's interest in and dedication to this matter [2,3].

As climate change poses a real and urgent threat to humanity and the environment [4 addressing the challenges facing architectural heritage thus becomes key to the future European architecture [5,6], and this can be accomplished by creating specific and loc guidelines, standards, and methodologies to address the modernization process and energ optimization of historic buildings [2].

In this paper, the term that is often used when referring to old building stock buildings with historical and architectural value, which is a description that include not only the buildings included in the National Register of Historic Monuments in eac member state but also refers to buildings that have historic and architectural value that ar not included in the above-mentioned lists. Currently, in Romania, there is no structure information regarding the actual number of buildings with historic and architectur value that exist; however, these buildings are widespread, and research in the field being developed.

Heritage buildings have been elaborated through a combination of design practice execution and design practices, techniques, and knowledge, and have been tested in well-defined framework and adapted to local, climatic, and sociological contexts. B the same principles, buildings (the architecture of today, ultimately) must serve curren needs through the development of procedures, methodologies, and practices to ensur the sustainability of architectural objects so that they can be used by future generation Hence, there is a need to create an intervention framework to modernize and optimiz the energy performance of buildings that are listed as historical monuments, which ar specific cultural buildings presenting traditional construction systems that are still widel used today.

In a European context, through the most recent communiqué of the European Unior titled "A Renovation Wave for Europe—greening our buildings, creating jobs, improvin lives" [2,5], a legislative framework has been proposed that supports the modernization o energy in existing building funds. Thus, in the current European context, the identificatio of customized solutions for energy efficiency [2,5] for the building envelopes of thos considered architectural heritage buildings is a key strategy for the modernization an energy optimization of the modernist building fund, and this can be achieved by studyin diagnosing, and implementing solutions to reduce general energy consumption and CO emissions for the existing building fund. Therefore, as directives and strategies to preven climate change and protect European heritage, the European Union has proposed the Gree Deal [2,6], which aims to accomplish the following goals:

- Reach the threshold of climate neutrality by 2050;
- Decouple economic growth from the use of resources;
- Ensure that no citizen or country is left behind;
- Ensure the efficient use of resources in a clean and circular economy;
- Restore biodiversity and reduce pollution.

In addition, on 14 July 2021, the European Commission adopted a set of proposals to make EU climate, energy, transport, and taxation policies adequate to reduce net greenhous gas emissions by at least 55% by 2030, compared to 1990 levels. Achieving these emissior reductions over the next decade is crucial for Europe to become the world's first climate neutral continent by 2050 and make the European Green Agreement a reality [6]. In thi context, the new proposals target sectors such as energy, transport, construction, anc renovation, with the aim of having 35 million buildings renovated by 2030.

2. Heritage Buildings in Romania

A number of European countries [7–20] have started to develop various nationa guidelines paired with research projects aimed at specific architecture within their border and to focus on the regeneration and energy optimization of the existing building fund

Therefore, the importance of proposing solutions specific to each case, i.e., on a case-by-case basis, has been acknowledged, given the multiple challenges that the global building fund presents (England, Hungary, Italy, Greece, Croatia, etc.) [7–18,20–40].

In Romania, roughly 31% of buildings date from before 1961, thus forming the stock of buildings with historical and architectural value [1]. Regardless of the fact that only a few of them are listed as historical monuments, they all carry importance from a historical and architectural point of view, being directly responsible for the characteristic of the country's various regions, thus keeping alive the local identity of local communities [1]. At the same time, even if the regulations for existing buildings impose limitations in terms of energy modernization interventions, many of the heritage buildings will benefit from individual optimization through customized solutions, which is necessary if we want the goal of reducing international greenhouse gas emissions by 80% by the year 2050 to be realistically attainable.

From a legislative point of view, in Romania, the process of authorizing energy efficiency interventions on buildings with historical and architectural value requires obtaining the consent of the Ministry of Culture, according to the legislative provisions in force [41–46].

Within the national legislation of Romania, there are various categories of buildings and areas of historical and architectural importance; e.g., they may be classified as historical monuments, according to the provisions of Law no. 422/2001 [41] (in the form of individual monuments, ensembles, or sites), or they may designated protected built-up areas, according to the provisions of Law no. 350/2001 [42] on territorial development and urban planning. In accordance with the legislative framework mentioned above, at present, in Romania, there are several strategies and programs funded at the European level with the aim of highlighting the gaps in the current legislative framework and, subsequently, of developing the legislative framework at the national level and beyond [1,44,45].

Regarding energy efficiency measures, the main regulatory act in Romania is Law no. 372/2005 [43] on the energy performance of buildings, with subsequent additions and amendments, though the requirements do not apply to buildings and monuments protected by heritage legislation

To be mentioned in this approach is a methodology developed within Reform R1.b, optimizing the legislative and normative framework to support the implementation of investments in the transition to green and resilient buildings and financed by PNRR, Pillar IV, and Component 5–Renovation Wave [2], titled "the intervention methodology for the non-invasive approach to energy efficiency in buildings with historical and architectural value" [1]. The main purpose of the methodology is to facilitate understanding of the behavior and characteristics of historic buildings in relation to energy efficiency interventions, the health of the spaces, and the increase in comfort level, as well as to establish the methodological framework for intervention in historic buildings, detailing the stages of analysis and choice of solutions within the projects in order to optimize and streamline costs in the medium and long term [1].

Currently, heritage buildings occupy an important share of the national building fund, with over 30,000 historical monuments declared in the list of historical monuments revised by the Ministry of Culture and National Identity in 2015 [46]. Their importance and value, and especially the general legal regime of the monuments' historical data and the need to protect them, require increased attention to monitor their behavior over time. Therefore, the impact of continuously increasing temperatures and climatic conditions must also be addressed, especially in the context of current climatic phenomena that can damage both the structure of a monument and any protected objects inside it.

2.1. Characteristics of Heritage Buildings in Romania

Heritage building stock, once modernized and optimized, can function as well as a new building without undergoing major interventions or incurring excessive costs. With the help of modern energy optimization technologies, as well as the involvement and education of the population regarding their impact on the general well-being of the natural and built environment, as already pointed out by European directives and by specialists in the field, we will be able to highlight and implement the essential aspects of the protection and effective use of the European architectural heritage.

In a study carried out in 2013, which analyzed the number of buildings with historic and architectural value in Europe [46], 19 European countries were analyzed. In the general ranking, Romania came in 15th place (see Figure 1) out of the 19 countries analyzed in the study. A key take away from the study was the knowledge that many of the European states that were included in the study had different criteria for classifying historical monuments taking into consideration the country's population and territory.

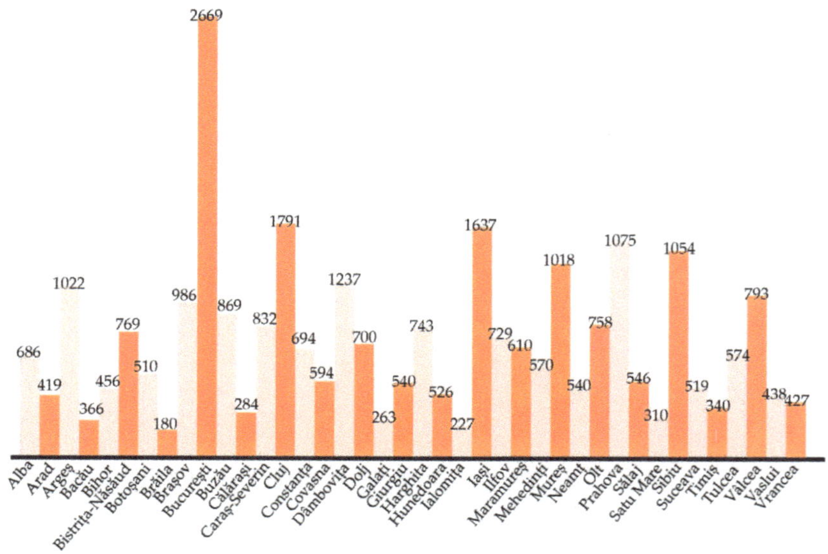

Figure 1. A visual representation of the current spread of national heritage buildings on Romanian territory. Information is provided the National Heritage Institute, Bucharest.

For example, Romania took over the wireframe method of categorizing and protecting heritage architecture monuments and sites from the French model, a state that has approximately 44,000 historical monuments, a relatively low number compared to the country's population and surface area.

As can be seen in the classification above [45], Romania has an extensive cultural heritage of great historic and architectural value, with the majority of these buildings being located in Bucharest, the country's capital city [47]. In total, there are 30,147 heritage sites included in the List of Historical Monuments and 8 sites included in the UNESCO World Heritage List [44,46].

The importance of rehabilitating these buildings thus becomes apparent, as a great number of these buildings are still in use and require urgent interventions in order to improve their interior conditions and the way in which the energy used for maintaining these conditions is employed.

The National Register of Historic Monuments in Romania [47] divided the monuments into two distinctive categories:

- Category A: Heritage monuments with national or universal value;
- Category B: Heritage monuments representative of the local cultural heritage;

As a country, Romania, on its territory, has a large number of heritage buildings (that have been indexed in the list of heritage buildings and are thus protected); however, there is also quite a high number of currently undefined buildings with historic and architectural value, i.e., buildings that have not been included in the National Register of Historic

Monuments in Romania but need to be protected and restored with the same degree of importance.

Heritage buildings are very diverse in terms of architectural style, most of them combining several styles, thus becoming unique; from fortresses to churches, from mansions to open-air museums, to administrative buildings, or private houses (residential buildings), the distribution of monuments with architectural and historical value is relatively balanced.

The counties with the largest number of buildings and sites in the category of monuments and architectural ensembles are Sibiu (840), Braşov (784), Argeş (765), Mureş (736), Prahova (737), and Iași (706) [44,46].

An important factor to be considered is that the above-mentioned buildings are not the only buildings in Romania with historic and architectural value; at present, there is an active endeavor to add more buildings to the list, and this is an ongoing process; more and more buildings are constantly being identified and added to the National Register of Historic Monument in Romania. An example of unique architecture on the Romanian territory is the Town Hall of Câmpulung town, as shown below (see Figure 2).

Figure 2. Town Hall of Câmpulung, Romania—main façade, showcasing the neo-Romanian style.

At the beginning of the 20th century, the interwar period marked an important moment in the development of Romanian architecture; through the drive of some renowned architects of the time, the quest for creating an original Romanian architectural style was on its way, and the results can be seen in the images above. Therefore, the restoration and inclusion in the National Register of Historic Monuments in Romania of these architectural and structural experiments was of great relevance and importance.

In Romania, heritage buildings are divided into several utility topologies, such as educational units (universities, schools, academies, institutes, and libraries), administrative buildings, restaurants, railway stations, hospitals, archives, and museums, each with its different regime of use and density of utilizers. Thus, it is equally necessary to study these types of buildings (schools/universities) in order to evaluate and propose energy efficiency measures since many of them are "energy-hungry". In this way, indoor climate/comfort conditions can also be improved (air quality, humidity, PMV, PPD, etc.).

The typologies of buildings with historic and architectural value serve a wide range of functions; e.g., individual housing, collective housing, administrative spaces, headquarters of institutions and social services, education and health spaces, and commercial or production spaces. If part of the existing building fund had been maintained in its original function, many heritage buildings would have been adapted for new uses, with a new set of needs and requirements for the use of space. If these transformations were made consistent with the structure and typology of historic buildings, these readjustments woul allow them to extend their life [1].

2.2. *Main Challenges in the Energy Optimization of Heritage Buildings in Romania*

When the topic of energy efficiency measures is considered for buildings with histori and architectural value, there are a multitude of causes of damage that come into discussio thus, the rehabilitation process needs to be conducted in a timely manner to buildings wit historical and architectural value, as follows [48]:

- Constructive causes: These are observable after a long period of time after the exec tion of the construction and are usually accompanied by other causes.
- Improper use of construction: This is one of the main causes resulting in damage to construction. Considering their very long lifespan, the human factor intervenes in th decision-making process with respect to intervention in historical buildings.
- Degradation of materials.
- The land connection solution.
- Humidity.
- Catastrophic actions.
- Lack of use.
- Lack of skills and experience and/or materials.
- Over-climatization.
- Lack of insulation and vapor barriers in historic buildings.
- Additional sources of moisture.
- Pollution.
- Inherent wear and tear.
- The light.
- Climatic changes.
- Effects of temperature and humidity fluctuations.
- Microbiological growth.
- Pests, i.e., vermin and insects.

As Romania is a four-season country, heritage buildings can be affected by the influ ence of temperatures, with major fluctuations due to low winter temperatures; in additior fires, excessive solar radiation, and excessive precipitation have also proven to be key fac tors that must be taken into consideration. The implications of thermal actions are diverse however, in the first phase of an energy efficiency project conducted on a building witl historic and architectural value, the focus should definitely be on the influence of humidit and very low temperatures, since the main priorities are ensuring interior comfort anc reducing energy consumption.

In addition, it is essential to consider the different structural and architectural ty pologies of heritage buildings, which can prove to be a good start in identifying the bes solutions for these types of buildings. For example, some buildings with historic and archi tectural value that have used massive walls as a structural solution are able to experience delays in receiving the variation of outside temperature due to their high thermal capacity and thermal inertia. This results in fewer thermal bridges; the indoor temperature is kep longer (depending on the variation of outdoor temperatures), and the risk of overheating or excessive cooling of the interior spaces is greatly reduced.

If the renovation processes and energy optimization steps do not take into consider ation the specific characteristics of the building's typology, the processes may lead to a decrease in the quality of their usage, create interior comfort problems for their inhabitants or irreparably destroy their historic and architectural value. Thus, it is extremely importan to take into consideration the built typology of the building during the pre-interventior process. The European policy regarding reducing energy consumption by 2050 makes it mandatory to optimize the energy of all buildings by 2050, including those in Romania This initiative has increased the importance of the existing building stock on a national level thus creating a positive context for the development of new solutions for interventions that ought to consider the specific requirements of public historic buildings.

3. Case Study: Câmpulung Town Hall Building

As case study, we have chosen a public building that is listed in the National Register of Historic Monuments in Romania, namely, the Town Hall building of Câmpulung, Romania, a B-category heritage building with the identification code: AG-II-m-B-13556. The structural typology of the building consists of massive walls that have good energetic values, with the shape of the building being a U form with an interior courtyard.

3.1. Description

The Town Hall of Câmpulung was originally created to function as the Muscel Prefecture and was built as part of the Ministry of Public Works program in order to equip the Old Kingdom with the necessary public buildings intended for county administrations. This project was carried out according to the plan of the architect Dimitrie Ionescu Berechet; it began in 1924, and was finalized in 1934, which is also the year in which the inauguration of the building took place (see Figure 3).

Figure 3. A visual representation of the Town Hall building at its inauguration in 1934.

The building was constructed with elements from the neo-Romanian, the neo-byzantine, and neo-gothic styles, and, as with many other buildings constructed in the same period, the result was a unique-looking building This was a period in Romanian history when a multitude of architectural experiments were taking place, mostly through the architects' desire to instill an architectural identity to the buildings of the time, a goal eventually achieved through several architectural experiments. Thus, in the case of the Câmpulung Town Hall building, we can observe neo-Romanian elements, such as the dimensions and decorations of the windows on the first floor, which present a neo-Gothic-style framing, as can be seen in the image below.

Architecturally, the Town Hall of Câmpulung (see Figure 4) features intricate façade carvings, high ceilings, and large windows typical of the period's style, which not only enhance its aesthetic value but also present unique challenges in terms of thermal insulation. The building's historical significance is underscored by its location in the heart of Câmpulung, making it a landmark of not only architectural but also sociocultural importance.

Figure 4. Town Hall of Câmpulung, Romania—side building façade.

3.2. *Identified Building Problems before the Intervention Process*

The main reason why the building needed to be optimized energetically was the fact that this heritage building has a public function, as this is the location of Câmpulung's Town Hall. In terms of problems, the building was dealing with high energy bills, unoptimized energy usage, thermal bridges, and interior spaces that could not be utilized as they had not been rehabilitated; hence, the administration decided to modernize the building, making it more accessible and user friendly while also being able to produce its own required energy.

Currently, the building faces significant energy efficiency challenges. It is primarily heated through an outdated central heating system, leading to high energy consumption and inefficiency. The building's heating system is centralized and relies on a single gas boiler located in the basement. This boiler operates at high temperatures and distributes heat through water radiators set at 80 °C. The conventional arrangement, albeit prevalent in older buildings, presents efficiency obstacles and sustainability issues due to its elevated energy usage and substantial heat dissipation.

The thermal performance of the building is compromised by several factors, including poor insulation in the attic, single-glazed windows, and significant air infiltration through aging window frames and doorways. These issues are enhanced by the building's large volume and high ceilings, which makes it difficult to maintain consistent indoor temperatures.

The thermal envelope of the structure is the main emphasis of the suggested energy optimization techniques. Among the main procedures meant to lower energy consumption while maintaining the architectural integrity of the building are replacing the window units with double-glazed panels, adding better sealing and insulating materials in the attic, and installing more-efficient heating systems. Along with enhancing the Town Hall's sustainability, these actions are supposed to act as a model for the preservation of other historic structures dealing with comparable issues.

The exterior walls of the building are made of 45 cm solid brick. The construction is provided with unheated space with a hipped roof. The floor above ground is made of concrete and does not have any thermal insulation in the soffit. The perimeter plinth is not thermally insulated. The joinery of the exterior windows and doors is made of wood. The heels are positioned on the inner face of the parapets.

The existing exterior finishes show mechanical wear at the level of the visible layers. Due to atmospheric agents, mechanical agents, and biological agents, as well as occasional rheological phenomena, the finishes have been affected up to now by dirt and discoloration caused by the action of ultraviolet rays, stains, etc. The building has no special shading elements on the façades.

As can be seen in the image above (see Figure 5), the state of the attic was unkept and no intervention was conducted to insulate the roof space, hence the occurrence of substantial heat loss and thermal bridges.

Figure 5. Image displaying the uninsulated attic.

However, because the attic had not been in its original form since the building's construction in 1934, this necessitated proper care and proper rehabilitation solutions so as not to lose the historic and architectural value of this interior space, while also enabling us to implement certain changes to the space to incorporate the HVAC elements to properly generate energy efficient results for the building.

3.3. Energy Analysis—A Pre-Intervention, Non-Invasive, Multicriterial Study

As the Town Hall of Câmpulung is a registered heritage building, the process of pre-intervention analysis had to be conducted in a non-invasive way. Thus, to analyze the current state of the building from an energy efficiency point of view, it was necessary to conducted multiple non-invasive studies:

- The energy audit of the building;
- An in situ study using measurement equipment, recording the humidity and temperatures on the south and north façade;
- A thermographic study of the building's envelope;

3.3.1. The Energy Audit

Part of the building issues presented were identified by studying the building's annual energy consumptions, as the annual heat consumption for space heating (discontinuous heating) is determined according to the Mc001/PII.1 Romanian methodology. The calculation temperature considered the fact that we have a daily variation, so the equivalent internal temperature was calculated to be 20 °C. Finally, the values based on which the building will be classified from an energetic point of view were determined. Adding up the entire energy consumption presented above results in a total annual energy consumption for heating of 714.7 MWh/year and a specific consumption of 326.7 kWh/m^2 year, respectively.

It can be observed in Table 1 above that the building's main elements do not meet the requirements needed for proper thermal insulation; thus, rehabilitation and thermal energy modernization solutions are needed in order to achieve appropriate thermal insulation and energy saving according to current requirements. It can also be observed that in this case, for example, even if the walls of the building are massive, they do not perform thermally as needed so as to form an optimal indoor climate. This highlights once again the importance of the project team's research before intervention to ensure that the right solutions have been selected in order to extend the building's (i.e., the Town Hall's) lifespan calculation.

Table 1. Results of a study conducted on the main construction elements of the studied heritage building in order to identify the current thermal resistance and determine if they meet the thermal insulation requirements for a building of this typology.

Construction Element	R' [m^{2}K/W] (Calculated)	R'min [m^{2}K/W] (Standard Value)	Meeting the Thermal Insulation Requirements
Exterior wall	17.1	1.7	No
Floor slope over basement/terrain	1.36	2.5	No
Floor slope terrace/Sky parlor/Attic	0.28	4	No
Exterior window frame	0.38	0.5	No

The highest consumptions are related to heating and, secondly, to the lighting system; thus, the energy efficiency measures tackled these parts (see Table 2). In Romania, within the legislation, the building is compared to a reference building (same architecture) but enhanced in terms of thermal resistance and systems. The reference building in this case is calculated as in Table 3.

Table 2. Energy consumption calculations—building Town Hall.

Consumption	Heating	Domestic Hot Water	Lighting	Total
Annual consumption [MWh/year]	714.7	1.6	50.6	766.9
Specific consumption [kWh/m^2/year]	326.7	0.7	23.1	350.6
CO_2 emissions [kgCO_2/m^2/year]	78.4	0.2	18.1	96.7
Energy class-Romanian legislation	E	A	A	**D**

Table 3. Energy consumption calculations—reference Town Hall.

Consumption	Heating	Domestic Hot Water	Lighting	Total
Annual consumption [MWh/year]	208.0	1.4	60.8	275.6
Specific consumption [kWh/m^2/year]	95.1	0.7	22.0	120.2
CO_2 emissions [kgCO_2/m^2/year]	22.8	0.2	21.8	46.7
Energy class-Romanian legislation	B	A	A	**A**

It must be mentioned that for historic buildings, there is no clear delimitation for reference buildings, and exterior walls are forbidden to be thermally insulated with external insulation. In this situation, the purpose is not to reach the reference values—this would be impossible—but rather to find the best solutions to the problem of reducing energy consumption while keeping the architectural value of the building.

3.3.2. Result from the In Situ Measurements of Temperature and Humidity

To better understand the state of the building and its indoor conditions during winter/spring season, testing the relative humidity and the temperature fluctuation was a key part of the study. Following the energy audit, we were able, with the help of the Testo 174 equipment, to record and analyze the date recorded as follows: we placed the equipment, that had been set to record, in an office facing the south façade and in an office facing the north façade; this positioning of the sensors was also meant to be used for comparison as the solar gains can severely impact the indoor temperatures.

As can be observed in Figure 6, the temperatures throughout March and April of this year, during both night and day, were different, with almost daily temperature drops.

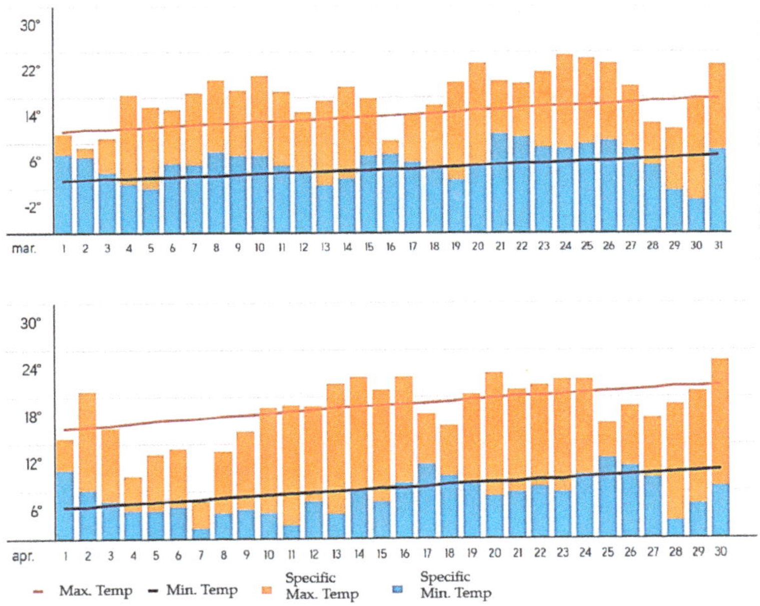

Figure 6. Exterior temperatures between March and April 2023, the period when the in situ measurements took place.

Romania is a country that experiences all 4four seasons during a year; however, in the passing years, the lines between winter and summer have been blurred and the changes are visible as little to no snow falls during the winter season. The reason why we chose to monitor the building during the end of winter and the beginning of spring was to better understand the state of a building with such characteristics.

- South Façade Office Measurements

The Testo equipment we set in one of the main offices of the building facing the south façade was positioned strategically in order to monitor the relation between the building's pre-intervention relative humidity and the interior temperatures registered. Thus, as the above Table 4 showcases, as this is the façade that was the most solar, the temperatures remained generally constant, i.e., between 21.7 and 24.4 °C, while the humidity throughout the period of the office monitoring (14 March 2023–11 April 2023), as can be seen in Figure 7, fluctuated from 17.1 to 39.1%rH. This can decrease the quality of the interior climate and reduce the work efficiency of the staff that work in the building, as the building is an administrative one.

Table 4. Results regarding the degree of relative humidity and temperature.

Parameter [1]	Min	Max	Mean Value
Humidity [%rH]	17.1	39.1	26.89
Temperature [°C]	21.7	24.4	23.308

[1] Parameters recorded between 14 March 2023 and11 April 2023, a transition period in between the cold seasc and a warmer one.

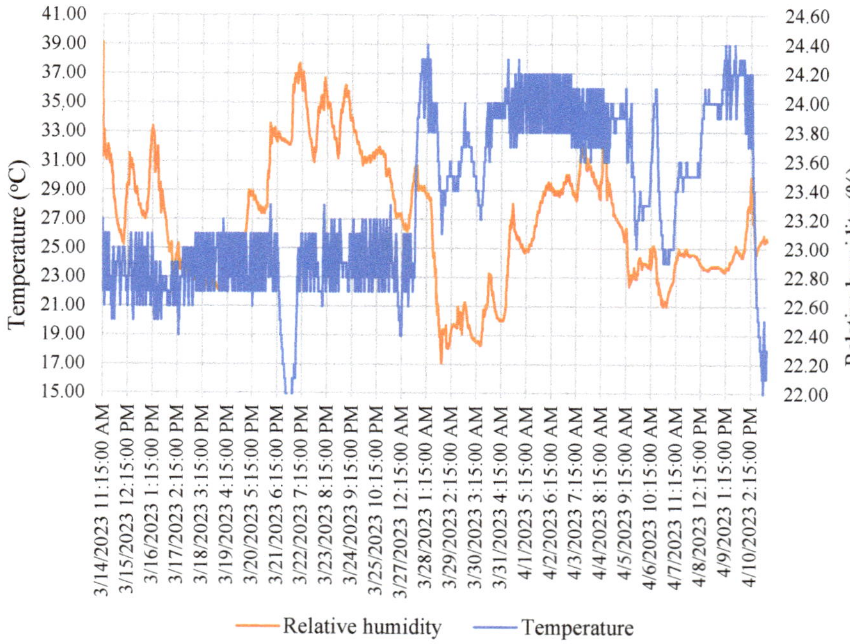

Figure 7. Results regarding the degree of relative humidity and temperature obtained from th equipment located in an office space facing the south façade of the building.

- North Façade Office Measurements

The equipment that was set to measure the relative humidity and temperatures in th main office facing the north façade was able to showcase, just as in the case of the sout façade, that the monitored temperature values were more constant (see Table 5) than th humidity levels; the latter were significantly more different, taking into consideration th climate conditions at that time in Câmpulung town, as there was more humidity in th exterior air.

Table 5. Results regarding the degree of relative humidity and temperature obtained from th equipment located in a room facing the north façade of the building.

Parameter [1]	Min	Max	Mean Value
Humidity [%rH]	19.5	41.7	29.83
Temperature [°C]	19.8	23.6	21.738

[1] Parameters recorded between 14 March 2023 and 11 April 2023, a transition period in between the cold seaso and a warmer one.

Moreover, the failure to address this issue, particularly in the context of Romania' seasonal climate variations, poses significant concerns. Summers in Romania are charac terized by high temperatures, while the transitions between colder and warmer season are protracted. During these transitional periods, there is a continuous exchange of ai between indoor and outdoor environments. This persistent state of flux can compromise the building's usability, rendering it an unsafe workspace for prolonged durations.

Figure 8 highlights that, because of temperature changes, the interior temperatures fluctuate considerably in the period during which the equipment monitored the building. Consequently, this means more chances of thermal bridge formations appearing, and even mold can occur over time, given the change in temperatures and the lack of necessary layers for thermal insulation. All this comes as a research validation tool for the energy audit conducted, showcasing with real-time data that the construction elements do not meet the requirements of thermal insulation and energy saving.

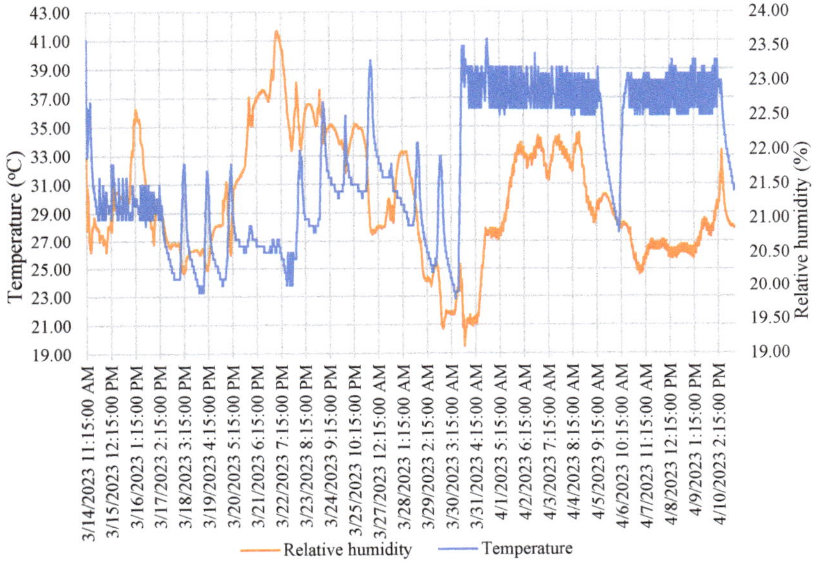

Figure 8. Results regarding the degree of relative humidity and temperature obtained from the equipment located in an office room facing the south façade of the building.

The south office room has a wider temperature range, with the lowest and highest recorded readings being 21.7 °C and 24.4 °C, respectively, with an average temperature of around 23.31 °C. In comparison, the north office room maintains a temperature range between 19.8 °C and 23.6 °C, with an average temperature of approximately 21.74 °C, which is lower. This suggests that the temperature in the south office is consistently higher than that in the north office. The humidity levels in the north office room typically exceed those in the south office room. The humidity levels in the north office range from 19.5% to 41.7%, with an average humidity of 29.83%. In comparison, the south office has humidity levels ranging from 17.1% to 39.1%, with a mean humidity of 26.89%. This indicates that the air in the north office consistently has a higher level of humidity. The north office room reveals a little more noticeable fluctuation in temperature and humidity. The north office exhibits a wider spectrum of temperature and humidity variations, which could have a more pronounced impact on thermal comfort and air quality, compared to the relatively milder south office.

3.3.3. Results from the Thermographic Study on the Building's Envelope

A thermographic study was also undertaken on the Câmpulung Town Hall, specifically examining the north façade of the structure. We have used a TESTO 872s thermal imaging camera to conduct thermographic observations of the Town Hall of Câmpulung. The weather conditions at the time of measurement were overcast, with a morning temperature of around 3 °C. The prevailing conditions were optimal for performing thermal imaging as they effectively reduced the influence of solar radiation on the thermal measurement enabling more-precise identification of heat losses from the building's exterior. This stud employed thermal imaging technology to detect regions experiencing heat losses, thu highlighting the energy efficiency problems of the building's exterior and better provin the necessity of implementing solutions. The outcomes of this assessment are crucia for planning future conservation and restoration endeavors, especially due to the Tow Hall's historical significance and unique architectural requirements. The heat loss throug the windows was particularly significant, with temperatures reaching up to 14.1 °C (se Figure 9). The significant difference in temperature clearly demonstrates that the window are vulnerable areas in the building's envelope. The outdated timber-framed windows c this historic structure are evidently insufficient in mitigating energy dissipation.

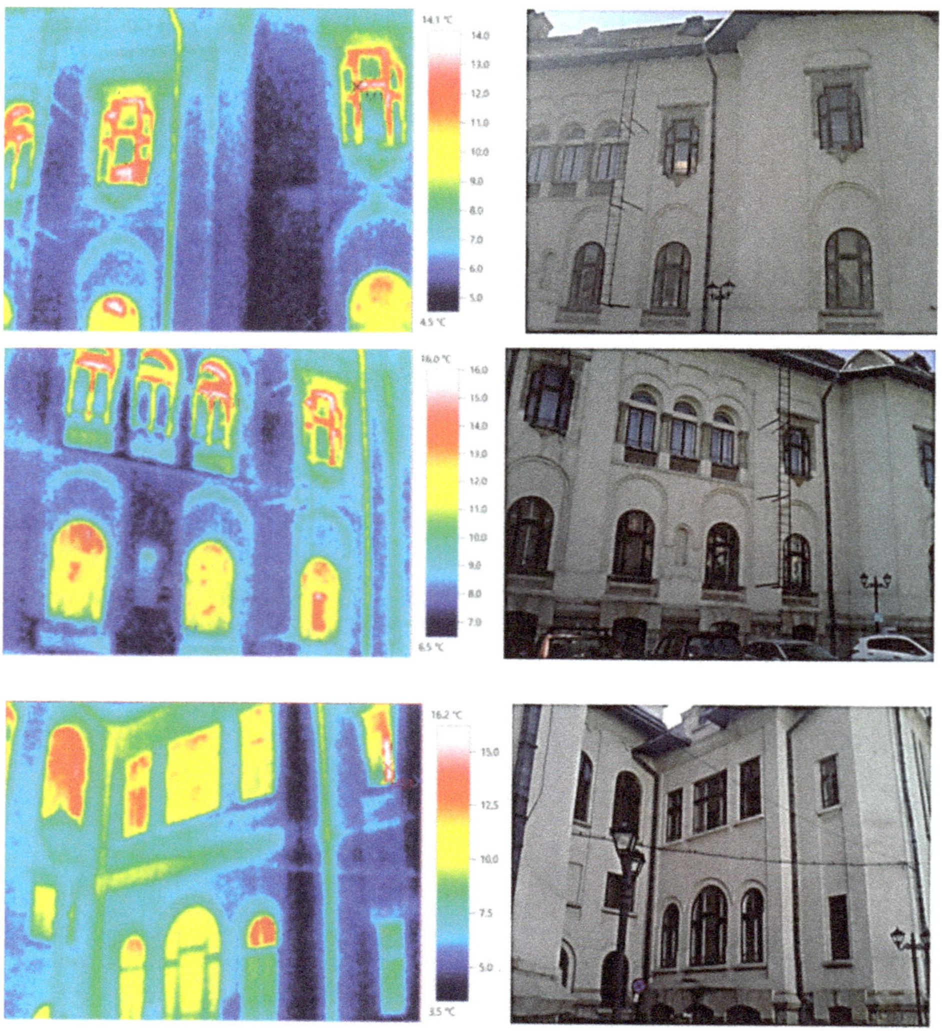

Figure 9. Multiple thermal images of the north façade of the building.

The thermographic research highlights a significant issue of energy inefficiency at the Câmpulung Town Hall, mainly caused by outdated construction materials and deterio rating architectural components. By replacing or updating the historical windows with thermally efficient alternatives that preserve the building's beauty and historical signifi cance, energy consumption can be reduced, and the sustainability of the property can be improved. This short thermographic study provides essential evidence to support the need for investing in energy efficiency measures at the envelope level.

4. Strategies and Solutions

The strategy implemented in this project was first to study the building in a non-invasive way via the use of the above-mentioned studies (thermal calculations, indoor measurements, and thermal camera photos).

Identified Solutions

Taking into consideration the energy audit and the in situ studies, a series of solutions were identified as proper for the intervention. They are as follows:

- Proposed solution 1: Changing the entire window as a result of the minimum thermal resistances provided for the external window frame (R'min > 0.9 m^2K/W);
- Proposed solution 2: Keeping the wood frame with restoration measures but replacing the glass with a high-performance example, namely, 4-12-4 type with low-e.

After changing the windows, the following must be taken into account:

- The sealing of cold air infiltrations in the joints of carpentry contours between the skirting board and the wall gap with a sealing film on the outside (width 29 cm), and filling the remaining spaces after installing the new windows with polyurethane foam and closing the joints with plaster;
- The water-repellent sealing of the joints on the outer contour of the frame with special materials (silicone putty, exterior sealing film, hydrophobic mortars, etc.), as well as covering the joints;
- Where necessary, replacing the existing galvanized sheet joists on the external horizontal gable at the bottom of the wall gaps with aluminum gables; the slope, the existence and shape of the teardrop, the sealing against the frame (nails with a wide head at small distances), the sealing against the wall (the edge of the board raised and covered with plaster on the upper part), etc., will be ensured;
- The unclogging (or creating, if there are none) of the holes at the bottom of the skirts intended for the removal of condensed water between the sashes.

The thermal modernization of the exterior window frame is proposed to be carried out in the following manner: changing the entire glazed surface with glass of type 4-12-4 energy performance to prevent the building's cooling requirement from increasing during the hot season; the solar coefficient of the glass will be $g < 0.35$.

Adopting the solution of the total replacement of the existing windows involves sealing the interior space and drastically reducing the number of air exchanges below the value necessary to dilute the CO_2 concentration and indoor humidity. Thus, before the rehabilitation, the air exchange was partially achieved through the leaks in and around the window frames.

In addition, by providing sealing gaskets, air freshening must be conducted in other ways, namely, by installing hygro-adjustable grid systems in the color of the wooden frame. The solutions identified for the doors at the entrance of the building are to be equipped with automatic, mechanical, or electric closing systems. For the door at the main entrance, it is recommended to choose a configuration similar to the existing one, consisting of two successive doors between which a buffer space from the outside environment is created.

The rehabilitation solutions chosen for the attic slab include insulating the floor towards the unheated bridge; the heat-insulating layer will be applied to the outer face of the support layer after uncovering the ballast and/or waterproofing layers as appropriate. The hydrothermal insulation solution will be made with a layer of fireproof mineral wool of a minimum 30 cm and a maximum conductivity of 0.04 W/mK.

Regarding the modernization solutions proposed for the interior lighting, the propose strategy was to replace the interior lighting luminaires, which currently have fluorescer lamps, with efficient LED lighting luminaires.

The increased efficiency of LED luminaires would lead to energy savings and therefor the achievement of another desired outcome. Also, their average lifetime is substantiall longer than any classic source, operating for up to 30,000 h without the luminous flu diminishing. LEDs are also able to withstand variations in supply voltage without affectin their lifetime.

The main solution to the problem of creating new energy sources was the installatio of a complete photovoltaic system of the "ON-GRID" type, a system with monocrystallin photovoltaic panels with a power of 25 kWp and a total area of 200 m^2. The system will b installed on the back part of the roof without affecting the building's aesthetic value.

The PV panels are installed to ensure the production of electricity for the building' own consumption, being connected to the external network, and they will consist o the following:

Monocrystalline photovoltaic panels with a power between 350 and 500 W and a tota nominal power of 25 kW, mounted on a support structure made of aluminum profile elements with a west orientation and an inclination of 30–40° compared to the incline plane and voltage inverters with a minimum efficiency of 95%.

Regarding the mounting system for panels, the solution was to use MC4-type con nectors for photovoltaic panels and a solar electric cable with a two-way energy mete (recording of energy consumed from the network and energy delivered to the network The surface available for mounting the panels is approx. 855 m^2, and the total area o the panels is 200 m^2. The total installed power of the system is 25 kWp, with an annua electricity production of approx. 32,500 kWh (from renewable sources).

As future research directives for energy optimization projects for buildings wit historic and architectural value, analyzing and creating a new category of similar building that are not listed in the monument category would prove useful not only to better addres their problems in a controlled way but to also unveil their real number and their actua spread in the country.

5. Results and Perspectives

The results identified in the case study further highlight the need for a pre-stud multicriterial analysis regarding the current state of the building alongside a heritage stud In situ studies of the building, conducted in a non-invasive way, have facilitated mor in-depth knowledge of the building, thus making sure that the solutions proposed wil extend the life cycle of the building. The solutions for energy reduction are presente in Table 6.

Table 6. Proposed energy efficiency solutions.

Solution	Description
S1	Keeping the wood frame with restoration measures but replacing the glass with a performant 4-12-4 type with low-e
S2	The hydrothermal insulation solution for the attic floor will be made with a layer of fireproof mineral wool of a minimum of 30 cm and a maximum conductivity of 0.04 W/mK
S3	Replacing the interior lighting luminaires, which currently have fluorescent lamps, with efficient LED lighting luminaires
S4	Installing thermostatic valves on all heating radiators
S5	Mounting 25 kW solar photovoltaic panels on the west façade
S1 + S2 + S3 + S4	Analysis of solutions 1–4
S1 + S2 + S3 + S4 + S5	Analysis of solutions 1–5–complete

The results of the simulations are summarized in Table 7.

Table 7. Results of annual consumption and energy reduction for each proposed solution.

Consumption	Heating	Domestic Hot Water		Lighting	Total
Solution S1					
Annual consumption [MWh/year]	575.4	1.6		50.6	627.6
Specific consumption [kWh/m^2/year]	263.0	0.7		23.1	286.9
Energy reduction			18.16%		
solution S2					
Annual consumption [MWh/year]	591.2	1.6		50.6	643.4
Specific consumption [kWh/m^2/year]	270.3	0.7		23.1	294.1
Energy reduction			16.1%		
solution S3					
Annual consumption [MWh/year]	714.7	1.6		31.9	748.1
Specific consumption [kWh/m^2/year]	326.7	0.7		14.6	342.0
Energy reduction			2.44%		
solution S4					
Annual consumption [MWh/year]	664.1	1.6		50.6	716.3
Specific consumption [kWh/m^2/year]	303.6	0.7		23.1	327.5
Energy reduction			6.59%		
solution S5					
Annual consumption [MWh/year]	714.7	1.6		18.1	734.4
Specific consumption [kWh/m^2/year]	326.7	0.7		8.3	335.7
Energy reduction			4.24%		
solution S1 + S2 + S3 + S4					
Annual consumption [MWh/year]	451.9	1.6		50.6	504.1
Specific consumption [kWh/m^2/year]	206.6	0.7		23.1	230.4
Energy reduction			43.29%		
solution S1 + S2 + S3 + S4 + S5					
Annual consumption [MWh/year]	419.9	1.6		0.0	421.5
Specific consumption [kWh/m^2/year]	192.0	0.7		0.0	192.7
Energy reduction			47.53%		
Final energy class	D	A		A	B

The energy efficiency solutions recommended for a building, including window upgrades and solar panel installation, have undergone comprehensive analysis to determine their individual and collective effects on energy usage. The initial solution (S1) entails enhancing the windows by installing high-performance glass, while preserving the original wooden frames. This approach results in a substantial 18.16% decrease in energy use, highlighting the crucial importance of window insulation in enhancing the overall energy efficiency of buildings, especially in historic buildings.

The second method (S2) improves thermal insulation by installing fire-resistant mineral wool on the attic floor. This method decreases energy usage by 16.1%, thus lowering the amount of heat lost via the roof, which is a prevalent source of inefficiency in older structures. By enhancing the insulation of the roof, this solution improves the building's thermal envelope and helps maintain stable inside temperatures, resulting in a more comfortable life and work environment.

The third solution (S3) centers on enhancing the lighting system by replacing outdated fluorescent lamps with energy-efficient LED luminaires. The change leads to a moderate reduction of 2.44% in overall energy consumption. However, it has a significant impact on the energy used just for lighting, which highlights the efficiency of LEDs in comparison to older lighting choices.

The fourth solution (S4) incorporates thermostatic valves on all heating radiators, enhancing the heating system's efficiency by optimizing heat distribution management. This modification results in a 6.59% decrease in energy consumption, demonstrating the advantages of contemporary heating controls in minimizing needless heating and improving the comfort of occupants.

The fifth option (S5) integrates renewable energy by installing 25 kW solar photovoltaic panels on the west façade. This not only reduces the building's dependence on external power sources by 4.24% but also promotes sustainability efforts by decreasing the carbon footprint linked to conventional energy sources.

When these ideas are implemented all together, the outcomes are notably remarkabl The integration of solutions S1 through S4 leads to a significant decrease in energy usag by 43.29%, demonstrating the synergistic impact of comprehensive building retrofits. Th addition of solar panels (S1–S5) further amplifies this effect, resulting in a 47.53% reductic in energy use and raising the building's energy rating to "B". This holistic strategy nc only greatly reduces operational expenses but also establishes the building as an exampl of sustainable retrofitting in historical structures, highlighting the feasibility of combi ing contemporary technologies with preservation principles. This technique not onl advances environmental objectives but also preserves the cultural and historic integrity c heritage structures.

6. Conclusions

This paper presented the framework for pre-intervention studies on a heritage buil ing, that is a building with historic and architectural value. The framework incorporate several primary non-invasive techniques, presented in detail in the paper. A key takeawa from the study was the importance of studying buildings with historic and architectur value in situ, monitoring them in periods of seasonal changes—with climate condition that are inconsistent—in order to better assess how the buildings handle these transitio periods and how to better assist them with strategic energy efficiency measures.

The proposed framework can help with providing a better understanding of th characteristics of buildings with historic and architectural value before starting the energ optimization processes, as the proposed pre-intervention studies can highlight importar aspects about the building's current state and intervention possibilities.

Buildings with historic and architectural value, once modernized and energy opt mized, can function as well as new buildings, meaning they will be able to accommodat proper interior comfort values for the inhabitants. As in the case of the Town Hall buildin there are plenty of older buildings that are still being intensively used to this day. Th proposed solution of integrating a solar panel system on the roof so as not to disrupt th general architecture of the façades but still to be able to generate energy locally could prov to be a great solution for these buildings. Ultimately, the positive outcome of consistentl modernizing this building stock is that we can live more sustainably and in healthie interior climates.

Author Contributions: Conceptualization, S.E.Ș. and T.C.; methodology, S.E.Ș. and T.C.; softwar T.C.; validation, T.C.; formal analysis, T.C.; investigation, S.E.Ș., T.C., R.P. and L.P.; resources, S.E. and T.C.; data curation, S.E.Ș. and T.C.; writing—original draft preparation, S.E.Ș.; writing—revie and editing, S.E.Ș., T.C., R.P. and L.P.; visualization, S.E.Ș. and T.C.; supervision, T.C.; proje administration, T.C. All authors have read and agreed to the published version of the manuscript.

Funding: This paper was funded by the National Research Grants of the UTCB (ARUT 2023), proje number UTCB-31.

Institutional Review Board Statement: Not applicable.

Informed Consent Statement: Not applicable.

Data Availability Statement: Data are contained within the article.

Conflicts of Interest: The authors declare no conflicts of interest.

References

1. *The Intervention Methodology for a Non-Invasive Approach to Energy Efficiency Measures for Buildings with Historical and Architectura Value/22.12.2022 2022*; Romanian Ministry Order 3568/2022; Romanian Ministry of Culture: Bucharest, Romania, 2022.
2. Communication from the Commission to the European Parliament, the Council, the European Economic and Social Committe and the Committee of the Regions. In *A Renovation Wave for Europe—Greening Our Buildings, Creating Jobs, Improving Lives* European Commission: Brussels, Belgium, 2020.

Communication from the Commission to the European Parliament, the Council, the European Economic and Social Committee and the Committee of the Regions. In *A Policy Framework for Climate and Energy in the Period from 2020 to 2030*; European Commission: Brussels, Belgium, 2014.

World Meteorological Organization. *Annual Report Highlights Continuous Advance of Climate Change*; World Meteorological Organization: Geneva, Switzerland, 2022.

European Council. *2030 Climate and Energy Policy Framework*; European Council: Brussels, Belgium, 2014.

Delivering the European Green Deal on the Path to a Climate-Neutral Europe by 2050; European Commission: Brussels, Belgium, 2021.

Stanica, D.-I.; Karasu, A.; Brandt, D.; Kriegel, M.; Brandt, S.; Steffan, C. A methodology to support the decision-making process for energy retrofitting at district scale. *Energy Build.* **2021**, *238*, 110842. [CrossRef]

Nägeli, C.; Farahani, A.; Österbring, M.; Dalenbäck, J.-O.; Wallbaum, H. A service-life cycle approach to maintenance and energy retrofit planning for building portfolios. *Build. Environ.* **2019**, *160*, 106212. [CrossRef]

Blázquez, T.; Ferrari, S.; Suárez, R.; Sendra, J.J. Adaptive approach-based assessment of a heritage residential complex in southern Spain for improving comfort and energy efficiency through passive strategies: A study based on a monitored flat. *Energy* **2019**, *181*, 504–520. [CrossRef]

Dornieden, T.; Gorbushina, A.A.; Krumbein, W.E. Biodecay of cultural heritage as a space/time-related ecological situation—An evaluation of a series of studies. *Int. Biodeterior. Biodegrad.* **2000**, *46*, 261–270. [CrossRef]

Godwin, P.J. Building Conservation and Sustainability in the United Kingdom. *Procedia Eng.* **2011**, *20*, 12–21. [CrossRef]

Rodrigues, C.; Freire, F. Building retrofit addressing occupancy: An integrated cost and environmental life-cycle analysis. *Energy Build.* **2017**, *140*, 388–398. [CrossRef]

Huijbregts, Z.; Kramer, R.; Van Schijndel, J. Computational modeling of the impact of climate change on the indoor environment of a historic building in Netherlands. In Proceedings of the 9th Nordin Symposium on Building Physics, Tampere, Finland, 29 May–2 June 2011.

Coillot, M.; El Mankibi, M.; Cantin, R. Heating, ventilating and cooling impacts of double windows on historic buildings in Mediterranean area. *Energy Procedia* **2017**, *133*, 28–41. [CrossRef]

Litti, G.; Khoshdel, S.; Audenaert, A.; Braet, J. Hygrothermal performance evaluation of traditional brick masonry in historic buildings. *Energy Build.* **2015**, *105*, 393–411. [CrossRef]

Caro, R.; Sendra, J.J. Evaluation of indoor environment and energy performance of dwellings in heritage buildings. The case of hot summers in historic cities in Mediterranean Europe. *Sustain. Cities Soc.* **2020**, *52*, 101798. [CrossRef]

Menassa, C.C. Evaluating sustainable retrofits in existing buildings under uncertainty. *Energy Build.* **2011**, *43*, 3576–3583. [CrossRef]

Murgul, V.; Pukhkal, V. Saving the Architectural Appearance of the Historical Buildings due to Heat Insulation of their External Walls. *Procedia Eng.* **2015**, *117*, 891–899. [CrossRef]

Sugár, V.; Talamon, A.; Horkai, A.; Kita, M. Energy saving retrofit in a heritage district: The case of the Budapest. *J. Build. Eng.* **2020**, *27*, 100982. [CrossRef]

Ruggeri, A.G.; Calzolari, M.; Scarpa, M.; Gabrielli, L.; Davoli, P. Planning energy retrofit on historic building stocks: A score-driven decision support system. *Energy Build.* **2020**, *224*, 110066. [CrossRef]

Biseniece, E.; Žogla, G.; Kamenders, A.; Purviņš, R.; Kašs, K.; Vanaga, R.; Blumberga, A. Thermal performance of internally insulated historic brick building in cold climate: A long term case study. *Energy Build.* **2017**, *152*, 577–586. [CrossRef]

Walker, R.; Pavía, S. Thermal performance of a selection of insulation materials suitable for historic buildings. *Build. Environ.* **2015**, *94*, 155–165. [CrossRef]

González, A.G. Keeping the Historical Heritage Alive: Methodology for the energy renovation of the historic residential stock of the east extension in Brussels. In Proceedings of the PLEA2013—29th Conference, Sustainable Architecture for a Renewable Future, Munich, Germany, 10–12 September 2013.

Bernardi, A.; de Carli, M.; Pockelé, L.; Poletto, F.; Galgaro, A.; Di Sipio, E.; Castelruiz, A.; Urchueguía, J.; Badenes, B.; Pasquali, R.; et al. Shallow Geothermal Energy for existing buildings—Overview and status of project GEO4CIVHIC 2022. In Proceedings of the European Geothermal Congress 2022, Berlin, Germany, 17–21 October 2022.

Corrado, V.; Ballarini, I.; Paduos, S.; Primo, E. Refurbishment of the Residential Building Stock toward the Nearly-Zero Energy Target Through the Application of the Building Typology. *Energy Procedia* **2016**, *101*, 208–215. [CrossRef]

Ascione, F.; Bianco, N.; De Masi, R.F.; Mauro, G.M.; Vanoli, G.P. Cost-Effective Refurbishment of Italian Historic Buildings. In *Cost-Effective Energy Efficient Building Retrofitting*; Elsevier: Amsterdam, The Netherlands, 2017; pp. 553–600.

Smiles, M.J.; Law, A.M.; Urwick, A.N.; Thomas, L.; Irvine, L.A.D.; Pilot, M.T.; Bowman, A.R.; Walker, A.B. Next steps in the footprint project: A feasibility study of installing solar panels on Bath Abbey. *Energy Sci. Eng.* **2022**, *10*, 892–902. [CrossRef]

Chen, X.; Qu, K.; Calautit, J.; Ekambaram, A.; Lu, W.; Fox, C.; Gan, G.; Riffat, S. Multi-criteria assessment approach for a residential building retrofit in Norway. *Energy Build.* **2020**, *215*, 109668. [CrossRef]

Johansson, P.; Wahlgren, P.; Eriksson, P. *Interior Super Insulation in Heritage Buildings. Challenges and Possibilities to Conserve Heritage Values and Increase Energy Performance*; Chalmers University of Technology Report ACE 2020:2; Chalmers University of Technology: Gothenburg, Sweden, 2020.

30. Dulnewa, J.; Elvisto, T.A.; Hahn, T.; Harten, J.; Menniks, M.; Prahm, J.; Saleh, S.; Scherz, D.; Simpson, S. *Improving the Energ Efficiency of Historic Buildings: The Four Pilot Projects of Co$_2$olBricks, Stand: October 2013 Co$_2$olBricks—Climate Change, Cultur Heritage and Energy Efficient Monuments*; Denkmalschutzamt: Hamburg, Germany, 2013; Volume 7.
31. GEO4CIHIC Consortium, Adriana Bernardi, Francesa Bampa, Historic Buildings and Buildings from the Universal Heritag GEO4CIVHIC Most Easy, Efficient and Low Cost Geothermal Systems for Retrofitting Civil and Historical Buildings. 20 Available online: https://geo4civhic.eu/ (accessed on 7 May 2024).
32. *4RinEU Reliable Models for Deep Renovation Guidelines and Technology Concepts for Managing Building End of Life*; Europea Commission: Brussels, Belgium, 2019.
33. Örn., T.; Nilsson, K. Identifying Cultural Building Values—Methododlogy review for energy efficiency alterations. Part c Cultural Heritage Preservation. In Proceedings of the EWCHP 2013. 3rd European Workshop on Cultural Heritage Preservatio Bozen/Bolzano, Italy, 18 September 2013; pp. 219–225.
34. Widström, T. *Enhanced Energy Efficiency and Preservation of Historic Buildings Methods and Tools for Modeling*; Architecture and th Built Environment; Royal Institute of Technology (KTH): Stockholm, Sweden, 2012.
35. *Energy Efficiency and Historic Buildings Application of Part L of the Building Regulations to Historic and Traditionally Construct Buildings*; HIstoric England (Ed.) English Heritage: Swindon, UK, 2011.
36. Galatioto, A.; Ricciu, R.; Salem, T.; Kinab, E. Energy and economic analysis on retrofit actions for Italian public historic building *Energy* **2019**, *176*, 58–66. [CrossRef]
37. *EUR 28055 EN*; Economidou, M.; Labanca, N.; Serrenho, T.; Bertoldi, P.; Zancanella, P.; Paci, D.; Panev, S.; Gabrielaitiene, Assessment of the first National Energy Efficiency Action Plans under the Energy Efficiency Directive. Publications Office of th European Union: Luxembourg, 2016. [CrossRef]
38. Sun, X.; Gou, Z.; Lau, S.S.-Y. Cost-effectiveness of active and passive design strategies for existing building retrofits in tropic climate: Case study of a zero energy building. *J. Clean. Prod.* **2018**, *183*, 35–45. [CrossRef]
39. Sesana, E.; Gagnon, A.S.; Ciantelli, C.; Cassar, J.; Hughes, J.J. Climate change impacts on cultural heritage: A literature revie *WIREs Clim. Chang.* **2021**, *12*, e710. [CrossRef]
40. Franzen, C.; Dahl, T.; Wedebrunn, O.; Bräunlich, K.; Esposito, E.; Colla, C.; Pulcini, F.; Gabrielli, E.; Khalil, M.; Helbig, C et al. D 2.5 Report on Methodology and Checklist | Efficient Energy for EU Cultural Heritage. 2010. Available online: htt //www.3encult.eu/en/project/workpackages/builtheritageanalysis/Documents/3ENCULT_2.5.pdf (accessed on 7 May 2024
41. *Romanian Law, No. 422/2001 on the Protection of Historical Monuments, Republished, Amended and Supplemented*; Romanian Ministr of Culture: Bucharest, Romania, 2006.
42. *Romanian Law, No. 350/2001 Regarding Territorial Development and Urban Planning*; Romanian Ministry of Culture: Buchares Romania, 2006.
43. *Romanian Law, No. 372/2005 Regarding Modernization and Energy Optimization Interventions for Existing Buildings*; Romanian Ministr of Culture: Bucharest, Romania, 2023.
44. *Romanian Sectorial Strategy in the Field of Culture and National Heritage for the Period 2014–2020—Analysis of Capacity and Fundin Needs*; Romanian Ministry of Culture: Bucharest, Romania, 2014.
45. *Romanian Strategy for Culture and National Heritage 2016–2022*; Romanian Ministry of Culture: Bucharest, Romania, 2016.
46. *Territorial Development Strategy of Romania: Foundational Studies—Services for the Development of Studies for the Implementation c Project Activities Titled 'Development of Tools and Models for Strategic Territorial Planning to Support the Upcoming Programming Perio Post-2013'*; Romanian Ministry of Culture: Bucharest, Romania, 2013.
47. The National Register of Historic Monument in Romania. Available online: https://patrimoniu.ro/en/articles/lista monumentelor-istorice (accessed on 7 May 2024).
48. Crișan, R. *Reabilitarea Patrimoniului Construit Teorie si Practica/Heritage Rehabilitation Theory and Practice*; OZALID: Buchares Romania, 2023.

applied sciences

MDPI

Article

Validation of Dynamic Natural Ventilation Protocols for Optimal Indoor Air Quality and Thermal Adaptive Comfort during the Winter Season in Subtropical-Climate School Buildings

Antonio Sánchez Cordero [1], Sergio Gómez Melgar [2,*] and José Manuel Andújar Márquez [2]

[1] Programa de Doctorado Ciencia y Tecnología Industrial y Ambiental, ETS Ingeniería, University of Huelva, 21004 Huelva, Spain; antonio.sanchez443@alu.uhu.es

[2] TEP192 Control y Robótica, CITES, ETS Ingeniería, University of Huelva, 21004 Huelva, Spain; andujar@uhu.es

* Correspondence: sergomel@uhu.es; Tel.: +34-687-88-07-14

Featured Application: The introduction of effective natural ventilation controls can provide acceptable indoor air quality and thermal comfort for school buildings in a subtropical climate.

Abstract: The need for energy-efficient buildings must be based on strong effective passive-design techniques, which coordinate indoor air quality and thermal comfort. This research describes the principles, simulation, implementation, and monitoring of two different natural cross-ventilation algorithm scenarios applied to a school-building case study affected by a subtropical climate during the winter season. These ventilation protocols, the steady and dynamic versions, can control the carbon dioxide concentration and actuate the window openings according to pre-defined window-to-wall ratios. The implementation of the monitoring process during three non-consecutive days in the winter of 2021 validates the opening strategy to maintain carbon dioxide below 800 ppm, described by the protocol Hygiene Measures Against COVID-19, and the temperature within the comfort ranges suggested by the adaptive UNE-EN 16798. The study shows that a steady opening of 2.16% window-to-wall equivalent ratio can be enough to maintain the requested comfort and carbon dioxide conditions. The use of the dynamic window ratios, from 0.23% to 2.16%, modified according to the measured carbon dioxide concentration, can partially maintain the carbon dioxide below the required limits for ASHRAE 62.1, Hygiene Measures Against COVID-19 and UNE-EN 16798 between 48.28% to 74.14% of the time. However, the carbon dioxide limit proposed by RITE, 500 ppm, is only achieved for 15.52% of the time, which demonstrates the inadequacy of the natural ventilation to fulfil the standard. Further improvements in the dynamic control of the openings in these buildings could lead to lower carbon dioxide concentrations while maintaining the thermal comfort in mild winter climates.

Keywords: indoor air quality; thermal adaptive comfort; Designbuilder^{TM}; monitoring; school buildings; natural ventilation

Citation: Cordero, A.S.; Melgar, S.G.; Márquez, J.M.A. Validation of Dynamic Natural Ventilation Protocols for Optimal Indoor Air Quality and Thermal Adaptive Comfort during the Winter Season in Subtropical-Climate School Buildings. *Appl. Sci.* **2024**, *14*, 4651. https://doi.org/10.3390/app14114651

Academic Editors: Daniel Sánchez-García and David Bienvenido Huertas

Received: 1 May 2024
Revised: 23 May 2024
Accepted: 24 May 2024
Published: 28 May 2024

1. Introduction

1.1. Energy Efficiency, Thermal Comfort and Indoor Air Quality

The built environment is responsible for considerable impacts on our present and future lives, considering both construction and the in-use stage. For example, in the year 2009, the construction sector was responsible for 23% of the total CO_2 emissions produced by the global economic activities [1], and so on, up to the present date.

Energy efficiency (EE), thermal comfort (TC) and indoor air quality (IAQ) are related key performance indicators to measure the sustainability assessment of buildings. This has been considered as the EE-TC-IAQ Dilemma [2]. The most recognized sustainabilit assessment tools, like Level(s), include them in their most relevant criteria [3]. Extensiv research has been carried out lately to demonstrate the relationship and the proper balanc between them [4,5].

A successful IAQ requires an adequate rate of air renovation, which necessarily implie a partial substitution of the indoor air [2]. Since it depends on the season and the loc climatic conditions [6], this air substitution can strongly influence the TC of any buildin type [7]. Temperature, relative humidity, and air movement are the basic indicators th affect TC. The proper correction of these hygrothermal conditions provides, by definitio additional heating/cooling potential energy demand from mechanical systems. Recen research has focused on this issue by considering the use of setpoint temperatures based o adaptive TC models [8]. Furthermore, the expected climate change scenarios will provid a certain increase in the cooling loads, as well as an uncertain reduction in the heatin loads [9].

In addition, the improvement of mechanical systems and the introduction of renewab energies play an important role in energy reduction with a proper energy productio demand management [10]. However, the introduction of passive-design principles i considered to be a more effective method to cut down energy demand [11,12].

1.2. Indoor Air Quality

Buildings are the places where humans spend almost 90% of our lives, including hom leisure and work [13]. If time at home is not considered, adults spent most of their time a work, whilst children and youngsters spent it in school buildings.

Recent trends in social sustainability assessment, such as Level(s), are boosting th development of better strategies to improve the IAQ of buildings [3]. However, the pace o implementing these improvements in buildings is slow, especially in existing ones, du to their age and lack of investment [14]. These buildings usually have deficiencies in TC IAQ and EE. For a robust and transparent assessment, IAQ can be defined by the Worl Health Organization as the absence of air pollutants [15]. The most common pollutants i buildings are the following: biological, contaminants, off-gassing emissions, carcinogen and particulate matter. Among these, CO_2 is the most relevant for most authors [16].

Among all types of buildings, the study of IAQ in school buildings is particularl relevant because their occupants have specific needs. Children can be more affected b pollutants than adults [17]: their lungs are not fully developed, due their low weigh they can suffer greater exposures than adults, and those exposures remain for longer i their lungs. Specifically, CO_2, affects human cognition and concentration capacity when i exceeds established limits [18].

Considering that schools have regular occupancy and design typology, it is convenien to anticipate design strategies that can be applied to buildings with similar conditions [19,20 Previous studies have described experimental approaches to identify the most suitabl window opening configuration, and their application in schools worldwide to achieve a good IAQ, TC and EE. Furthermore, urban air-quality climatic conditions are extremel important in describing an effective natural ventilation strategy.

Schools located in polluted areas have restrictions in applying outdoor air renova tions. In such locations, the forecast of air quality can be monitored through real-tim meteorological stations, or though interpolation models [21].

Field measurement and building energy simulation can be used to suggest optima strategies for balancing energy use and indoor environment quality, simultaneously. Fo those schools in hot climates with mild winters the strategies must be oriented to summe overheating risks [14]. Although the performance of the envelope is not as important as th ventilation rate, when natural ventilation is applied, it still needs to be included in a dee analysis of the IAQ-TC-EE Dilemma [2]. Then, thermal transmittance of the envelope is th most relevant indicator to be considered [6]. For schools in warm and moderate climate the design strategies must be oriented both to the summer and winter season, based on im proved ventilation schemes and well-designed energy-conscious building [2]. Empowering occupants as agents actively engaging in their own comfort can be an interesting solution when mechanical systems are not available [22]. Finally, studies in schools comparing both natural and mechanical ventilation have demonstrated a better performance for IAQ, but worst results for TC with natural ventilation for all the teaching hours [20].

1.3. Recent Evolution of the Indoor Air Quality Standards

Since the SARS-CoV-2 airborne transmission was demonstrated [23], IAQ indicators have occupied a major role in overcoming the disease. For those enclosed environments, the CO_2 concentration became the only indicator for providing effective advice on the airborne transmission [24]. With the need to maintain school buildings open, after the declaration of the disease, the national government established a set of protocols to reduce the infection risk [25]. A maximum CO_2 concentration of 800 ppm was established as a reference safety guide, which must be addressed by natural ventilation. Unfortunately, most of the school buildings did not have CO_2 monitoring devices at the beginning, and those with them failed to address the TC criteria [7]. Since them, the monitoring of CO_2 has become recently a very popular easy-to-use tool for IAQ evaluation [26].

Several ventilation standards, such as EN 16798-1 [27], ASHRAE 62.1 [28], RITE [29], and some others like WELL [30], have established limits for the most relevant IAQ indicators. These include CO_2, particulate matter, and volatile organic compounds, but CO_2 is the most relevant of them because of its usability as an airborne infection indicator. Their limits for CO_2 concentrations (between 500 and 1000) are shown in Table 1.

Table 1. Most relevant IAQ indicators by standard.

Indicator	Protocol [2]	EN 16798-1	ASHRAE 62.1	WELL [30]	RITE [29]
CO_2	800 ppm	900 ppm	1000 [1] ppm	900 ppm	500 ppm

[1] recommend considering assumption [31], [2] hygiene measures against COVID-19 [25].

In the case of RITE, buildings are classified into categories of buildings from 1 to 4. There, school buildings are included in category IDA 2, in which the CO_2 limit is 500 ppm. Table 2 summarizes the IAQ levels with the corresponding CO_2 concentrations applied in this research.

Table 2. Classroom simulation parameters.

Main Parameters		Thermal Transmittance		
		Element	Components	U (W/(m^{2}K))
Area	49.60 m^2	Roof boards	Asbestos + air cavity	3.077
Height	2.85 m	Ceiling	Concrete slab	1.505
Volume	141.36 m^3	Flooring	Concrete slab	1.505
Occupancy	26 pupils + teacher	External wall	2 brick leaves	1.390
Occupancy ratio	0428 p/m^2	Openings	Aluminum frame + 2 panes of glass (3 + 6 + 3)	1.960
Area	49.60 m^2	Interior partition	2 brick leaves	1.425

1.4. Minimum Natural Ventilation to Ensure Indoor Air Quality

There is a vast body of literature that recognizes how natural ventilation plays a vital role in IAQ, TC and EE [4,5,32–34]. On sites where urban pollution is not an issue, IAQ can be considered as a result of the balance between the indoor pollutants produced against the outdoor renovation rate introduced [32]. The first element of the balance is provided by the number, the type of activity and the age of the occupants, while the second element of the balance depends mostly on natural/mechanical ventilation. In most of the cases, the infiltration through the building's envelope can be avoided due to its lack of relevance in comparison with the other elements [33]. Polluted sites are not considered adequate for air renovation without previous filtration.

Natural ventilation is driven by different thermodynamic airflow issues. That airflow is provided due to different air pressures and temperatures or by wind flows [34]. The basic classification describes two kinds of natural ventilation: single or cross [35]. The the effectiveness of the natural ventilation depends on the size, the proportions, the shap and the position of the openings and the window-to-wall ratio (WWR) of each facade [36 Traditionally, the WWR has been extensively used in research due to the fact that it is easy t use and understand, and influences the way of combining different building performanc objectives: daylighting, thermal, and acoustic comfort, as well as cooling/heating energ demand [37]. Elements for defining a cross-ventilation system are graphically described i Figure 1 below: occupancy (O), window inlet (Wi), window outlet (Wo), and room size i depth (D), length (L), and height (H).

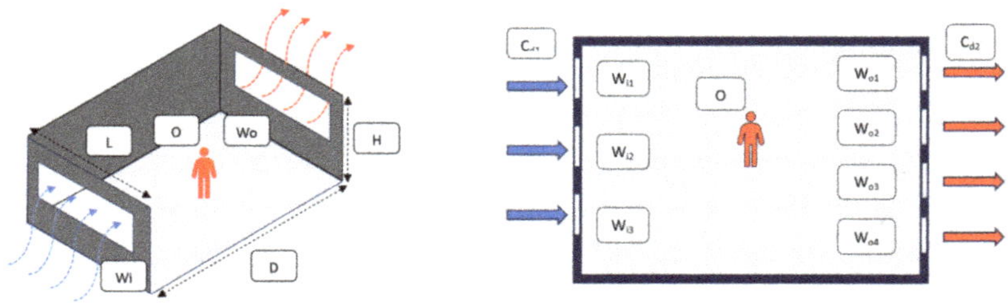

Figure 1. Main parameters affecting cross ventilation.

According to the literature, cross ventilation works when local winds pass throug the Wi in the windward side, cross the room and are expelled out through the Wo i the leeward façade [36,38]. For most of the airflow studies, the Venturi effect cannot b fully presumed, and an additional computational fluid dynamic must be considered t obtain the assumptions. If the window is the most relevant factor to manage, then ther are several options to consider, such as size, wide-to-height ratio, height position, an the number of single/multi windows per façade [39–41]. As described by these studie different configurations in the Wi-to-Wo ratio provide an internal pressure rise when th Wi/Wo is bigger than 1. If it is smaller, then the internal pressure decreases, providin better airflow conditions. Therefore, it results in a more effective ventilation configuratio for reducing the Wi-to-Wo ratio [36,39]. For those rooms with a different window size a the windward and leeward façades, an average combined-window area can be defined, a described in Equation 1 [36]. The equation includes a summation of areas at each of th opposite sides of the room. The result is expressed in Aeq as window-to-wall equivalen ratio ($WW_{eq}R$), useful for rooms with a different opening size at each side of the room.

$$Aeq = \frac{\sum_i Wii * \sum_i Woi}{\sqrt{\left(\left(\sum_i Wii\right)^2 + \left(\sum_i Woi\right)^2\right)}} \qquad \begin{aligned} Aeq &= Equivalent\ opening\ area \\ \sum Wii &= Windward\ opening\ summation \\ \sum Woi &= Leeward\ opening\ summation \end{aligned} \quad (1$$

1.5. Adaptive Thermal Comfort Model

The adaptive comfort model was firstly described by De Dear and Brager in 1998 [40 The model describes the influence of local and behavioral factors on the perception o TC. The model is based on TC field studies in countries worldwide, and established th foundations of the thermal models for the ASHRAE 55 [41]. Since then, several authors hav introduced variations in the original comfort algorithm to better approximate particula local conditions. The *Smart Controls and Thermal Comfort* project from the European Unior has been considered for the development of the UNE-EN 16798-1 [27], with both passiv and mechanical operations. The passive UNE-EN model equation described in the Standard Section A.2.2 [27], is used in this research to validate the TC.

1.6. Aims and Novelty of This Research

Based on the results obtained in previous works [7], the aim of this research is to demonstrate the validity of a proper $WW_{eq}R$ schedule to better achieve both IAQ and TC, using different ventilation protocols. As described by other authors, there are interesting possibilities for achieving IAQ without mechanical ventilation in school buildings [6,14,20], but they may compromise TC and EE. This depends on the climatic conditions [6,14,20], the envelope performance [6], set-point temperatures [14], the consideration of adaptive thermal comfort [42] and the efficiency of the mechanical systems [20]. It is necessary to elaborate window-opening management protocols that are based on natural ventilation and low-cost sensors in schools where mechanical ventilation is not an option. Then, the validation of these protocols is necessary to understand how to solve the IAQ-TC-EE Dilemma within specific climatic conditions and building configurations.

The research aims to carry out the following:

- Using Designbuilder simulations of this case study, obtain the $WW_{eq}R$ that provides enough air renovation to fulfill the CO_2 levels under the required limits:
    - For the steady natural ventilation protocols (SNVPs), it describes a $WW_{eq}R$ that allows the achievement of the required CO_2 all the time.
    - For the dynamic natural ventilation protocol (DNVP), it describes the functioning of the dynamic $WW_{eq}R$ according to the indoor CO_2 sensors, to minimize the outdoor air infiltration.
- Using the monitoring data from this case study, validate the proper functioning of the $WW_{eq}R$ in buildings with manual operation, providing enough air renovation to fulfill the CO_2 levels requested.
- Considering adaptive models, anticipate the TC outcomes that may arise due to the introduction of outdoor air without heat-recovery units.
- Provide a summary of design-and-management rules of thumb for scholar buildings in a subtropical climate to achieve acceptable IAQ during the operational time as an alternative to installing mechanical ventilation systems.

The paper contains the following: Section 1 introduces the problem and the perspective of the research, with the objective and aims to be achieved. Section 2 describes the materials and methodology used in the research. It includes the simulation, the case study description and the proceedings and materials for the data monitoring. Section 3 shows the results for the different scenarios described in Section 2. Thus, Section 4 discusses the outcomes from the data of the preceding section. Finally, Section 5 provides the main conclusions obtained and suggests future works. Some relevant material is included in Sections A and B.

2. Materials and Methods

This section contains the materials and methods used within this research to cover the different ventilation protocols proposed to improve the IAQ and TC of a case study in the city of Sevilla, Spain. The study includes two research stages: first, a software simulation, and second, an experimental installation and the measuring equipment for real-time monitoring devices.

2.1. Monitoring Materials

The real-time monitoring systems are shown in Figure 2, and their specifications in Table 3:

- A TEMTOP M2000C 2nd CO_2 Air Quality Monitor (orange). This includes a Sensair nondispersive infrared CO_2 sensor, a sensirion SHT31-ARP sensor for temperature and relative humidity, and an NDIR CO_2 sensor.
- A meteorological station, Froggit HP 1000SE PRO (white), includes a set of sensors to provide a full description of weather data in real time, as well as an indoor hydrothermal sensor able to measure humidity and temperature, and a central console. All of them are connected via radio at 868 MHz.

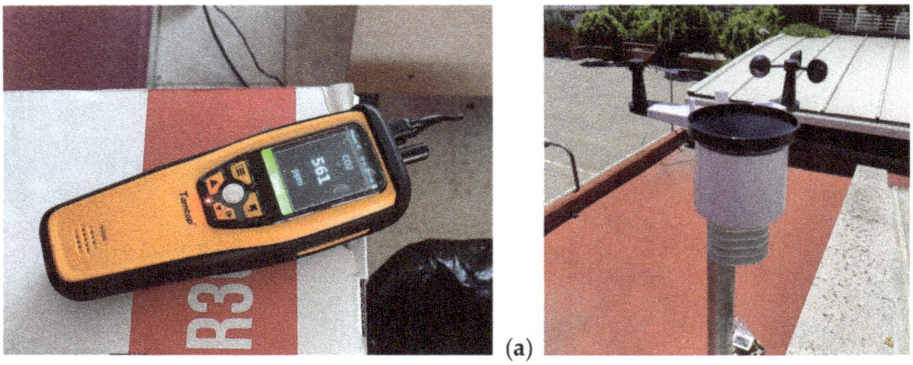

(a) (b)

Figure 2. Digital monitoring systems: TEMTOP M2000C 2nd air-quality monitor on the left (**a**) an the Froggit HP1000 SE wireless meteorological station on the right (**b**).

Table 3. TEMTOP M2000C 2, most relevant specifications.

Specifications	TEMTOP M2000C 2nd	Froggit HP 1000SE PRO	
	Indoor	Meteo station	Indoor sensor
Operation frequency		868 Mhz	868 Mhz
Temperature range	0–50 °C	−40–60 °C,	−40–60 °C,
Temperature accuracy	±0.3 °C	±1 °C	±1 °C
Relative humidity range	0–90%	1–99%	1–99%
Relative humidity accuracy	±3%	±5%	±5%
CO_2 measuring range	400–5000 ppm		
CO_2 accuracy	±40 ppm		

2.2. Simulation Software

The software chosen for the building simulation is *DesignBuilder*^{TM} V7.0.2.006 (DB) [43 which runs *EnergyPlus*^{TM} as the data calculation engine. It can simulate current and im proved conditions to compare different scenarios in terms of CO_2 emissions, indoor hy drothermal conditions, and energy consumption, among others. The software is recognize by several of the most influent standards, like the ANSI/ASHRAE, and it is commonl used in academia [44,45].

2.3. Methodology

The methodology proposed in this paper is based on two different approaches: a CO concentration *DesignBuilder*^{TM} simulation and onsite IAQ data monitoring to evaluate th validity of the ventilation protocols, the SNVP and DNVP. Both focus on a real case stud in a school building in the city of Sevilla, Spain. All the windows have been scheduled wit the same opening, whether they are windward or leeward.

The study considers $WW_{eq}R$ in measuring the average ratio, including all the window in all facades, instead of the traditional WWR. The *DesignBuilder*^{TM} simulation is used t obtain the most appropriate $WW_{eq}R$ to be used in the opening protocol to be followed b the school staff during the on-site monitoring process.

The duration of the monitoring process was reduced to a few days, due to the restric tions imposed by the school principal and the COVID-19 protocols. The authors decided t choose a set of non-consecutive days to evaluate different weather conditions, coverin different wind and temperature conditions.

Considering the monitoring devices described in Table 3, the following experimenta error is expected: ±0.3 °C indoor temperature, ±3% indoor relative humidity, ±40 ppm indoor CO_2, ±1.0 °C outdoor temperature, and ±5% outdoor relative humidity.

2.3.1. Case Study

The study focuses on a classroom in a school building in the city center of Sevilla (Spain). It is placed at 37.39 north latitude, and 5.98 east longitude, and influenced by a subtropical climate, Csa in the Köppen–Geiger classification [46]. The average temperatures are 11.3 °C in December, 25.7 °C in June, and 18.2 °C for the whole year. The average humidities are 75% in December and 42% in June. The prevailing winds are southwest in summer and northeast in winter.

This classroom has been selected because it is representative of the average school buildings in this region, but also because it has already been used in previous studies of IAQ and TC [7]. The distribution and position of the different elements within the classroom are presented in Figure 3.

- Figure 3a provides a layout with the position of the openings and the seats for the occupants. It also contains the position of the radiators (in red) and the heat pump (in blue). There is only 1 entrance to the classroom, which was always closed during the experimentations. The graphic scale and the north compass sign are included to classify the window openings as north/south. The north façade can be described as windward, while the south façade is considered leeward for the purposes of cross-ventilation analysis.
- Figure 3b provides an axonometric 3D view of the room, which adds extra information. Both windward and leeward openings include a set of solar-protection louvres, but during winter they are always set at a horizontal position. Therefore, it is considered that they have no influence on the ventilation process. Blue boxes describe the split heat-pump system. The yellow dot describes the approximate position of the IAQ monitoring devices (Figure 4a), while the yellow line describes the wireless connection with the central console (Figure 4b).

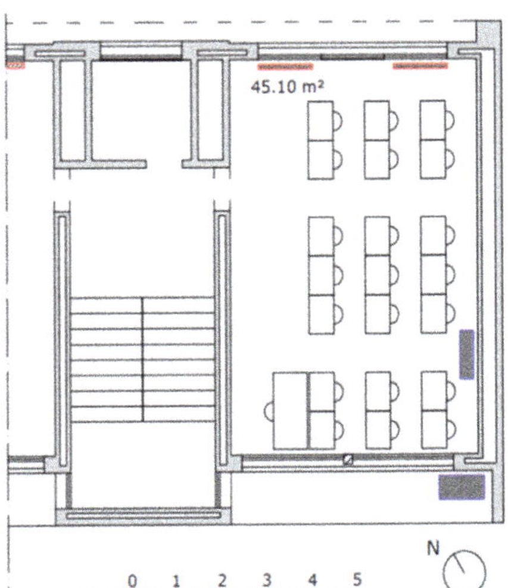

(a)

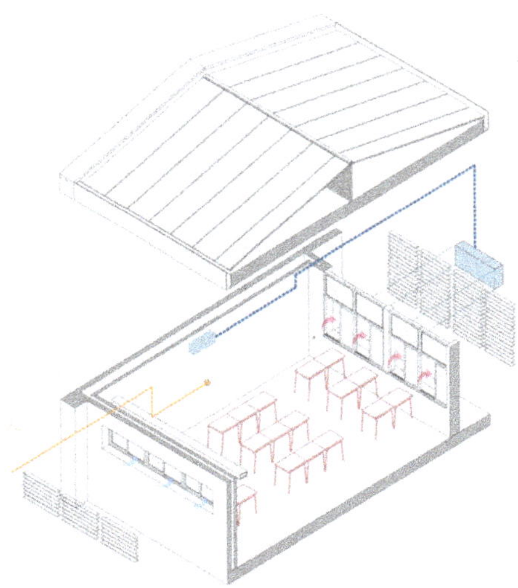

(b)

Figure 3. Opening elevation layout (**a**) and *DesignBuilder™* model 3D view (**b**).

The classroom volume is 141.36 m^3. It is usually occupied by 25 pupils of about 11 years old and a teacher, which makes one person every 1.90 m^2 (see Table 2), for the main parameters. The classroom activity schedule is organized from Monday to Friday. The attendance hours are from 9:10 to 12:05 and from 12:30 to 13:55. Students have a playtime out of the room between 12:05 and 12:30.

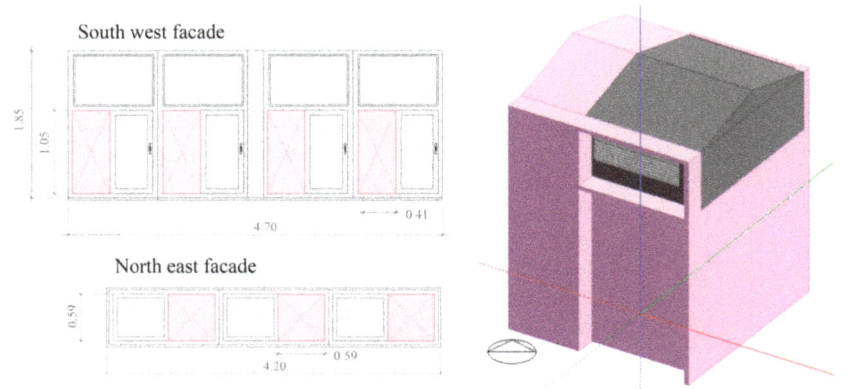

(**a**) (**b**)

Figure 4. Classroom layout (**a**) and north 3D axonometric view of classroom (**b**).

Construction elements were reviewed on-site and their thermal transmittance and other physical properties were calculated using *DesignBuilder*^{TM}. No insulation materials were found during the visual inspection, as would be expected for a 1970s building. The roof is composed of two separate layers with a non-ventilated interior air cavity of 0.60 m average thickness. Ceiling and flooring both include a concrete slab. The external walls are composed of two bricks sheets with an interior variable-thickness air cavity (cavity thickness from 5 to 15 cm). The windows in the south and north façades include aluminum frames plus double-glazing panes with air-gap insulation, as described in Table 2 (thermal transmittance), which gives a thermal summary of these construction elements.

The classroom is cross-ventilated through the windows in both the windward and leeward facades, as shown in Figure 4a. Both show, in red, the maximum effective opening to be considered for natural ventilation following the definition of the $WW_{eq}R$.

2.3.2. Natural-Ventilation Protocols

After an intense debate in academia, described in Section 1, the worldwide scientific community established natural ventilation as the most reliable source of improvement of the IAQ and the consequence reduction in the airborne transmission of several infections such as COVID-19. In the case of the Andalucía region (Spain), the regional Government prescribed the need to provide windows which open fully to guarantee natural ventilation. However, this was a temporary solution that causes additional heating/cooling demands which must be addressed at some point. To solve this issue, in a second step, the authors prosed a natural ventilation protocol that seeks the minimum air renovation to guarantee a maximum CO_2 concentration of 800 ppm. Ideally this will help to maximize the comfort periods while reducing the heating/cooling energy demand. This was called the COVID-19 natural ventilation protocol, as it mainly addressed the infection issue. The protocol described ideal periods to provide constant air renovation through manual window operation but without any control of its effectivity.

As an improvement of these early natural ventilation protocols, this paper suggests different approaches, based on the following:

- CO_2 concentration that can be obtained through CO_2 sensors. They are cheap and easy to find and allow any worker in an educational building to have a real-time IAQ assessment. Even more, they can be connected remotely to cloud control via an app so anyone can access the IAQ data report. In *DesignBuilder*^{TM} the CO_2 is obtained through virtual sensors, as described in Section 3.1.

- The opening factor described as a $WW_{eq}R$, which can be manually or mechanically operated on site, following any of the suggested algorithm rules. It is also possible to model it in *DesignBuilder*^{TM} through EMS scripting, as described in Section 3.1.

The Steady Natural-Ventilation Protocol

The SNVP can increase the natural ventilation while being easily implemented in the building. It describes an ideal steady-state opening WWR for all the operational time, according to the classroom configuration. See Figure 5 for details. For scenarios where the outdoor CO_2 is circa 400 ppm, the SNVP can be described as follows:

- The $WW_{eq}R$ is calculated via *Designbuilder*^{TM} simulations to maintain the CO_2 concentration under 800 ppm.
- The $WW_{eq}R$ obtained is then applied on-site to the case study in scenario 1. If the CO_2 concentration is below 800 ppm during this time, then the objective is accomplished and the $WW_{eq}R$ can be maintained while the rest of the conditions remain the same.
- On the contrary, if the CO_2 concentration is not below 800 ppm, then the $WW_{eq}R$ must be revised in several steps, until it finally decreases to under 800 ppm.

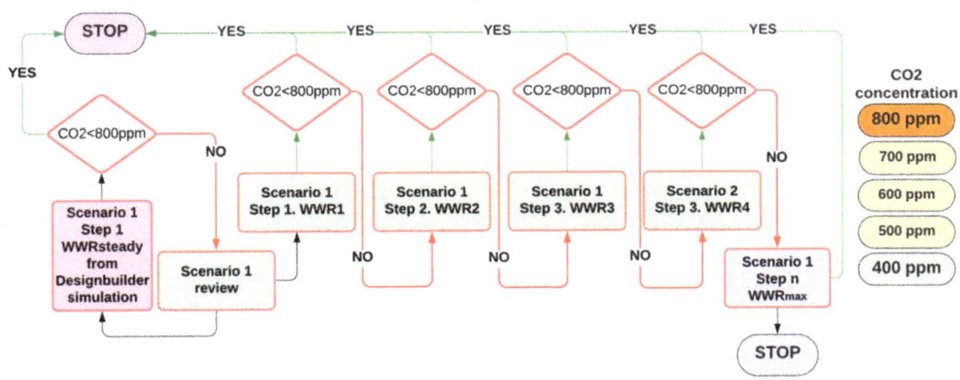

Figure 5. SNVP for optimal IAQ during winter season.

The Dynamic Natural-Ventilation Protocol

The DNVP has been designed as an improvement on the SNVP to increase air renovation control while minimizing the negative effect of thermal loads in extreme seasons. The ease of the DNVP will provide the opportunity for a successful implementation of natural ventilation policies in those school buildings where complex management systems are not available, for some reason. The DNVP can be seen in Figure 6. For scenarios where the outdoor CO_2 is circa 400 ppm, the summary of the DNVP can be described as follows:

- In the first step, if the CO_2 concentration is above 800 ppm, then windows must be fully opened to maintain the CO_2 as low as possible.
- In the next step, when CO_2 is below 800 ppm, the windows on both sides of the room will be opened according to a rising scale of $WW_{eq}R$. The $WW_{eq}R$ is the average number of the WWR in the case of cross ventilation with different window sizes in the windward and leeward facade. For each of these openings, the CO_2 concentration is measured. If the value remains below 800 ppm, then the process is stopped.
- If the CO_2 rises above 800 ppm, then a full opening must be applied to quickly remove pollutants until the CO_2 approaches the CO_2 min. The CO_2 is the result of the natural ventilation according to the maximum $WW_{eq}R$ of each case. Then, the CO_2 concentration can be considered as around 550 ppm. The $WW_{eq}R$ is then set to the next higher value from the previous step. Again, if the value remains below 800 ppm, then the process is stopped. On the contrary, if the CO_2 rises above 800 ppm, then the full opening scenario must be applied to repeat the flush process and the $WW_{eq}R$ applied for the next corresponding value.

- The process must be repeated in several steps until the CO_2 remains below 800 pp
or, alternatively, the CO_2 is above 800 ppm, but the windows remain fully open.
- In the next step, when CO_2 is below 800 ppm, the windows in both sides are opene according to the increasing scale of $WW_{eq}R$. The $WW_{eq}R$ is the average number of th WWR in the case of cross ventilation with different window sizes in the windwar and leeward side. To ease the process for the school buildings management team, th $WW_{eq}R$ used came from an easy-to-remember opening width: 2-4-8-10-15-22 cm i each of the windows at the same time.

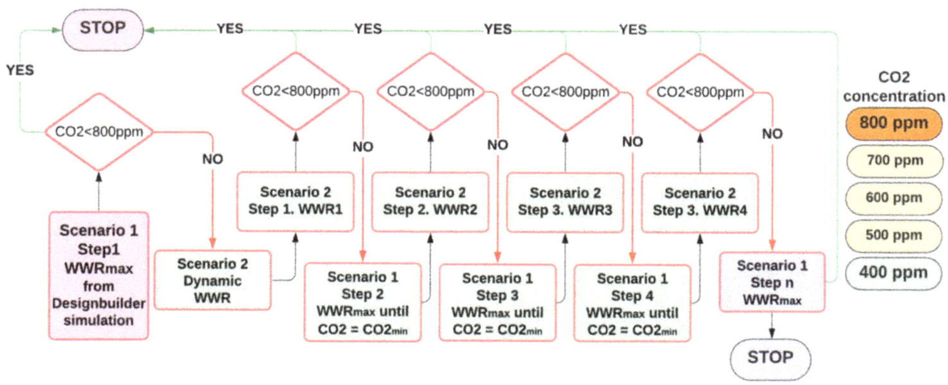

Figure 6. DNVP for optimal IAQ during the winter season.

2.3.3. Opening Considerations for This Case Study

According to previous experiments [36], if the size of the windows on the leeward an windward side are different, then the most effective opening configurations include th smaller windows in the windward side, which is the configuration of this case study. Th equivalent opening area is calculated through Equation (1) and shown in Table 4.

Table 4. WWR and $WW_{eq}R$ depending on opening configuration.

Windward/Northeast Windows 59 cm Height × 134 cm Wide			$WW_{eq}R$ [1] as Equation (1)	Leeward/Southwest Windows 105 cm Height × 107.5 cm Wide		
Opening Conditions				Opening Conditions		
Wide (cm)	Wide (%)	WWR (%)		WWR (%)	Wide (%)	Wide (cm)
2	1.5%	0.23%	0.23%	0.54%	1.9%	2
4	3.0%	0.46%	0.45%	1.07%	3.0%	4
8	6.0%	0.92%	0.87%	2.14%	6.0%	8
15	7.5%	1.72%	1.54%	4.02%	11.2%	15
22	11.2%	2.53%	2.16%	5.90%	16.4%	22
30	15.2%	3.45%	2.80%	8.04%	22.3%	30
51	47.7%	6.78%	4.10%	10.99%	44.0%	41

[1] equivalent ratio calculated with Equation (1) considering windows at both sides (wind and lee).

2.3.4. Calculation of Indoor Carbon Dioxide Concentration

This research uses the calculated natural ventilation mode in *Designbuilder*^{TM} wit a discharge coefficient of 0.65. The CO_2 concentration equations are described in th EnergyPlus^{TM} Engineering Reference [47]. *The Carbon Dioxide Predictor-Corrector* is base on the following Transient Mass Balance Equation (see Equation (3)). Each element o Equation (2) is detailed in Appendix A.

$$\rho_{air}\, V_Z\, C_{CO2}\, \frac{dC_Z^t}{dt} = \sum_{i=1}^{N_{si}} kgmass_{sched\ load} * 1.0^{\,6} + \sum_{i=1}^{N_{zones}} m_i\,(C_{zi} - C_z^t) + (C_\infty - C_z^t) + m_{sys}(C_{SUP} - C_Z^t) \quad (2$$

As described in Appendix A, the CO_2 concentration depends on internal CO_2 load the zone air density and volume, CO_2 due to outdoor infiltration and ventilation, whic are present in this simulation, and also on CO_2 transferred from other zones and CO_2 from the air supply stream, which are not present.

The standard *EnergyPlus*^{TM} *epw weather file* [48] has been modified to establish the average wind speed and direction for a typical morning in December, with direction 260°, speed 2.6 m/s.

The *DesignBuilder*^{TM} simulation includes a scheduled ventilation protocol which is not within the current options of the software capabilities. To ensure the proper function of the DNVP, the authors used the EMS runtime scripting tools described in *The Application Guide for EMS* [47]. The scripting code is presented in Appendix B. The EMS script is acting over the openings to ensure the CO_2 remains under 800 ppm, as described in the DNVP in Figure 4. The EMS window-opening actuator is connected to two variables/sensors: *Air_CO2_Concentration* and *TrendDirection. TrendDirection* evaluates whether the CO_2 concentration is rising or decreasing.

It is expected that slight variations will be found between the simulation forecast and the real-time monitoring data. This is provided by the expected delay between the software simulation and the human management of the windows.

Finally, the monitoring process introduces an evaluation TC according to the method described in the UNE-EN 16798-1:2020 [27], Section A.2.2. UNE. The adaptative comfort is based on the following Equation (3):

$$\theta_{rm} = (1-\alpha) \times \left\{\theta_{ed-1} + \alpha \cdot \theta_{ed-2} + \alpha^2 \cdot \theta_{ed-3}\right\} \quad (3)$$

θ_{rm} Outdoor running mean temperature for the considered day
θ_{ed-1} Daily mean outdoor air temperature for the i-th previous day

3. Results

This section provides a comprehensive review of IAQ in two different scenarios, according to the methodology explained in Section 2.

3.1. Simulation Stage

At the simulation stage, the validity of the DNVP is evaluated for each $WW_{eq}R$ against the SNVP, from 0.23% to 2.16% $WW_{eq}R$.

The first set of simulations were run in *DesignBuilder*^{TM} with a constant wind speed of 2.6 m/s, direction 260°, and a constant $WW_{eq}R$. These simulations are plotted in Figure 7. Additionally, the figure includes the occupancy level and the recommended CO_2 concentration limit defined by the World Health Organization and the RITE [29]. The occupancy defined is 21 people and the CO_2 limit is 800 ppm.

The results shown in Figure 7 describe how a 2.16% $WW_{eq}R$ is the only simulation remaining below the established CO_2 limit of 800 ppm. All the other configurations of $WW_{eq}R$ provide simulation results between 800 and 2600 CO_2 ppm. In all run simulations, the CO_2 increases exponentially until it reaches a flat line, which remains constant, except in the simulations with the $WW_{eq}R$ under 0.45%. In these simulations, the CO_2 is still rising until the occupancy is zero, between 12:05 and 12:30. The simulations show the $WW_{eq}Rmax$ for this case study, which is 4.10%. For the $WW_{eq}Rmax$ the average CO_2 is circa 544 ppm.

For the next set of simulations run in *DessignBuilder*^{TM}, all the parameters remain the same except the $WW_{eq}R$, which is controlled by the DVNP EMS script, as explained in Appendix B. The $WW_{eq}Rs$ that provided a CO_2 concentration below 800 ppm are not considered in this simulation, following the DNVP algorithm. The results of the CO_2 concentration obtained through the calculations via Equation (2), explained in Appendix A, are shown in Figure 8. CO_2 limits and wind conditions remain the same as those in Section 3.1.

The results shown in Figure 8 describe the running loops of the different $WW_{eq}Rs$. Each of the $WW_{eq}R$ loops rises until 800 ppm, then they change to 4.10% $WW_{eq}R$ until the CO_2 concentration decrease to 400 ppm or as low as possible. As expected, for thos opening configurations with a lower $WW_{eq}R$, the number of loops is higher.

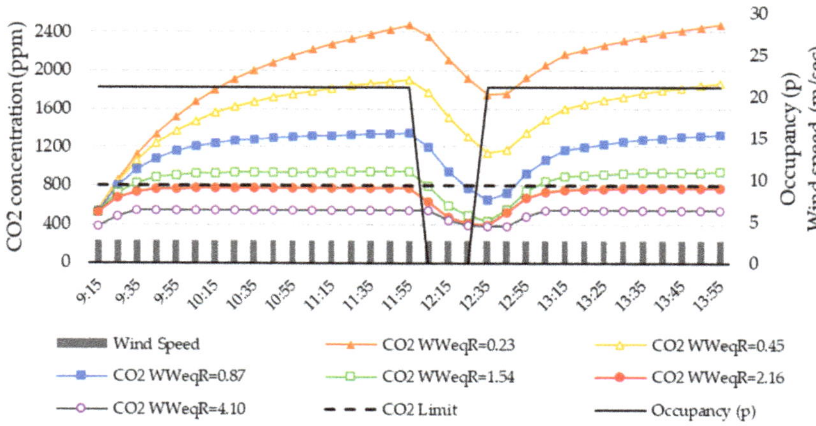

Figure 7. CO_2 concentration for different $WW_{eq}Rs$ with SNVP.

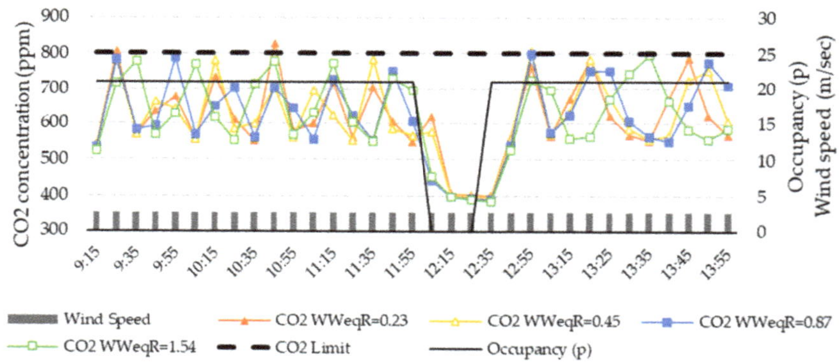

Figure 8. CO_2 concentration for different $WW_{eq}Rs$ with DNVP activated.

Figure 9 summarize the results of simulations for 0.23% $WW_{eq}R$, but comparin both the SNVP and the DNVP. Regarding the selected 0.23% $WW_{eq}R$, the results of CO concentration, when the DNVP is on, are below 800 ppm, while they achieve circa 250 when the SNVP is on.

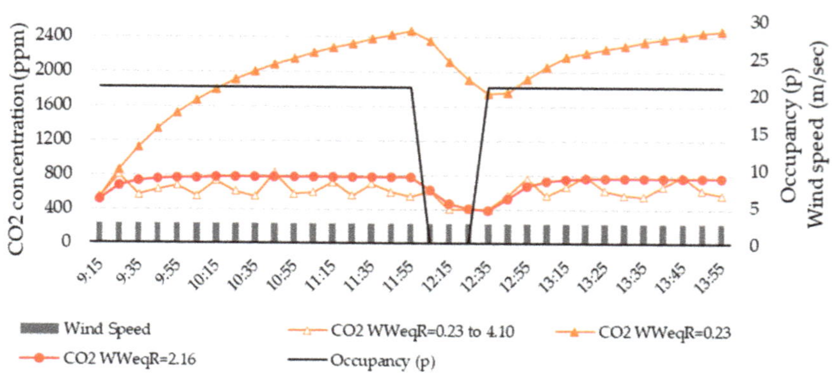

Figure 9. CO_2 concentration at 0.23% $WW_{eq}R$ for simulation with DNVP vs. SNVP.

Data shown in Figures 7–9 are organized and presented in Table 5.

Table 5. Simulation summary of average CO_2 concentration in ppm for several $WW_{eq}Rs$.

$WW_{eq}R$	Protocol	09:10 09:30	09:30 10:00	10:00 10:30	10:30 11:00	11:00 11:30	11:30 12:00	12:00 12:30	12:30 13:00	13:00 13:30	13:30 14:00	Peak
0.23 (%)	DNVP	608	643	637	641	634	620	485	579	627	640	641
	* SNVP	623	1274 *	1764 *	2063 *	2261 *	2414 *	2136 *	1827 *	2191 *	2416 *	2416 *
0.45 (%)	DNVP	599	642	632	628	630	632	477	581	632	642	642
	* SNVP	613	1180 *	1532 *	1707 *	1807 *	1880 *	1539 *	1236 *	1618 *	1813 *	1880 *
0.87 (%)	DNVP	593	652	645	647	644	642	427	573	644	649	649
	* SNVP	596	1037 *	1229 *	1292 *	1320 *	1340 *	984 *	790 *	1173 *	1302 *	1340 *
1.54 (%)	DNVP	565	653	667	680	668	662	431	574	636	636	680
	* SNVP	569	861 *	927 *	937 *	940 *	943 *	641 *	602 *	893 *	937 *	940 *
2.16 (%)	DNVP	543	729	750	752	755	757	469	544	745	757	757
	SNVP	547	749	773	774	774	774	525	551	758	771	774
4.10 (%)	DNVP	-	-	-	-	-	-	-	-	-	-	-
	SNVP	544	544	544	544	544	544	481	418	532	544	544

The results of CO_2 concentration are grouped by a set of rows for the same $WW_{eq}R$. Each column describes a 30 min time frame to show the average CO_2 concentration simulated every 30 min period. Values over the 800 ppm CO_2 limit are shown in red *.

The results shown in Table 5 confirm that all $WW_{eq}R$ configurations maintain the CO_2 concentration under the limit when the DNVP is active, while only a 2.16% $WW_{eq}R$ configuration maintains the CO_2 concentration objective when the SNVP is active.

3.2. Monitoring Results

At the monitoring stage, the most efficient $WW_{eq}Rs$ for the DNVP are validated during several days in 2021: 18th and 26th November and 17th December. There, real variable occupancy and weather applies instead of the constant ones considered in Section 3.1.

3.2.1. Monitoring of CO_2 Concentration through Different $WW_{eq}Rs$ with DNVP

The $WW_{eq}R$ applied on site did not match totally with the $WW_{eq}R$ described in the DNVP algorithm. The school staff suggested reducing some $WW_{eq}Rs$ to avoid unwanted wind gusts from strong winds from 4.10% to 2.80%, and from 1.54% to 1.07%.

The results gathered after several weeks of IAQ data monitoring during November to December 2021 can be seen in Figures 10–12. Figures 10 and 11 shows the evolution of the CO_2 concentration in ppm during the DNVP on-site monitoring, from the lowest (0.23%) to the highest (1.07%) $WW_{eq}R$, including the maximum ventilation flux at 2.80% $WW_{eq}R$. Figure 11 shows the CO_2 concentrations when $WW_{eq}R$ is 2.16% with the SNVP. Figure 10 shows the CO_2 concentration evolution for different $WW_{eq}Rs$, from 0.23% to 1.07%. At each CO_2 concentration peak, the school building staff modify the $WW_{eq}R$ to 2.80%, until the CO_2 falls below 400 ppm or to the lowest possible CO_2 concentration. The wind speed shown in Figure 10 indicates a great deviation from the average statistical conditions used in Section 3.1, which has a notable effect on the CO_2 concentration. In accordance with the strong wind and the promising results of the lower $WW_{eq}R$, the school building staff decided to slightly increase the $WW_{eq}R$ to 1.07%, to reduce the TC unsatisfaction of the occupants. Therefore, in the last monitoring stint, between 12:30 and13:55, the CO_2 remains between 800 and 1000 ppm.

Figure 11 shows the CO_2 concentration evolution for different $WW_{eq}Rs$, from 1.07% to 1.82%, with additional variations up to 2.80%, according to the DNVP algorithm.

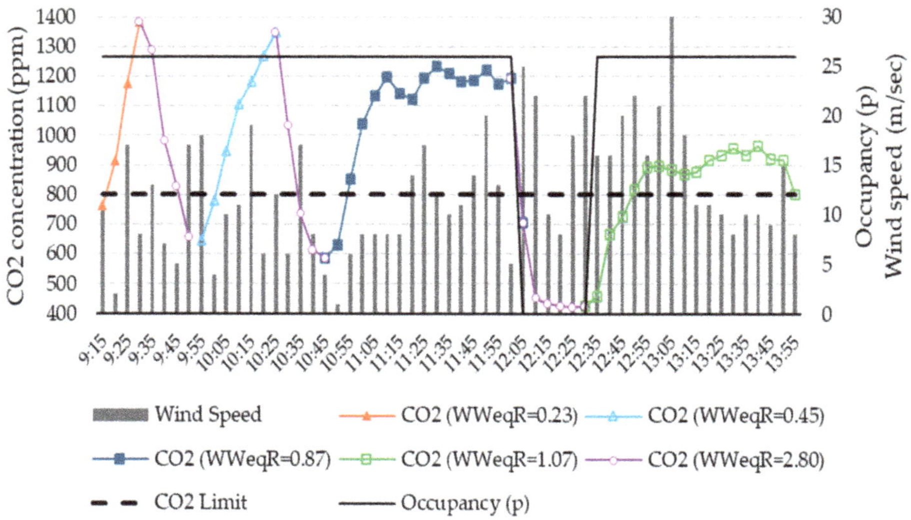

Figure 10. CO_2 concentration for different $WW_{eq}Rs$ with DNVP on the 18th of November 2021.

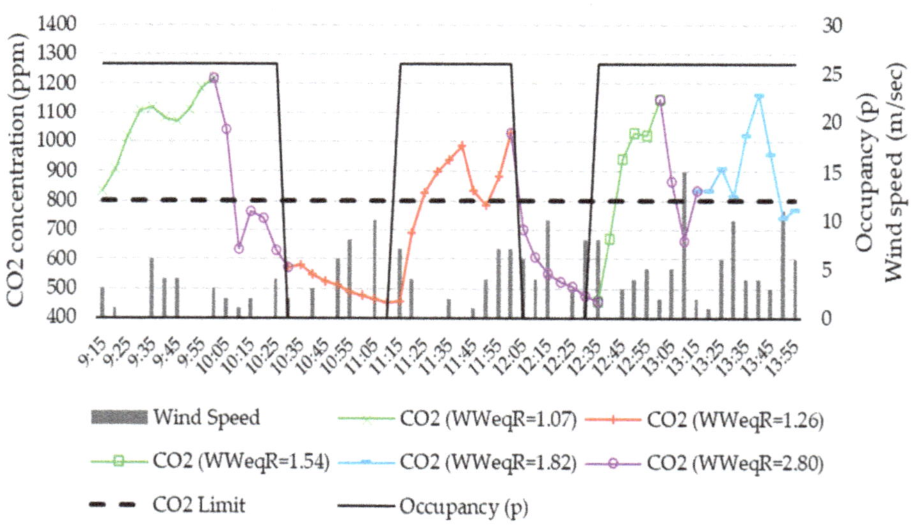

Figure 11. CO_2 concentration for different $WW_{eq}Rs$ with DNVP on the 26th of November 2021.

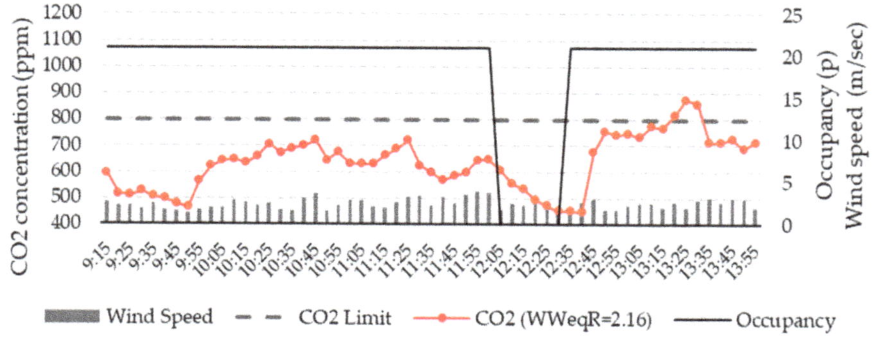

Figure 12. CO_2 concentration for 2.16% $WW_{eq}R$ with SNVP on the 17th of December 2021.

The wind speed decreases from Figure 10 to Figure 11, which seems to affect the CO_2 concentration. Although the $WW_{eq}R$ was changed between 1.54% and 1.82%, the CO_2 concentration rises quickly, circa 1200 ppm, as it did with the 0.87% $WW_{eq}R$ in Figure 10. In both cases, in Figures 10 and 11, the occupancy remained the same and there were no other incidents during the experiment.

3.2.2. Monitoring of CO_2 Concentration through 2.16% $WW_{eq}R$ with SNVP

A new monitoring was carried out on the 17th of December using a 2.16% $WW_{eq}R$, as suggested by the simulations and the previous monitoring processes. The results are shown in Figure 12.

During the whole process, the SNVP was not activated because the CO_2 concentration remained all the time below 800 ppm, except for a 10 min period from 13:20 to 13:30. Therefore the $WW_{eq}R$ selected (2.16%), achieved the maintaining of adequate levels of CO_2 concentration. There were some modifications in the key environmental parameters: the occupancy was 21 instead of 26 and the wind speed remain constant, between 2 and 3 m/s. Both parameters have a major influence on CO_2 concentration, as described in Equation (2).

Figure 13 shows a comparison of occupancy and wind conditions between the 18th and 26th of November and the 17th of December 2021. In all of them, the wind direction trend is quite similar, a northwest direction, but the wind speed is quite different. The first graph in Figure 13a shows peak speed values over 20 m/s, while the graph in Figure 13b shows lower wind speeds. Results in Figure 13c are similar to those average values described in Section 2.3.4., as used in the simulations. Figure 13a–c also includes the outdoor temperature, which is also relevant in comparing the TC with the climatic conditions. Figure 13a,c show outdoor temperatures from 10 °C to 20 °C, while Figure 13b shows a 5 °C-lower outdoor temperature.

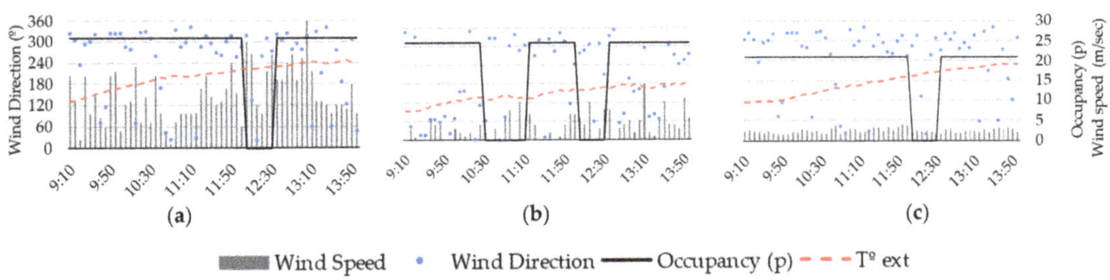

Figure 13. Wind conditions on 18th and 26th of November (**a**,**b**) and 17th of December 2021 (**c**).

Table 6 includes a summary of the average values for each 30 min time frame for the three different days monitored. Each day starts with a specific $WW_{eq}R$, from 0.23% to 2.16%. The table also includes the peak value for 2.16%. The table also includes the peak value for each period. During all the time frames, the concentrations of CO_2 remain below the suggested limits when the DNVP activates a $WW_{eq}R$ bigger than 2.16%, for a typical windy winter day in Seville.

Table 6. Monitorization summary of average CO_2 concentration in ppm for several $WW_{eq}Rs$.

$WW_{eq}R$	Protocol	09:10 09:30	09:30 10:00	10:00 10:30	10:30 11:00	11:00 11:30	11:30 12:00	12:00 12:30	12:30 13:00	13:00 13:30	13:30 14:00	Peak
0.23–1.07 18th-Nov.	DNVP	964 *	880 *	1106 *	570	1139 *	1201 *	607	666	896 *	915 *	1201 *
1.07–1.82 26th-Nov.	DNVP	880 *	1114 *	839 *	538	562	888 *	655	767	877 *	912 *	1114 *
2.16 17th-Dec.	DNVP	529	503	652	684	661	603	552	588	787	711	787

3.2.3. Summary of CO_2 Assessment for DNVP and SNVP

Figure 14 shows the time during monitoring when CO_2 concentration is below th recommended limits according to the most relevant IAQ standards.

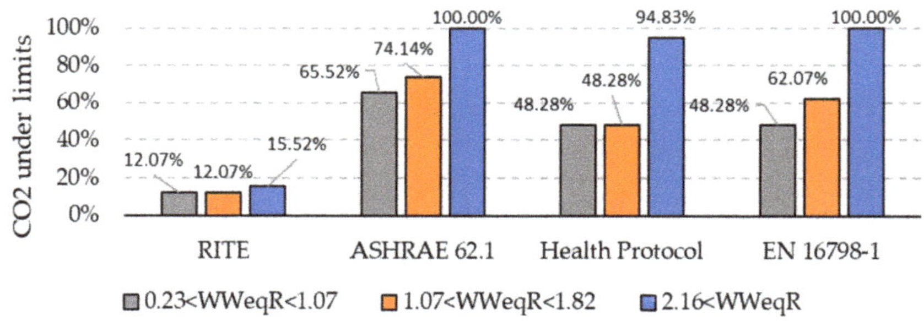

Figure 14. Percentage of time with CO_2 under limits described in Section 1 for several standards.

Figure 14 plots the results obtained in the three monitored days and groups them int color bars (gray, brown, and blue) to present a summary of time when the CO_2 is below th required limits: the 18th and the 26th of November, and the 17th of December.

When the $WW_{eq}R$ is 2.16%, the 17th of December (blue bar), the CO_2 concentratio is adequate during 100% of the time according to the EN 16798-1 and ASHRAE 62.1, an during 94.83% of the time according to the Health Protocol, but only during 15.52% of th time according to the RITE. However, the RITE is mandatory for new buildings and/o thermal facilities' major renovations, but not for existing buildings.

When the $WW_{eq}R$ is between 0.23% and 1.82%, the 18th and the 26th of Novembe (gray and brown bars), then the time within an adequate CO_2 concentration is reduce During the time of data monitored on the 18th of November, with a $WW_{eq}R$ between 0.2 and 1.07 (gray bar), the IAQ was adequate for the standards between 12.07% and 65.52% o the time. During the time of the data monitored on the 26th of November, with a WW_{eq} between 1.07 and 1.82 (brown bar), the IAQ was adequate for the standards only betwee 12.07% and 74.14% of the time.

According to the IAQ monitoring realized, if RITE is not considered in the analysi the CO_2 concentration was appropriate during all the time when the $WW_{eq}R$ was at leas 2.16%, on the 17th of December. If the $WW_{eq}R$ used is between 0.23% and 1.82%, then th IAQ can be considered appropriate between 48.28% and 74.14% of the time.

3.2.4. Temperature Monitoring during the Activation of the DNVP

The TC has also been monitored during this experiment, to achieve the necessar balance with IAQ. In Figure 15, the temperatures and CO_2 concentration have been plotte with their respective limit levels. The CO_2 concentration limits are set at 800 ppm, whil the temperature limits are set as described in the UNE-EN 16798 for the adaptive model o passive operation buildings.

Figure 15a includes the data obtained on the 18th of November, with $WW_{eq}R$ fro 0.23% to 1.07% according to the DVNP algorithm. The interior temperature remains for a the monitored time within the requested limits for adaptive comfort.

Figure 15b includes the data obtained on the 26th of November, with $WW_{eq}R$ fro 1.07% to 1.82% according to the DVNP algorithm. The interior temperature achieves th TC band only at the end of the monitored time (from 12:35).

Figure 15c includes the data obtained on the 17th of December, with $WW_{eq}R$ at 2.16 according to the DVNP. The interior temperature remains for all the monitored time withi the requested limits for adaptive comfort.

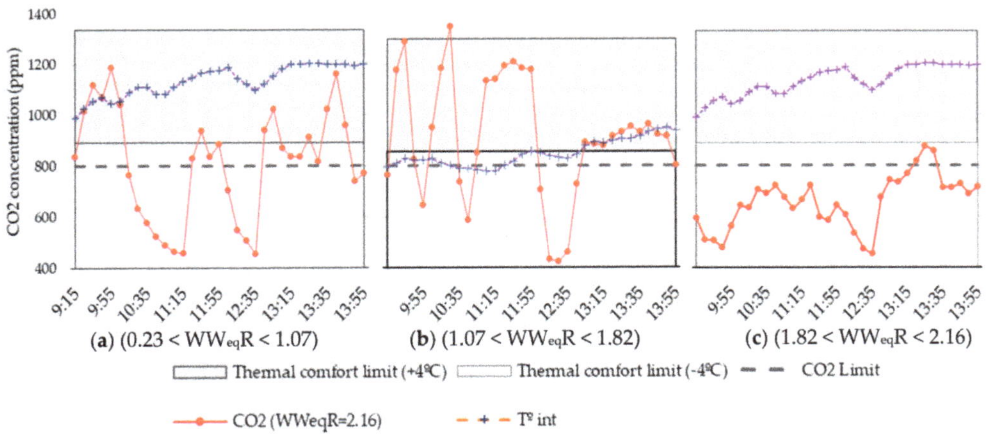

Figure 15. CO_2 concentration and TC for a DNVP with $WW_{eq}R$ from 0.23% to 2.16%.

4. Discussion

4.1. The Convenience of Natural vs. Mechanical Ventilation in Achieving Adequate IAQ

The results shown in Section 3 describe how a natural ventilation protocol can provide acceptable IAQ in cross-ventilated school buildings. The adequate management of the $WW_{eq}R$ as proposed by the DNVP described in Figure 6 helps to achieve IAQ and TC. The alternative implementation of the SNVP can be equally successful in achieving the IAQ objective, but it may bring unwanted effects in TC.

Different IAQ standards have been considered in this study. If the standard applied is RITE [29], considering category IDA 2 with a 500 ppm CO_2 concentration limit, then the period below the CO_2 objective will be small. However, the RITE standard is only applicable for new construction buildings and major renovations of thermal facilities, which is not the focus of this research. The rest of the buildings that are not affected by the RITE, mostly in-use buildings, are focused on achieving IAQ with the objectives described in ASHRAE 62.1 [28], EN 16798 [27] and The COVID-19 Health Protocol [25]. For these standards, the effectiveness of the natural ventilation strategies can be achieved for circa 100% of the classroom period for buildings with cross ventilation in a subtropical climate with the proper $WW_{eq}R$ implementation regarding the DNVP.

Results shown in Section 3.1 are based on the hypothesis of an outdoor CO_2 concentration of 400 ppm. The results shown in Section 3.2, for periods with no occupancy, demonstrate the validity of the outdoor CO_2 concentration close to 400 ppm. Future studies may include additional analysis comparing the effectiveness of natural ventilation when the outdoor air quality is above 400 ppm. For those buildings, in cities with high pollution levels, the DNVP should include connection to outdoor sensors [21].

This study is focused on those indicators that affect exclusively thermal comfort. However, high-noise urban areas may be inadequate for natural ventilation. Future versions of the DNVP should include noise-level outdoor sensors to prevent unwanted effects.

4.2. Dynamic vs. Steady Ventilation Controls

The results provided in Section 3.1. show interesting results for maintaining the CO_2 concentration below 800 ppm with constant modifications of the $WW_{eq}R$. The design and implementation of the DNVP seeks to minimize the discomfort provided by the air renovations but maintaining the requested IAQ objective. However, the results shown in Section 3.2 demonstrate no great difference between the application of the SNVP and DNVP algorithms. Figure 15a–c indicates no benefit with the implementation of a DNVP, but according to Figure 13a–c there was a significant modification to the weather conditions compared with those considered as average in the simulations in Section 3.1. Each scenario shown in Section 3 provides comparable, but not identical, results about the convenienc of using the DNVP or the SNVP to achieve 800 ppm as the maximum CO_2 concentratior

The results provided by other authors [20], in similar climate conditions, indicate 60 hours with no TC when natural ventilation is the only source for obtaining IAQ. Howeve the results, provided in Figure 15a–c demonstrate a higher period of TC. Figure 15c, wit 2.26% $WW_{eq}R$, achieves 94.38% of IAQ according to the Health Protocol CO_2 requiremen and a 100% TC according to the adaptive comfort model [40]. Future studies should focu on the need to extend the monitoring period to include the effects of climate change [42]

4.2.1. Simulation Results

Most of the authors describe the IAQ with natural ventilation as a discussion on th need to maintain a constant or variable $WW_{eq}R$ [6,14]. The results shown in Section 3. indicate that a variable $WW_{eq}R$, alternating a higher and lower $WW_{eq}R$, can maintain a acceptable IAQ while reducing periods with high $WW_{eq}R$. That is useful to avoid constar unwanted wind gusts and potential thermal discomfort.

When the SNVP is active, it is necessary to have a minimum $WW_{eq}R$ of 2.16% t achieve the CO_2 objective. The use of the maximum $WW_{eq}R$ available for this classroor may provide a 545 CO_2 concentration, which does not achieve the requirements of RIT (IDA 2 category) [29]. However, the requirements from all the other standards coul be fulfilled.

When the DNVP is applied to the simulation (Appendix B), each of the WW_{eq} considered, from 0.23% to 1.54%, achieves the CO_2 concentration objective. However, thi anticipates the need to maintain a constant control on the openings and the $WW_{eq}R$.

The results are based on a 0.65-discharge coefficient as standard value for *Designbuider*^{TI} simulations, and an average 2.6 m/s wind speed, in the northeast direction. It is nc expected that climate conditions can be included in the DNVP algorithm, but futur simulation works may include different discharge coefficients according to different schoc shapes and orientations [6,14].

4.2.2. Monitoring Results

The results shown in Section 3.2 verify the simulation data obtained in Section 3.1 i many ways. However, due to physical restrictions, some parameters had to be change during the monitoring process: people occupancy, and wind conditions. The monitorin described in Section 3.2 applies the DNVP to a real scenario. In both cases, the occupanc was 26 instead of the 21 used in Section 3.1, which provided an increase in the CO concentration. Due to mechanical restrictions in opening the windows, caused by a lack c maintenance, the staff in charge of the experiment had to adapt different $WW_{eq}R$ from th ones used in the simulations.

The research achieves the CO_2 objective requested by the ASHRAE 62.1, the Healt Protocol and EN 16798-1 when the $WW_{eq}R$ is 2.16%, but its success is reduced to 50% whe the $WW_{eq}R$ is reduced. This confirms the strategies described in previous experiments [7 and the adequacy of this $WW_{eq}R$ for this case study, as adequate for the SNVP. The result in Figure 14 demonstrate the difficulties in achieving the CO_2 required by RITE for natura ventilation, due to its extreme request.

However, the monitoring results for validating the DNVP provide some slight diffe ences from those obtained in Section 3.1, due several factors:

First, the variations steps of $WW_{eq}R$ for adjusting the IAQ were smaller than expecte in the simulation results in Figure 8. Human management of the opening process cannc be fully followed by the school staff. Future versions of the DNVP must consider huma limitations or introduce a mechanical control.

Second, wind and occupancy changes and variations could not be anticipated from th simulation to the monitoring process. Future versions of the DNVP could include low-cos outdoor air-quality sensors and a meteorological station to anticipate changes in the trend of CO_2 concentration.

4.3. Adaptive Comfort with Natural Ventilation

The improvement in the IAQ of the buildings by adding natural ventilation [7] may have some comfort flaws [48]. IAQ can potentially affect TC and consequently the EE of the buildings [12,33]. However, there are passive-design measures that may combine good IAQ with TC to maintain a low energy demand. Some of these measures are based on adaptive TC [9], and they seek the combination of natural ventilation with climatization in a mix-mode approach. As shown in Figure 15a,c, for subtropical climates with mild winters, the TC can be achieved for 100% of the time if the adaptive comfort approach [27] is considered. However, the results shown in Figure 15b demonstrate that significant variations in outdoor temperatures produce a reduction in the time within the TC limits. It has been demonstrated that the introduction of control patterns in existing buildings with natural ventilation such as the DNVP, whether they are manual or mechanical, can improve the IAQ while maintaining adaptive TC on mild winter days.

The DNVP must be revised to be used in school buildings in different climatic regions, to demonstrate its convenience. Other authors suggest overheating as the main barrier for TC introducing natural ventilation to achieve IAQ in a hot, humid climate [22]. There, the DNVP may introduce an additional input temperature sensor.

5. Conclusions

The research describes the principles, simulation, implementation, and monitoring of two different scenarios of natural cross ventilation applied to a school building affected by a subtropical climate. The findings are summarized in the following points:

- Two different ventilation control algorithms, the SNVP and DNVP, were developed by the contribution of the software simulation to guarantee an acceptable IAQ, according to the most-accepted standards.
- For this case study simulation, the SNVP provides a 2.16% $WW_{eq}R$ to maintain CO_2 below 800 ppm for a typical winter day in a subtropical climate, with 2.6 m/s wind and a northeast direction, when the occupancy is twenty-one students.
- The simulation of the DNVP anticipates that any $WW_{eq}R$ below 2.16% can maintain CO_2 below 800 ppm if it increases the $WW_{eq}R$ to remove the unwanted CO_2.
- The simulations, run in *Designbuilder*^{TM}, obtain a different grade of validation for the SNVP. The results, with a steady 2.16% $WW_{eq}R$, shown in Figure 9, are similar to the monitored results in Figure 12. However, the accuracy of the simulated DNVP is reduced because of a lack of accuracy in the opening management. The DNVP simulation describes double the $WW_{eq}R$ modifications compared to the DNVP monitoring. The implementation of manual proceedings applied by the building staff seem to be less effective than mechanical controls. This may be considered in future studies.
- According to the results obtained in the monitoring process, the application of the SNVP maintains the CO_2 concentration under 800 ppm during 100% of the operational time when the $WW_{eq}R$ was 2.16%, but failed during part of the study when the DNVP was applied. That suggests the need to implement better $WW_{eq}R$ controls to increase the efficiency of the protocols, as described in the simulation process.
- The CO_2 limits proposed by RITE, 500 ppm, were only achieved for 15.52% of the time with the maximum $WW_{eq}R$ applied, which demonstrates the inadequacy of the natural ventilation to fulfil the standard.
- When the DNVP is used with a $WW_{eq}R$ below 2.16%, it can partially maintain CO_2 under the required limits for those standards different to RITE. The analysis demonstrates compliance for 48.28% to 65.52% of the monitored time if the $WW_{eq}R$ is between 0.23 and 1.07, and a compliance for 48.28% to 74.14% of the monitored time if the $WW_{eq}R$ is between 1.07 and 1.82.
- During the monitoring process, the TC was also monitored to demonstrate the adequacy of the natural ventilation protocols with the adaptive TC described in the UNE-EN 16798 [27], for category III with a ±4 °C of adaptation. The application of the DNVP on a mild winter day achieved 100% TC during the full monitoring period.

For schools in colder climatic zones, the TC could be compromised, if the rest of th conditions remain the same. However, passive-design principles could be used.

- Future versions of these ventilation protocols may include temperature sensors t improve the $WW_{eq}R$ control. Therefore, it could lead to an increase in energy efficienc to become nearly zero-energy buildings by means of passive-design techniques.

Author Contributions: Conceptualization, methodology, software, validation, formal analysis, inve tigation, data curation, writing—original draft preparation, and visualization, A.S.C.; supervisio project administration, writing—review and editing, S.G.M. and J.M.A.M. All authors have read an agreed to the published version of the manuscript.

Funding: This work has been funded by the Research Center for Technology, Energy and Sustainab ity (CITES) at the University of Huelva (Spain).

Institutional Review Board Statement: Not applicable.

Informed Consent Statement: Not applicable.

Data Availability Statement: Data are contained within the article.

Conflicts of Interest: The authors declare no conflicts of interest.

Appendix A

The Transient Mass Balance Equation. As briefly described in Section 2.3, th *DesignBuilder*^{TM} CO_2 prediction is based on the calculation provided by *EnergyPlus*^{TI} in the Engineering Reference, Section 2.5 [47]. The full equation and all its elements ar described here:

$$\rho_{air}\, V_Z\, C_{CO2}\, \frac{dC_Z^t}{dt} = \sum_{i=1}^{N_{si}} kgmass_{sched\ load} * 1.0^{\ 6} + \sum_{i=1}^{N_{zones}} m_i\,\left(C_{zi} - C_z^t\right) + \left(C_\infty - C_z^t\right) + m_{sys}\left(C_{SUP} - C_Z^t\right) \quad \text{(A1}$$

where

$\sum_{i=1}^{N_{si}} kgmass_{sched\ load}$ = sum of each sum of scheduled internal carbon dioxide load The zone air density is used to convert the volumetric rate of carbon dioxide generatio from user input into mass generation rate [kg/s]. The coefficient of 10^6 is used to conver the units of carbon dioxide to ppm.

$\sum_{i=1}^{N_{zones}} m_i\,\left(C_{zi} - C_z^t\right)$ = carbon dioxide transfer due to interzone air mixing [ppm-kg/s

C_{zi} = carbon dioxide concentration in the zone air being transferred into this zone [ppm

$m_{inf}\left(C_{a??} - C_z^t\right)$ = carbon dioxide transfer due to infiltration and ventilation of outdoo air [ppm-kg/s]

$C_{a??}$ = carbon dioxide concentration in outdoor air [ppm]

$m_{inf}\left(C_{sup} - C_z^t\right)$ = carbon dioxide transfer due to system supply [ppm-kg/s]

C_{SUP} = carbon dioxide concentration in the system supply airstream [ppm]

m_{sys} = air system-supply mass flow rate [kg/s]

$\rho_{air}\, V_Z\, C_{CO2}\, \frac{dC_Z^t}{dt}$ = carbon dioxide storage term in zone air [kg/s]

C_Z^t = zone air carbon dioxide concentration at the current time step [ppm]

ρ_{air} = zone air density [kg/m^3], V_Z = zone volume [m^3]

C_{CO2} = carbon dioxide capacity multiplier [dimensionless]

Appendix B

EMS script for DNVP. The *DesignBuilder*^{TM} simulation includes a ventilation protoco schedule, which is not within the current options of the software capabilities. To ensur the proper function of the DNVP, the authors used the EMS runtime scripting tools. Th scripting code for north windward windows is presented here (the same has been used fo the south leeward side, but with the corresponding opening factors):

```
<ForAllExternalWindows> {Tag=north}
EnergyManagementSystem:Sensor,
 <LoopWindowVariableName>Air_CO2_Concentration,
 <LoopWindowZoneIDFName>,
 Zone Air CO2 Concentration;
EnergyManagementSystem:Actuator,
 Venting_Opening_Factor_<LoopWindowVariableName>,
 <LoopWindowIDFName>,
 AirFlow Network Window/Door Opening,
 Venting Opening Factor;
EnergyManagementSystem:TrendVariable,
CO2_trend_log_<LoopWindowVariableName>, !- Name
<LoopWindowVariableName>Air_CO2_Concentration, !- EMS Variable Name
120; !- Number of Timesteps to be Logged
EnergyManagementSystem:GlobalVariable,
CO2_previous_state_dir_<LoopWindowVariableName>;
EnergyManagementSystem:OutputVariable,
CO2 2tstep dir <LoopWindowVariableName>, ! Name
CO2_previous_state_dir_<LoopWindowVariableName>, ! EMS Variable Name
Averaged, ! Type of Data in Variable
ZoneTimeStep; ! Update Frequency
Output:Variable, *, CO2 2tstep dir <LoopWindowVariableName>, Timestep;
Output:Variable, *, Window_current_state_<LoopWindowVariableName>, Timestep;
<LoopNextWindow>
! extra outputs for viewing in the results viewer
<If BuildingAttribute HourlyOutput = 1 Then>
Output:Variable, *, Zone Air CO2 Concentration, hourly;
<Endif>
<If BuildingAttribute TimesteplyOutput = 1 Then>
Output:Variable, *, Zone Air CO2 Concentration, timestep;
<Endif>
EnergyManagementSystem:ProgramCallingManager,
 CO2 Window Control,
 InsideHVACSystemIterationLoop,
 CO2WindowControl;
EnergyManagementSystem:Program,
 CO2WindowControl,
<ForAllExternalWindows> {Tag=north}
 ! on/off control of window opening factor
Set CO2_previous_state_dir_<LoopWindowVariableName> = @TrendDirection CO2_trend_log_<LoopWindowVariableName>
2,
if <LoopWindowVariableName>Air_CO2_Concentration < 400,
        Set Venting_Opening_Factor_<LoopWindowVariableName> = 0.015,
Elseif <LoopWindowVariableName>Air_CO2_Concentration < 800,
        if CO2_previous_state_dir_<LoopWindowVariableName> > -100,
                Set Venting_Opening_Factor_<LoopWindowVariableName> = 0.015,        else,
                Set Venting_Opening_Factor_<LoopWindowVariableName> = 0.477,        endif, else,
        Set Venting_Opening_Factor_<LoopWindowVariableName> = 0.477, endif,
<LoopNextWindow>;

Output:Variable, *, AFN Node CO2 Concentration, Timestep;
Output:Variable, *, Zone Air CO2 Concentration, Timestep;
```

eferences

Huang, L.; Krigsvoll, G.; Johansen, F.; Liu, Y.; Zhang, X. Carbon emission of global construction sector. *Renew. Sustain. Energy Rev.* **2018**, *81*, 1906–1916. [CrossRef]

Becker, R.; Goldberger, I.; Paciuk, M. Improving energy performance of school buildings while ensuring indoor air quality ventilation. *Build. Environ.* **2007**, *42*, 3261–3276. [CrossRef]

Cordero, A.S.; Melgar, S.G.; Márquez, J.M.A. Green building rating systems and the new framework level(s): A critical review of sustainability certification within Europe. *Energies* **2019**, *13*, 66. [CrossRef]

Pisello, A.L.; Castaldo, V.L.; Taylor, J.E.; Cotana, F. The impact of natural ventilation on building energy requirement at inter-building scale. *Energy Build.* **2016**, *127*, 870–883. [CrossRef]

Rodrigues, A.M.; Santos, M.; Gomes, M.G.; Duarte, R. Impact of natural ventilation on the thermal and energy performance of buildings in a Mediterranean climate. *Buildings* **2019**, *9*, 123. [CrossRef]

6. Hwang, R.L.; Huang, A.W.; Chen, W.A. Considerations on envelope design criteria for hybrid ventilation thermal management school buildings in hot–humid climates. *Energy Rep.* **2021**, *7*, 5834–5845. [CrossRef]
7. Melgar, S.G.; Cordero, A.S.; Rodríguez, M.V.; Márquez, J.M.A.; Control, T.E.P.; Técnica, E.; De Ingeniería, S.; De Huelva, U Huelva, C.P. Influence on indoor comfort and energy demand due to the application of COVID-19 natural ventilation protoco for schools at subtropical climate during winter season. In Proceedings of the 3rd International Symposium on Architectu Research Frontiers and Ecological Environment (ARFEE 2020), Guangzhou, China, 18–20 December 2020; p. 6.
8. Sánchez-García, D.; Bienvenido-Huertas, D.; Rubio-Bellido, C. Computational approach to extend the air-conditioning usage adaptive comfort: Adaptive-Comfort-Control-Implementation Script. *Autom. Constr.* **2021**, *131*, 103900. [CrossRef]
9. Sánchez-García, D.; Rubio-Bellido, C.; Tristancho, M.; Marrero, M. A comparative study on energy demand through the adaptiv thermal comfort approach considering climate change in office buildings of Spain. *Build. Simul.* **2020**, *13*, 51–63. [CrossRef]
10. Melgar, S.G.; Cordero, A.S.; Rodríguez, M.V.; Andújar Márquez, J.M. Matching energy consumption and photovoltaic productio in a retrofitted dwelling in subtropical climate without a backup system. *Energies* **2020**, *13*, 6026. [CrossRef]
11. Melgar, S.G.; Bohórquez, M.Á.M.; Márquez, J.M.A. UhuMEB: Design, construction, and management methodology of minimur energy buildings in subtropical climates. *Energies* **2018**, *11*, 2745. [CrossRef]
12. Melgar, S.G.; Martínez Bohórquez, M.Á.; Andújar Márquez, J.M. Uhumebr: Energy refurbishment of existing buildings i subtropical climates to become minimum energy buildings. *Energies* **2020**, *13*, 1204. [CrossRef]
13. Soares, N.; Bastos, J.; Pereira, L.D.; Soares, A.; Amaral, A.R.; Asadi, E.; Rodrigues, E.; Lamas, F.B.; Monteiro, H.; Lopes, M.A.R.; et a A review on current advances in the energy and environmental performance of buildings towards a more sustainable bui environment. *Renew. Sustain. Energy Rev.* **2017**, *77*, 845–860. [CrossRef]
14. Hwang, R.L.; Liao, W.J.; Chen, W.A. Optimization of energy use and academic performance for educational environments i hot-humid climates. *Build. Environ.* **2022**, *222*, 109434. [CrossRef]
15. World Health Organization; Regional Office for Europe. *WHO Guidelines for Indoor Air Quality: Selected Pollutants*; World Healt Organization: Geneva, Switzerland, 2010.
16. Mannan, M.; Al-Ghamdi, S.G. Indoor air quality in buildings: A comprehensive review on the factors influencing air pollution i residential and commercial structure. *Int. J. Environ. Res. Public Health* **2021**, *18*, 3276. [CrossRef]
17. Bateson, T.F.; Schwartz, J. Children's response to air pollutants. *J. Toxicol. Environ. Health Part A Curr. Issues* **2008**, *71*, 238–24 [CrossRef] [PubMed]
18. Bogdanovica, S.; Zemitis, J.; Bogdanovics, R. The Effect of CO_2 Concentration on Children's Well-Being during the Process c Learning. *Energies* **2020**, *13*, 6099. [CrossRef]
19. Stabile, L.; Dell'Isola, M.; Russi, A.; Massimo, A.; Buonanno, G. The effect of natural ventilation strategy on indoor air quality i schools. *Sci. Total Environ.* **2017**, *595*, 894–902. [CrossRef]
20. Alonso, A.; Llanos, J.; Escandón, R.; Sendra, J.J. Effects of the COVID-19 pandemic on indoor air quality and thermal comfort c primary schools in winter in a mediterranean climate. *Sustainability* **2021**, *13*, 2699. [CrossRef]
21. Sofia, D.; Giuliano, A.; Gioiella, F.; Barletta, D.; Poletto, M. Modeling of an air quality monitoring network with high space-tim resolution. *Comput. Aided Chem. Eng.* **2018**, *43*, 193–198. [CrossRef]
22. Domínguez-Amarillo, S.; Fernández-Agüera, J.; González, M.M.; Cuerdo-Vilches, T. Overheating in schools: Factors determinin children's perceptions of overall comfort indoors. *Sustainability* **2020**, *12*, 5772. [CrossRef]
23. Morawska, L.; Milton, D.K. It Is Time to Address Airborne Transmission of Coronavirus Disease 2019 (COVID-19). *Clin. Infec Dis.* **2020**, *71*, 2311–2313. [CrossRef]
24. Peng, Z.; Jimenez, J.L. Exhaled CO_2 as a COVID-19 infection risk proxy for different indoor environments and activities. *Enviro Sci. Technol. Lett.* **2021**, *8*, 392–397. [CrossRef]
25. Ministerio de Sanidad Gobierno de España. Medidas de Prevención, Higiene y Promoción de l Salud Frente a COVID-19 Par Centros Educativos en el Curso 2020–2021. Available online: http://www.educacionyfp.gob.es/dam/jcr:7e90bfc0-502b-4f18-b 06-f414ea3cdb5c/medidas-centros-educativos-curso-20-21.pdf (accessed on 1 November 2021).
26. Chatzidiakou, L.; Archer, R.; Beale, V.; Bland, S.; Carter, H.; Castro-Faccetti, C.; Edwards, H.; Finneran, J.; Hama, S.; Jone R.L.; et al. Schools' air quality monitoring for health and education: Methods and protocols of the SAMHE initiative and projec *Dev. Built Environ.* **2023**, *16*, 100266. [CrossRef]
27. *EN 16798-1:2020*; Energy Performance of Buildings—Ventilation for Buildings—Part 1: Indoor Environmental Input Parameter for Design and Assessment of Energy Performance of Buildings Addressing Indoor Air Quality, Thermal Environment, Lightin and Acoustic. European Standard: Newark, DE, USA, 2020. Available online: https://standards.iteh.ai/catalog/standards/cen b4f68755-2204-4796-854a-56643dfcfe89/en-16798-1-2019 (accessed on 1 November 2021).
28. *ANSI/ASHRAE Standard 62.1-2019*; Ventilation for Acceptable Indoor Air Quality. American Society of Heating, Refrigerating an Air-Conditioning Engineers, Inc.: Atlanta, GA, USA, 2019.
29. RITE. Reglamento de Instalaciones Térmicas en los Edificios. *Bol. Estado* **2021**, *71*, 33748–33793. Available online: https://www.bo es/eli/es/rd/2021/03/23/178 (accessed on 1 November 2021).
30. International Well Building Institute. *The Well Building Standard V2*; International WELL Building Institute (IWBI): New York, N USA, 2021.
31. Kapsalaki, M. *ASHRAE Position Document on Indoor Carbon Dioxide*; ASHRAE: Atlanta, GA, USA, 2022.
32. Persily, A.; de Jonge, L. Carbon dioxide generation rates for building occupants. *Indoor Air* **2017**, *27*, 868–879. [CrossRef]

- Gilani, S.; O'Brien, W. Natural ventilation usability under climate change in Canada and the United States. *Build. Res. Inf.* **2021**, *49*, 367–386. [CrossRef]
- Daemei, A.B.; Limaki, A.K.; Safari, H. Opening Performance Simulation in Natural Ventilation Using Design Builder (Case Study: A Residential Home in Rasht). *Energy Procedia* **2016**, *100*, 412–422. [CrossRef]
- Jiang, Z.; Kobayashi, T.; Yamanaka, T.; Sandberg, M. A literature review of cross ventilation in buildings. *Energy Build.* **2023**, *291*, 113143. [CrossRef]
- Tecle, A.; Bitsuamlak, G.T.; Jiru, T.E. Wind-driven natural ventilation in a low-rise building: A Boundary Layer Wind Tunnel study. *Build. Environ.* **2013**, *59*, 275–289. [CrossRef]
- Pathirana, S.; Rodrigo, A.; Halwatura, R. Effect of building shape, orientation, window to wall ratios and zones on energy efficiency and thermal comfort of naturally ventilated houses in tropical climate. *Int. J. Energy Environ. Eng.* **2019**, *10*, 107–120. [CrossRef]
- Zhang, X.; Weerasuriya, A.U.; Wang, J.; Li, C.Y.; Chen, Z.; Tse, K.T.; Hang, J. Cross-ventilation of a generic building with various configurations of external and internal openings. *Build. Environ.* **2022**, *207*, 108447. [CrossRef]
- Yoon, N.; Piette, M.A.; Han, J.M.; Wu, W.; Malkawi, A. Optimization of Window Positions for Wind-Driven Natural Ventilation Performance. *Energies* **2020**, *13*, 2464. [CrossRef]
- de Dear, R.; Schiller Brager, G. The adaptive model of thermal comfort and energy conservation in the built environment. *Int. J. Biometeorol.* **2001**, *45*, 100–108. [CrossRef] [PubMed]
- *ANSI/ASHRAE Standard 55-2013*; Thermal Environmental Conditions for Human Occupancy. American Society of Heating, Refrigerating and Air-conditioning Engineers, Inc.: New York, NY, USA, 2010.
- Bienvenido-Huertas, D.; de la Hoz-Torres, M.L.; Aguilar, A.J.; Tejedor, B.; Sánchez-García, D. Holistic overview of natural ventilation and mixed mode in built environment of warm climate zones and hot seasons. *Build. Environ.* **2023**, *245*, 110942. [CrossRef]
- DesignBuilder Software Ltd. DesignBuilder v7. Available online: https://designbuilder.co.uk/download/documents (accessed on 1 November 2021).
- Berkeley, L.; Ridge, O.A.K.; Ut-battelle, M.B.Y.; For, A.; Energy, S.; Or, D.; In, T.; Form, A.N.Y.; By, O.R.; Means, A.N.Y.; et al. *Application Guide for EMS*; U.S. Department of Energy: Washington, DC, USA, 2020.
- Schulze, T.; Eicker, U. Controlled natural ventilation for energy efficient buildings. *Energy Build.* **2013**, *56*, 221–232. [CrossRef]
- Kottek, M.; Grieser, J.; Beck, C.; Rudolf, B.; Rubel, F. World map of the Köppen-Geiger climate classification updated. *Meteorol. Zeitschrift* **2006**, *15*, 259–263. [CrossRef]
- U.S. Department of Energy. Engineering Reference—EnergyPlus™ Version 22.1.0 Documentation. 2022. 1774p. Available online: https://energyplus.net/assets/nrel_custom/pdfs/pdfs_v22.1.0/EngineeringReference.pdf (accessed on 1 November 2021).
- Department of Energy's (DOE) Building Technologies Office (BTO). Weather Data Download—Sevilla 083910 (SWEC). 2021. Available online: https://energyplus.net/weather-location/europe_wmo_region_6/ESP/ESP_Sevilla.083910_SWEC (accessed on 1 November 2021).

applied sciences

MDPI

Article

Experimental Studies and Performance Characteristics Analysis of a Variable-Volume Heat Pump in a Ventilation System

Anton Frik, Juozas Bielskus, Rasa Džiugaitė-Tumėnienė and Violeta Motuzienė *

Department of Building Energetics, Vilnius Gediminas Technical University, 10230 Vilnius, Lithuania; anton.frik@vilniustech.lt (A.F.); juozas.bielskus@vilniustech.lt (J.B.); rasa.dziugaite-tumeniene@vilniustech.lt (R.D.-T.)

* Correspondence: violeta.motuziene@vilniustech.lt

Featured Application: The results of the study may be useful in exploring the extended capabi ties of controlling the operational cycle of a heat pump.

Abstract: Air-to-air heat pumps are used in today's ventilation systems increasingly often as the provide heating and cooling for buildings. The energy transformation modes of these units are subje to constant change due to the varying outdoor air state, including temperature and humidity. Whe choosing how to operate and control energy transformers, it is important to be able to adapt effectivel to the changing outside air conditions. Nowadays, modern commercial heat pumps offer two leve of control flexibility: a compressor with a variable speed and an electronic expansion valve. Th combination of control elements has boosted the seasonal energy efficiency of heat pumps. For a lon time, cycle control possibilities have been dominated by electronic controls. The authors of this pape aim to present an additional element to the traditional heat pump controls, which provides a thir level of control over the cycle. To achieve the objective, experimental investigations of a heat pum integrated into a ventilation unit have been carried out under real-life conditions. The experimen involved varying the operating modes of the unit by adjusting the compressor speed, the positio of the expansion valve, and the volume of the system loop. The study examined the performanc characteristics of the heat pump and found that the performance of a variable-volume heat pum is comparable to that of a conventionally operated typical constant-volume heat pump system. I addition, the study found that by adding a third level of volume control to the active heating circu in combination with conventional controls, the heat pump's heat output range could be extende by 69.62%. The study determined the variation of the heat pump cycle in the *p-h* diagram with th variation of the loop volume. The benefits and drawbacks of a heat pump with a variable-volum loop are discussed in this study.

Keywords: variable volume air-to-air heat pump; ventilation unit; control level; efficiency; heat pum

Citation: Frik, A.; Bielskus, J.; Džiugaitė-Tumėnienė, R.; Motuzienė, V. Experimental Studies and Performance Characteristics Analysis of a Variable-Volume Heat Pump in a Ventilation System. *Appl. Sci.* **2024**, *14*, 3933. https://doi.org/10.3390/app14093933

Academic Editors: Daniel Sánchez-García and David Bienvenido Huertas

Received: 3 April 2024
Revised: 1 May 2024
Accepted: 2 May 2024
Published: 5 May 2024

1. Introduction

While most countries are implementing policies to increase the energy efficiency c buildings, the average energy consumption per person in the building sector has remaine practically unchanged since 1990 [1]. Progress has not been rapid enough to compensate fc the growth in the floor area (3% per year) and the increasing energy demand in building Building heating, ventilation and air conditioning (HVAC) systems account for half of th energy consumed in the European Union (EU), much of which is wasted through inefficien use [2].

The heat energy demand of new buildings has significantly decreased with the impl mentation of heat transfer reduction measures under the Energy Performance of Building Directive [3]. At this point, ventilation and air conditioning become the dominant energ use systems in buildings. Additionally, the number of offices is growing both in Lithuani and around the world, leading to a rise in energy consumption. About a quarter of energy use in the buildings sector is accounted for by office buildings, where more than 70% of final energy is consumed by HVAC systems [4]. HVAC systems achieve greater energy efficiency with new monitoring and control technologies [5]. Therefore, these systems specifically require new innovative technological solutions to maintain proper indoor air quality and to use energy efficiently.

New knowledge is needed to identify opportunities for creating more advanced measures that contribute to the EU and Lithuanian strategic goals of reducing energy demand and global greenhouse gas (GHG) emissions through the integration of renewable energy technologies. The primary energy from renewable sources, such as wind energy and solar energy (photovoltaic solar cells), is typically converted into electricity. Since heat pumps require electricity to operate, they become the potential thermal energy generators in HVAC systems, capable of making a contribution to solving the problem of increasing energy efficiency in the building sector. Regarding Zhu et al. [6], the future breakthroughs for integrated renewable energy systems are the use of Machine Learning (ML) with Artificial Intelligence (AI); improved battery technologies to reduce their pollution; a combination of heat pumps and phase change energy storages (PCES); and building design for lower lifecycle emissions. While heat pumps combined with thermal and electrical storages are still under development, further improvements in the efficiency and cost-effectiveness of heat pump technologies are needed. Olympios et al. [7] investigated the performance and cost of different compressor types for small-scale heat pumps (under 30 kW). The authors indicated that rotary vane and scroll compressors resulted in higher coefficient of performance (COP) values (around 3 for rotary vane) at an air temperature of +7 °C, while reciprocating pistons were more competitive at an air temperature of −15 °C (10% higher than scroll). The cost-effective solution for residential buildings is rotary vane compressors for all locations studied. However, for the commercial buildings scroll compressors were the better choice in warm and mild weather, while reciprocating pistons performed slightly better in colder regions. Experimental studies of Wen et al. [8] highlighted the potential of variable-speed control for single screw compressors in heat pumps. By adjusting the motor speed, the authors achieved a balance between efficiency and heating capacity. The study showed that single-screw compressors perform better at relatively high motor speeds. However, the optimal speed depends on the operating conditions, particularly the evaporation temperature. Therefore, for low evaporation temperatures (colder conditions), a combination of higher speed and liquid refrigerant injection can improve the overall performance of heat pump with a single-screw compressor. Wu et al. [9] analysed the optimal selection and design of air source heat pumps (ASHP) based on new parameters, such as the average unit load rate and the comprehensive nominal heating capacity during the heating season. The study showed that a careful selection of ASHP capacities based on building load and using a system with multiple units can significantly improve the overall COP (up to 6%) and energy efficiency of central heating systems. Other research by Koopman et al. [10] provides a detailed air-to-water heat pump simulation model by incorporating an evaporator and condenser pressure drop. The study highlights the influence of ambient conditions, relative humidity, condenser capacity, and refrigerant choice on the COP and pressure drop characteristics for the optimal efficiency of the heat pump. The authors found that the ambient temperature strongly influences the COP: at higher temperatures (20 °C vs. 7 °C), the COP increases by up to 35%; at lower temperatures (−10 °C), the COP decreases. Relative humidity can improve the COP if condensation occurs (up to 10.4% gain). A higher condenser capacity reduces the COP, especially at higher temperatures. The type of the refrigerant affects the pressure drop, which has an impact on the reduction of the COP.

The seasonal energy efficiency ratio is one of the criteria to reduce the energy demand in buildings by improving HVAC systems. In this case, their integral energy performance over the entire operating period depends on both high-rated and instantaneous efficiencies and the ability to maintain them across an extensive range of energy requirements [11].

During the heating season, the prevailing factor is the average outdoor air temperatur and its constant fluctuation affects the seasonal coefficient of performance ($COP_{HP\ SEZ}$) (the heat pump. Recent research by Pineda Quijano et al. [12] shows a growing interes in developing new heat pumping technologies, which assures higher seasonal energ performance. The authors found that magnetocaloric heat pumps with MnFePSi materia and dynamic control had a high potential for efficient building heating. Considerin motor/drive system efficiency, an estimated system $COP_{HP\ SEZ}$ of 4.5 is achievable.

The integration of heat pump technologies into ventilation systems is a less widesprea solution. In recent years, the situation has been changing, with more manufacturer introducing air handling units with integrated heat pumps [13]. However, the operatin range of heat pumps integrated into ventilation units is often limited by the outdoor ai parameters. And the operating mode ensuring the required supply air temperature i not always efficient and optimal. The control measures of heat pumps have remaine unchanged for many years, with the control performed by adjusting the compressor spee and the position of the expansion valve. It is necessary to increase the number of contro components to improve the energy efficiency of integrated heat pumps.

1.1. Control Methods of Heat Pumps Integrated into Ventilation Systems

One distinctive feature of HVAC systems is their technologically defined yet constantl stochastic energy demand throughout the year, the day, and sometimes even within hour The operation modes of heat pumps need to be adjusted depending on the continuousl changing ambient temperature. Additionally, decisions must be made regarding th outdoor air temperature at which the highest nominal indicators can be achieved whil maintaining system functionality at other temperatures.

Based on the thermodynamic dependencies of fluids, under steady-state condition all of the system's state points within the cycle of the heat pump can be defined on the basi of three values of enthalpy at the outlets of the evaporator, compressor, and condense along with one pressure or temperature value (corresponding to one of those three en thalpies). To calculate heat transfer quantities, the refrigerant mass flow rate is require This degree of freedom is a hydrodynamic rather than a thermodynamic variable an cannot be captured in the *p-h* diagram of the cycle. Therefore, the heat pump cycle has total of four thermodynamic degrees of freedom. Jain and Alleyne [14] presented a cas of optimizing these degrees of freedom to reduce the amount of destroyed exergy. It ha been theoretically proven that optimization could increase the COP of the refrigeratio system by 52.5%. However, optimal operational points of the cycle can be achieved b simultaneously adjusting the system's high and low pressures.

Jensen and Skogestad [15] examined the possible degrees of freedom in the vapo compression cycle from the component perspective. They defined five controllable degree of freedom: compression work, expansion valve permeability, optimal heat transfer i two heat exchangers, and the "active charge" in the cycle. "Active charge" refers to th mass accumulated in the process equipment, mostly in two heat exchangers. Jensen an Skogestad [15] suggested controlling the amount of "active charge," which indirectl determines the "pressure level" in the cycle. However, others, when optimizing an controlling vapor compression cycle systems, essentially ignored this degree of freedon Additionally, in this work [15], three out of the five identified degrees of freedom ar constrained for technical reasons, leaving the problem of 2 degrees of freedom, where th size of condensation sub-cooling and evaporation superheating is optimized.

In practice, most control methods present the refrigerant mass flow rate and th pressure or temperature of the evaporator to the control system operation. The enthalpies o the condenser and evaporator outlets depend on heat transfer indicators and environmenta conditions. The refrigerant flow rate can be controlled by a variable-speed compressor. Th state of the evaporator is controlled by a variable-opening expansion valve. This provide two control degrees of freedom. The position of the expansion valve that ensures th required evaporator conditions must depend on a specific operating mode of the refrigeran flow rate (or compressor speed). Most control methods studied by researchers aim to simulate one or both of the abovementioned degrees of freedom, while more advanced controllers focus on manipulating these variables through optimization of the objective function, such as reducing exergy destruction, maximizing the COP, or maintaining heating or cooling loads.

1.2. Control Measures

Traditional controls for heat pump operation involve a combination of a compressor and an expansion valve. The compressor operation/capacity can be controlled in several modes: on/off, cascading compressors with different capacities and modulating operation (variable-speed compressor). In modern heat pumps, the combination of a variable speed compressor and a variable position expansion valve is becoming increasingly common. This solution has at times enabled an increase in the seasonal $COP_{HP\ SEZ}$. Continuous advancements are pursued to enhance the $COP_{HP\ SEZ}$.

Modulating (continuous) control is the most appropriate control mode for the compressor in an integrated heat pump system. As there is no thermal accumulation in such systems, the heat pump used for air conditioning must respond immediately to the constantly changing heat demand. Therefore, on/off or stepwise control heat pumps are not suitable for integration into air conditioning equipment. Moreover, as shown by the works of other researchers [16,17], variable-speed operation is more efficient compared to stepwise and on/off operations. Aprea et al. [16] carried out a study of an air-to-water heat pump operating in on/off and modulating modes. The results showed a 20% reduction in electricity consumption for the heat pump using continuous control. Reducing the compression ratio (degrees) during the modulation phase was the main source of energy efficiency [16]. The degradation of efficiency during the on/off cycle operation, under partial load conditions of the heat pump, occurs due to the start-up losses, which depend on the device's characteristics and operating time [18]. Dongellini et al. [17] showed that the seasonal energy efficiency of a heat pump operating in a continuous control mode was 8.3% and 15.4% higher compared to pumps operating in step and on/off modes. In addition, modulating control minimises the number of compressor on/off cycles. This has a positive effect on the compressor's life.

1.3. Control Algorithms

Control algorithms for the heat pump cycle can be categorized into conventional, advanced, and smart. Conventional control algorithms are the most widely used type of control algorithm for heat pumps. They rely on fixed set points and simple on/off control logic, making them easy to implement and maintain. However, they may be less efficient and unable to adapt to changing conditions.

Advanced control algorithms are more complex but offer a wider range of features and functionalities. They can consider more variables, such as outdoor and indoor temperature and system operating conditions. They can also use more sophisticated control techniques, such as fuzzy model predictive control (MPC). Advanced control algorithms can improve the efficiency and performance of heat pumps, but they may be more expensive and complex to implement. Chae et al. [19] compared the performance of on/off control and proportional-integral (PI) control for air-to-air heat pumps. The results show that PI control can improve the system's efficiency and stability compared to on/off control. Wang et al. [20] implemented MPC on an air-to-air heat pump and demonstrated its ability to optimize energy consumption while maintaining comfort levels. Hlanze et al. [21] presented a model-based predictive control strategy to optimise the performance of phase change material (PCM) ceiling panels coupled to a multi-stage air-to-air heat pump. Putrayudha et al. [22] applied fuzzy logic control to a ground source heat pump system and showed its effectiveness in improving system performance and user comfort. Yang and Ge [23] presented the effectiveness of a PLC control system and optimized operation strategies in improving a desiccant-coated heat exchanger-based heat pump fresh air system's efficiency

(COP) and dehumidification rate. The study primarily used PID control methods. A PL control system was developed to manage the system's components. Optimized operatio strategies were established for the electronic expansion valve, compressor frequency an switchover period. The results showed an increase in COP of 25% in summer and 78% i winter compared to a non-optimised system.

Artificial intelligence (AI) control algorithms are the most advanced type of contr algorithm for heat pumps. They can learn from data and adapt to changing conditions i real-time. They can also integrate with other smart home devices and systems to optimiz energy use and comfort. AI control algorithms have the potential to significantly improv the efficiency, performance, and convenience of heat pumps; however, they are still i their early stages of development. Rohrer et al. [24] investigated using deep reinforcemer learning (Deep RL) for optimal control of air-source heat pumps, highlighting its potenti for significant energy savings.

1.4. Experiments on the Improvement of the Vapor Compression Cycle

The vapor compression cycle has thermodynamic losses in comparison to the idea reverse Carnot cycle. The compression of a single-phase gas and isenthalpic expansion ar related to these losses [25]. The first set of losses results in high refrigerant temperature high compression work, and significant heat release in the condenser. When considerin the vapour compression cycle for refrigeration purposes, these losses are seen as a negativ factor. However, when the cycle is used for heating, some of these drawbacks can be turne into advantages, converting energy losses into useful energy. Additionally, in cold climat zones where air-source heat pumps operate, a high compression ratio is necessary, as i determines the outdoor temperature at which the system can function. The second set c losses leads to significant throttling losses and low refrigeration capacity.

In examining the advancements in vapour compression cycle technologies aimed a reducing thermodynamic losses and improving efficiency/effectiveness, three directions c cycle improvement are distinguished: sub-cooling cycles, expansion loss recovery cycle and multi-stage cycles.

Research into sub-cooling cycles focuses on the internal heat exchanger of the suctio line [26,27], the thermoelectric cooler [28,29], and the mechanical secondary cooler [30,31 Expansion loss recovery cycles mainly concentrate on expanders (essentially, turbin use) [32,33] and ejectors [34,35] used in vapour compression cycles. Multi-stage cycl studies include vapour or liquid-phase refrigerant injection cycles [36], two-phase refrige ant injection cycles [37], and saturation cycles [38].

Furthermore, the heat pump cycle is examined from the perspective of the discharg temperature of the refrigerant. In one study [39], three cycle improvements were modelle to reduce the compressor discharge temperature. The impact of two-phase refrigeran pumping, liquid-phase injection, and two-phase injection on the compressor discharg temperature was analyzed at fixed condensation and evaporation temperatures. The re search shows that all three methods have the potential to reduce the discharge temperatur and thus increase the adiabatic efficiency of the compression process. However, two-phas injection outperforms liquid injection and two-phase suction for both heating and coolin efficiency by 4.8% and 11.8%, respectively.

Some researchers aim to improve the efficiency of the vapor compression cycle no through individual cycle measures (directly in contact with the refrigerant) but by alterin the parameters of the secondary fluid. In their work [40], the authors presented a heat pum operating in a refrigeration mode with air, pre-cooled by an indirect water vapour coolin condenser, passing through it. Thermodynamically, this solution reduces the refrigerant ga condensation temperature, consequently decreasing the work performed by the compresso In hot and dry climates, this method can reduce cooling energy consumption by abou 20%, but it requires a significant amount of water for the water vapour cooling condense to function.

Efficiency is sought through a combination of maximizing load selection with modulation and on/off control systems [41], where the use of storage tanks reduces the number of compressor on/off cycles.

An extractable air heat pump or extractable air heat recovery system is essentially an air-to-air heat pump. In cold climate zones, such traditional systems face many problems as the outdoor air temperature drops, including a high-pressure ratio, degradation of the mechanical lubrication properties in the compressor, a decrease in the refrigerant mass flow rate, and a decrease in the heating capacity [42,43].

The mass flow rate of the working fluid in the heat pump is a crucial aspect of the ventilation heat recovery system, affecting its efficiency. As the ratio of condensation and evaporation pressures increases during heat pump operation, the mass flow rate of the refrigerant decreases [44]. This problem is addressed in several ways. One option involves refrigerant vapour injection cycles. Experimental studies [44] show that using such a cycle helps maintain a nearly constant flow rate of the refrigerant passing through the condenser as the outdoor air temperature decreases.

Another option is to use multiple lower-capacity heat pump combinations. In this case, air passes through individual heat exchangers one after another, gradually changing its parameters. This allows for the reduction of the pressure ratio of each heat pump in-volved in the process. This method prevents excessive reduction of the mass flow rate of the working fluid. Additionally, this approach enables exercising control over the heating capacity. Moreover, the problem of a high-pressure ratio can be alleviated by using an ex-tractable air heat pump, as the air passing through the integrated heat pump evaporator in the ventilation system has a higher (indoor) temperature compared to the (outdoor) temperature of the air passing through the condenser. From the perspective of the vapour compression cycle, this means that the refrigerant's evaporation temperature can be higher, and the condensation temperature can be lower compared to a traditional heat pump, where the evaporator is located outdoors, and the condenser indoors. The isotherms of the vapour compression cycle approach each other, eliminating the high compression degree in the cycle, thereby reducing the work performed by the compressor.

Several studies are made to increase the performance of the heat pump compressor with programming. Lee et al. [45] showed a variable-speed control optimization model of the heat pump. Machine learning (ML) was used to predict the thermal load and outdoor temperature, and mixed-integer programming for compressor control optimization. This approach allowed proposing optimal schedules for a heat pump operation in a selected location, reducing its electricity consumption by 9% [45].

Clauß and Georges [46] studied the effect of the complexity and flexibility of the predictive control model of the heat pump on the energy demand of the building. Genetic algorithms (GA) are used to optimise the $COP_{HP\ SEZ}$ by combining the maximum heat pump load, compressor displacement and system operating parameters [47]. However, even in these recent studies, the focus is not on problem-solving by evaluating the possibilities of thermodynamic cycle control, which is the subject of a different set of studies [48]; however, these are specific and not directly related to the characteristics of HVAC systems.

As can be seen from this literature review, the majority of work on improving heat pump efficiency has focused on traditional, well-known and widely used heat pump control measures and their optimal operation strategies. A little attention is given to exploring the thermodynamic cycle of the heat pump and seeking new technological measures to modify and control it. This potential direction could improve the performance indicators of the heat pump integrated into ventilation systems.

The main benefit and novelty of this study is the experimentally proven potential of variable loop volume control technology to improve the performance of a variable-volume heat pump integrated into a ventilation system. The main findings of this study indicate that the control based on volume variation is the most sensitive to the heat output parameters of the heat pump compared to the control based on the compressor or expansion valve.

The variable-volume control technology under investigation may have the potentia
to extend the operating ranges of integrated heat pumps in the future, which woul
lead to energy savings in buildings. The goal of the article is to reveal and explore th
thermodynamic possibilities of optimally controlling the heat pump cycle integrated into
variable heat demand ventilation system. The research object is the air-to-air heat pum
integrated into an air handling unit.

2. Materials and Methods

The methodology section discusses the research object and tools.

2.1. Subject of the Study

This paper deals with an experimental investigation of an air handling unit (AHU
with an integrated air-to-air heat pump (HP). The theoretical background is also discusse
The schematic diagram of the air-to-air heat pump integrated into a ventilation system i
shown in Figure 1. The main components of the experimental bench can be divided int
two parts: AHU and HP. The AHU part consists of air ducts with the heat exchanger
(condenser and evaporator) and fans of fresh air supply (F_s) and air exhaust (F_e). The H
part consists of the typical main components as the compressor (CM), the evaporator (EV
the condenser (CN) and the expansion valve (EEV).

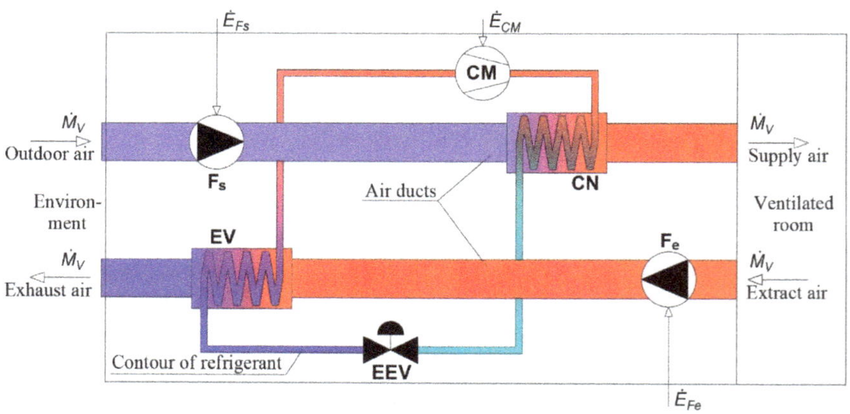

Figure 1. The schematic diagram of a ventilation unit with an integrated heat pump.

The theoretical studies were carried out to determine how the evaporation and co
densation temperatures vary with the heat pump's heating power (Figure 2). To that en
thermodynamic and parametric analysis were used. The results show the dependencies c
the heat pump parameters and their control guidelines.

Figure 2 shows that as the heating power of the HP increases, so does the conder
sation temperature ($T_{izo\ CN}$), while the evaporation temperature ($T_{izo\ EV}$) decreases, i.e
the isotherms move away from each other as the heating power increases. The increasin
temperature difference between the isotherms reflects the increasing pressure differenc
between the refrigerant in the evaporator and the condenser. The compressor must develo
a higher compression ratio. The horizontal dashed line (T_R) shows the air temperatur
supplied to the room as the heat output varies. In this case, the supply air temperatur
is constant.

In order to obtain the maximum efficiency of the heat pump, it is necessary to var
its parameters as shown in Figure 2. Therefore, in order to determine the possibilities o
controlling the heat pump in this mode, an experimental test bench was installed with a
additional extension of the control possibilities, i.e., volume variation.

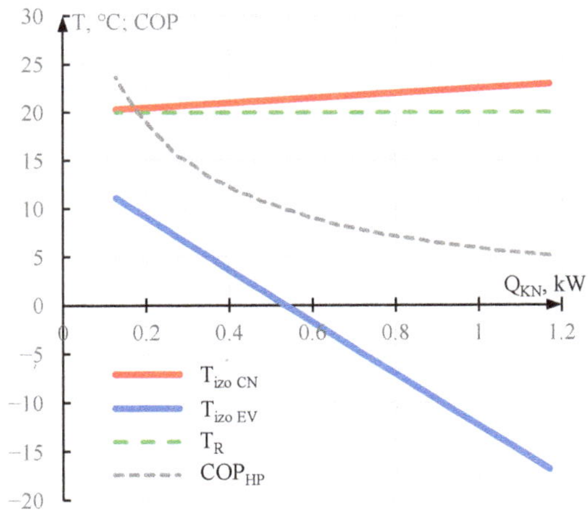

Figure 2. The dependence of condensation, evaporation temperatures and the COP on the heating power of HP [49].

2.2. Experimental Part

During the study, an experimental set-up was built where, in addition to standard means (compressor speed control, expansion valve position change), the power, condenser and evaporator parameters are also controlled with an introduction of volume change. The schematic diagram of the experimental set-up is shown in Figure 3.

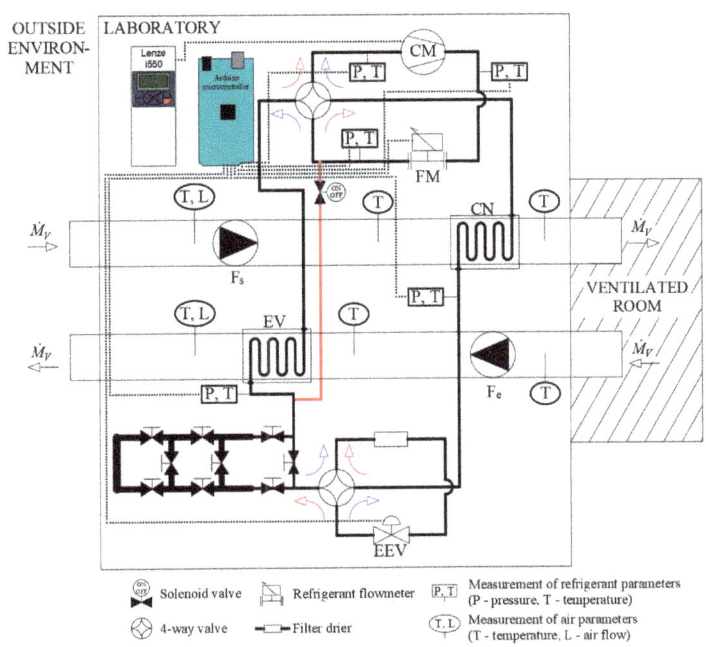

Figure 3. The schematic diagram of the experimental test bench with its main components.

Figure 3 shows that the main objective of the operation of the AHU is to provide the required indoor air quality. The purpose of an integrated HP is to recover heat and to preheat the outside air. Table 1 shows the technical characteristics of the main components of the experimental test bench.

Table 1. The technical characteristics of the main components of the experimental test bench.

The Component of the Experimental Bench	Marking in the Diagram	Technical Characteristics
Supply and exhaust fan RS-315L EC	F_s and F_e	Air flow rate: 430 m^3/h
EC-10 Fan motors potentiometer	-	Control signal: 0–10 V
Heat pump compressor GMCC (Foshan, China) YA281X3CS-4MT	CM	Refrigerant: R410A; voltage: 380–415 V; heating capacity: 6.92 kW
Frequency inverter Lenze i550	-	Voltage: 380–415 V
Expansion valve Carel (Breganze, Italy) E2V14 BSF01	EVV	Maximum differential pressure: 35 bars; full closing steps: 480.
Condenser	CN	DX coil heat exchange surface: 4.5 m^2; air flow rate: 400 m^3/h, pressure drop: 25 Pa
Evaporator	EV	DX coil heat exchange surface: 5.9 m^2; air flow rate: 400 m^3/h, pressure drop: 18 Pa
Microcontroller board Arduino Mega 2560	-	Operating voltage: 5 V; digital I/O: 54; analog input: 16; flash: 256 KB; SRAM: 8 KB.

The test bench is equipped with two four-way valves to allow the HP to be use for cooling. A hot steam line is installed to defrost the evaporator when it freezes. Thi paper presents a study addressing only the heating season. The speed of the compressor i changed using a Lenze (Hamelin, Germany) i550 frequency converter. The expansion valv position is changed using an Arduino (Ivrea, Italy) Mega 2560 microcontroller board.

This experimental bench has the advantage of being equipped with a multi-stag volumetric enlarger, which allows the system volume to be increased. The multi-stag volumetric enlarger is located between the evaporator (EV) and the electronic expansio valve (EEV). A schematic diagram of the multi-stage volumetric enlarger is shown i Figure 4.

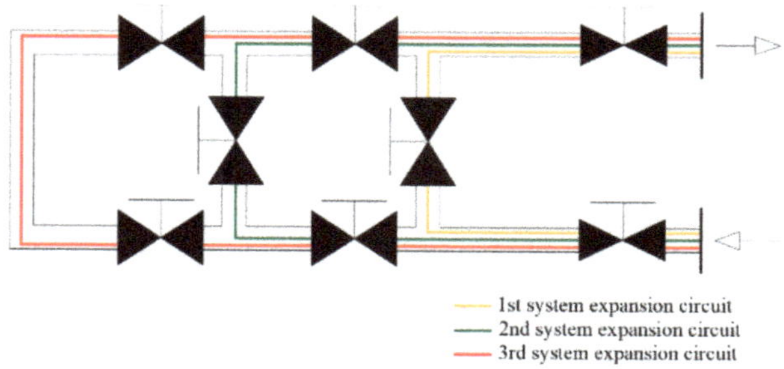

Figure 4. A multi-stage volumetric enlarger for increasing the volume of a heat pump.

The multi-stage volumetric enlarger consists of a number of shut-off ball valves an piping. The combination of these components allows the internal volume of the hea pump to be gradually increased by 17.0% (+0.81 dm^3) (V1), 26.4% (+1.26 dm^3) (V2), 34.4% (+1.64 dm^3) (V3) in relation to the initial volume (V0 = 4.76 dm^3). A test cycle is performe with the volume expansion. After the cycle, the volume is increased and the test is repeate

Figure 5 presents the images of the experimental test bench.

(a)

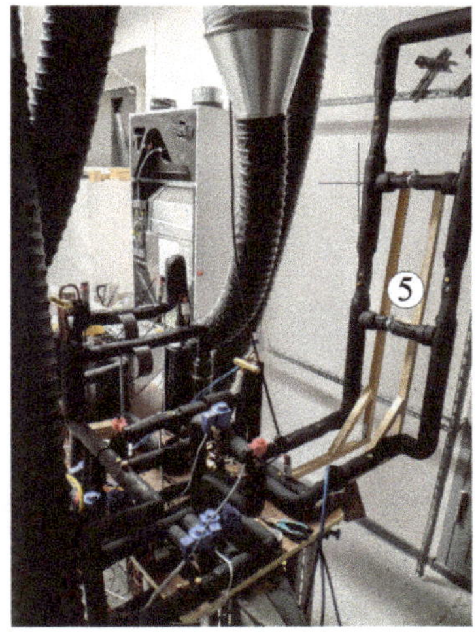

(b)

Figure 5. Digital photos of the experimental bench: (**a**) the experimental bench without thermal isolation and volume expansion; (**b**) the experimental bench with thermal isolation and a volume expansion unit.

Figure 5 shows digital photos of the experimental test bench where part (a) shows the experimental bench at the production stage, when there is no thermal insulation or volume expansion installed yet, and part (b) shows the bench, which is insulated and equipped with volume expansion and additional components, such as flexible ducts that supply and extract air. In addition, the main components are presented and marked in the pictures: (1) the compressor; (2) the evaporator; (3) the expansion valve; (4) the condenser; and (5) the volume expansion unit.

The following section discusses the measurement tool used to record the performance of the device.

2.3. Measuring Equipment

This study used a large number and variety of measuring equipment. During the study, a large number of values were recorded and an energy balance was constructed to determine whether the measurements were accurate. Below is a list of the measurement equipment used (Table 2).

Table 2. Used measurement equipment.

Name of Measuring Equipment	Properties	Notes
AHLBORN (Goettingen, Germany) ALMEMO 2890-9 measuring instrument and data logger	Measuring inputs: 9; Outputs: 2; Memory: 100,000 measured values	Used to record differential air pressure values.
Pressure measuring connector for differential pressure FDA 602 S1K	Measuring range: ±1250 Pa; Operating range: −10 to +60 °C, 10 to 90% RH; Accuracy: ±0.5% of the final value in the range of 0 to the final positive value	Measures the pressure difference across the airflow measurement diaphragm.
160 mm diameter airflow calibrated diaphragm	Measuring range: nominal air flow 400 m^3/h; Difference pressure: 400 Pa at nominal air flow; Accuracy: ±0.66% at 400 m^3/h	The length of the straight section upstream of the measurement point is 12 diameters; downstream, 6 diameters.

Table 2. *Cont.*

Name of Measuring Equipment	Properties	Notes
Multi-channel data logger Onset (Burlington, MA, USA) Hobo H22-001	Measuring inputs: 3 FlexSmart multi-channel modules and up to 6 Smart Sensors; Outputs: 2; Operating range: −40 to +60 °C; Memory: 512K nonvolatile flash data storage.	Used to record temperatures and electricity parameters
12-Bit temperature smart sensor	Measuring range: −40 to 100 °C; Accuracy: ±0.2 °C from 0° to 50 °C.	Measures air and freon temperature. Connects to Hobo H22-001
3-Phase AC kWh transducer sensor onset T-VER-8044-100	Measuring range: 0 to 100 A; Operating range: 0 to +60 °C, 0 to 95% RH; Input primary voltage: 480 Volts AC rms; Output: 4–20 mA; Accuracy: ±1% per ANSI (from 10 to 100% of CT rating).	Connects to Hobo H22-001
Pressure sensors Sanhua (Shaoxing, China) YCQC05L09	Measuring range: 0 bar to 44.8 bar; Operating range: −40 to +80 °C, max. working pressure 75 bar; Output: 4–20 mA; Accuracy: ±0.8% FS.	Used with Arduino Mega 2560

The following parameters were recorded during the experimental studies:

- refrigerant pressures, temperatures and flow rates;
- airflow (determined by using the calibrated diaphragms), temperature and relative humidity.

The tests were carried out under realistic climatic conditions and measurements were taken at 10 s intervals. Figure 3 shows the locations of the installed sensors and the recorded parameters.

An experiment is conducted to determine how the operating parameters of the heat pump change when it is controlled by three control measures: compressor, expansion valve, and circuit volume. Throughout the study, the heat pump operates under different combinations of these three control components: the compressor speed is varied between 1800, 1920, 2100, and 2220 rpm; the expansion valve permeability is varied between 30%, 50%, and 70%; the system volume is varied between V0 and V4. All combinations of control component parameters throughout the experiment are presented in Table 3.

Table 3. Combinations of heat pump control component parameters.

Volume	Speed of Compressor (rpm); Opening Degree of Expansion Valve (%)											
V0	1800; 30	1920; 30	2100; 30	2220; 30	1800; 50	1920; 50	2100; 50	2220; 50	1800; 70	1920; 70	2100; 70	2220; 70
V1	1800; 30	1920; 30	2100; 30	2220; 30	1800; 50	1920; 50	2100; 50	2220; 50	1800; 70	1920; 70	2100; 70	2220; 70
V2	1800; 30	1920; 30	2100; 30	2220; 30	1800; 50	1920; 50	2100; 50	2220; 50	1800; 70	1920; 70	2100; 70	2220; 70
V3	1800; 30	1920; 30	2100; 30	2220; 30	1800; 50	1920; 50	2100; 50	2220; 50	1800; 70	1920; 70	2100; 70	2220; 70

The experiment commences with the smallest system volume, compressor speed, and expansion valve permeability (V0, 1800 rpm, 30%), during which the operating parameters of the heat pump (refrigerant pressures, temperatures) are observed to remain constant for 10 min, indicating a steady-state process. Subsequently, the compressor speed is increased to 1920 rpm, and again, the process is allowed to stabilize. This approach is applied to all combinations of certain expansion valve and compressor speed settings. Once all compressor speeds have been examined, the expansion valve permeability is increased, and the compressor speed is varied again from 1800 rpm to 2220 rpm.

Meanwhile, when all combinations of compressor speed and expansion valve permeability have been explored, the system volume is increased from V0 to V1, and tests are repeated, varying the speed and the permeability. This process yields all operational results for the heat pump (48 data points). Parameters for further analysis of each operating combination are based on the average of the parameters measured during a steady-state period of 10 min.

2.4. Determining the Actual Performance Parameters of the HP from a Theoretical Model

The experimental study generates a large amount of data, which is processed and structured in the results section to produce the final result.

In this work, the compressor is a major focus, so its compression ratio is used in the data analysis, which determines its performance and efficiency. The compressor compression ratio is calculated as follows:

$$Compression\ Ratio = \frac{P_1}{P_2}, \tag{1}$$

where P_1 is the inlet pressure and P_2 is the outlet pressure.

A heat pump has more components than just the compressor. Therefore, the heat pump performance is characterized by other quantities, such as the amount of heat (or power) produced, the amount of electricity consumed and its COP.

The heating capacity provided by the heat pump is evaluated as follows:

$$\dot{Q} = \dot{M} \cdot c_p \cdot (T_{out} - T_{in}) = \dot{M} \cdot (h_{out} - h_{in}), kW \tag{2}$$

where M is the air mass flow rate, kg/s; c_p is the fluid specific heat, kJ/(kg·K); T_{out} and T_{in} are the end and start temperatures of the process, K; and h_{out} and h_{in} are the enthalpies of the process at the end and at the start of the process, kJ/kg.

The above formula is suitable for determining the heat quantities and capacities for different types of working material, i.e., both air and refrigerant.

The overall COP of the heat pump is determined as follows:

$$\text{COP} = \frac{\dot{Q}}{\dot{E}} \tag{3}$$

where Q is the quantity of heat produced by the heat pump (supplied to the room), kW (or kWh); E, the amount of electricity withdrawn to produce the amount of heat, in kW (or kWh). The electricity consumed by the installation shall be recorded in kWh according to the power factor.

3. Results and Discussion

In this study, a ventilation system with an integrated heat pump was tested. The ventilation unit does not use a heat recovery heat exchanger. The heat recovery function is performed by the heat pump. The main purpose of the heat pump is to extract heat from the exhaust air and to warm the cold air supplied from the outside. All tests are carried out under real-life climatic conditions and the control of the heat pump has been extended with additional functions.

A volume expansion device is implemented in the heat pump circuit, which allows the initial volume of the refrigerant circulating circuit to increase by 17.0% (+0.81 dm^3) (V1), 26.4% (+1.26 dm^3) (V2), and 34.4% (+1.64 dm^3) (V3) in comparison with the initial system volume V0 (4.76 dm^3). The tests are carried out at different speeds of the HP compressor and at different expansion valve capacities. The compressor speeds are 1800, 1920, 210 and 2200 rpm. The expansion valve flow positions are 30%, 50%, and 70%. The heat pum circuit uses 1.13 kg of R410A refrigerant.

3.1. General Characteristics

In this subsection, the typical performance of an experimental heat pump is present-e The purpose is to determine how the heat pump performs when controlled by standar means in combination with the system volume change and to make a comparison with th characteristics of several compressors provided by the manufacturers.

3.1.1. The Degree of Compression and Flow Rate

One of the typical characteristics of heat pump compressors is the dependence of th degree of compression (the ratio of the refrigerant pressure upstream and downstream the compressor) on the mass flow rate of the refrigerant. Figure 6 shows the operatin points obtained during the experiment and Figure 7 shows the operating points of differer compressors as declared by the equipment manufacturers.

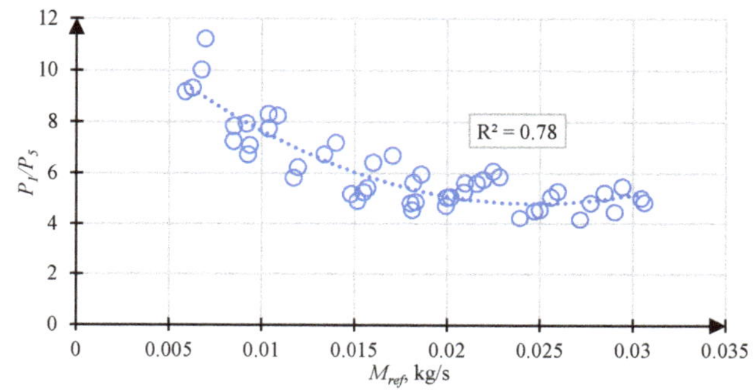

Figure 6. The relationship between the compressor pressure ratio P_1/P_5 and the flow rate (exper mental data).

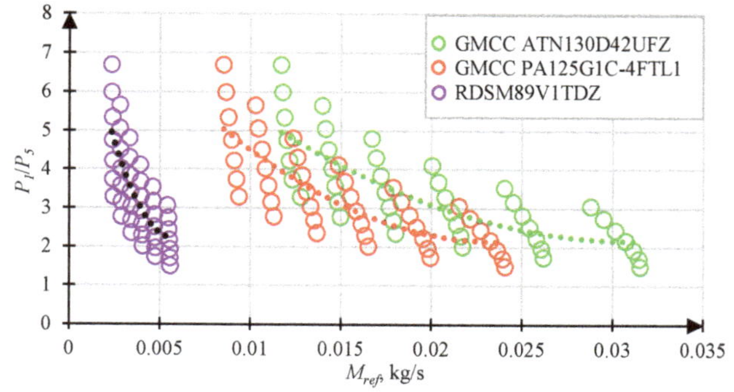

Figure 7. The relationship between the compressor pressure ratio and the flow rate (manufacturer data).

As mentioned above, Figure 6 shows, in circles, the points of dependence of th refrigerant compression ratio on the flow rate for heat pump compressors from differen manufacturers, and the dotted line shows the regression equation for these compresso specific points, which is generated by a second-degree polynomial.

It can be seen that the trend of the compression ratio reported by the compresso manufacturers (Figure 7) is in line with the values obtained during the experiment (Figure 6

Observably, as the refrigerant flow rate increases, the degree of compression generated decreases. A similar trend was observed during the experiment.

3.1.2. Flow Rate and Evaporation Temperature

Another characteristic of heat pumps is the dependence of the refrigerant flow rate on the evaporation temperature. Figure 8 shows this characteristic for an experimental heat pump obtained during testing. It can be seen that a higher evaporation temperature leads to a higher flow rate through the system.

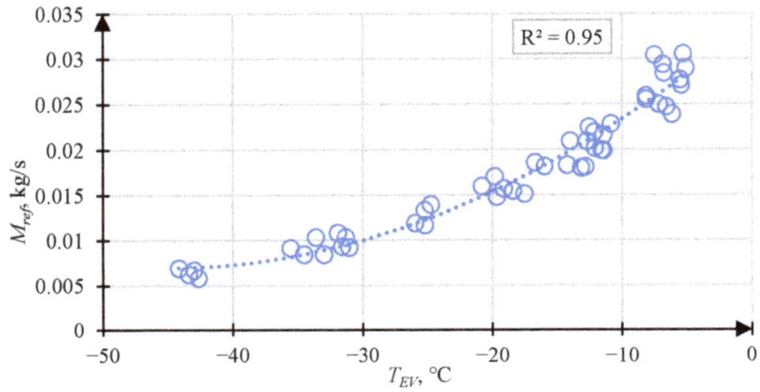

Figure 8. The relationship between the refrigerant flow rate and the evaporation temperature (experimental data).

Figure 8 shows that as the evaporation temperature increases, so does the flow rate. These points are joined by a regression line, which is a second-degree polynomial and the resulting regression rate is greater than 95%. This value indicates that the regression equation reproduces the measurement points with sufficient accuracy to allow the equation to be used in place of the measurements in further studies. Figure 9 shows the curves provided by the compressor manufacturers.

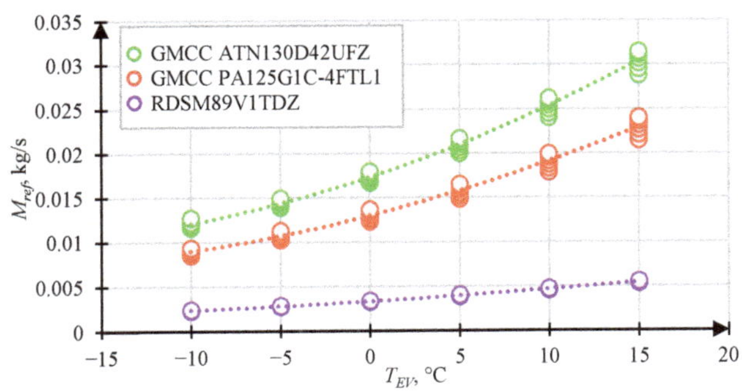

Figure 9. The relationship between the refrigerant flow rate and the evaporation temperature (manufacturers' data).

Figure 9 shows a trend similar to the experimental results. This similarity indicates that the results of the experimental studies are sufficiently accurate. The power and flow rate dependencies are then presented in the next subsection.

3.1.3. Power and Flow Rate

Another characteristic parameter of a heat pump is the dependence of the heating power on the mass flow rate. Figures 10 and 11 present these characteristics obtained from experimental tests and data from different manufacturers, respectively.

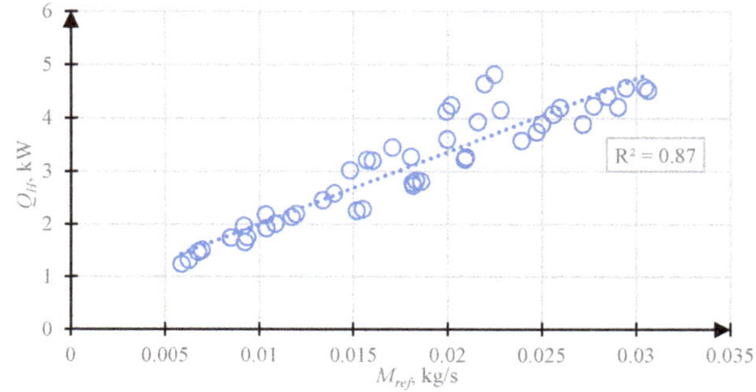

Figure 10. Heating power vs. flow rate (experimental data).

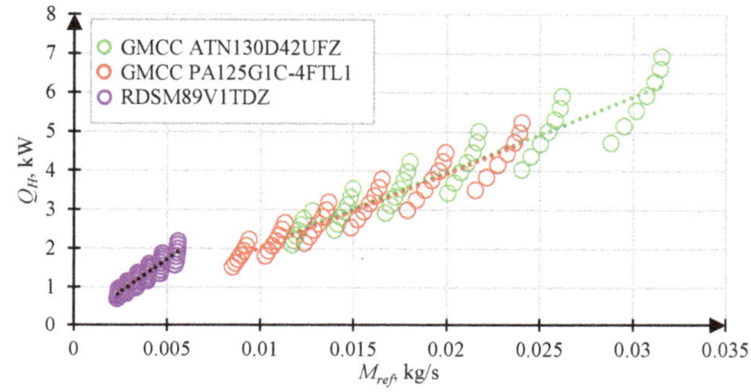

Figure 11. Heating power vs. flow rate (manufacturers' data).

Figure 10 shows that as the flow rate increases, so does the power of the heat pum and when the points are connected by a regression line, a regression coefficient of ove 87% is obtained. This coefficient shows that the performance can be calculated wit sufficiently high accuracy. The figures below show the performance data from compressc manufacturers.

From Figure 11, it can be seen that the compressor performance characteristics ar similar to Figure 10, which again shows that the experimental studies repeated by th authors are logical and can be trusted.

3.2. *The Effect on the Performance Parameters of a Change in the Heat Pump Circuit Volume (V V1, V2, V3)*

This section presents the results of experimental tests where the volume of the syster is changed but the mass of the refrigerant is kept constant. The refrigerant pressure i 11 bar when the heat pump system is not operating. The tests start from the smallest syster volume. The test results are presented starting with the variation of the heating capacity c the heat pump at different system loop volumes.

3.2.1. Heating Power

One of the most common descriptors of how heat pumps work is the heating capacit In this case, the figure below shows how the heating power varies at different compressc speeds, throttle positions, and volumes of the HP system. The heating power was evaluate by measuring the air flow rate and the temperatures before and after the condenser.

Figure 12 shows that by adjusting the heat pump operation using conventional meth ods in the baseline case with a loop volume of V0, a range of supplied air heating capacit from 3.89 kW to 4.82 kW is achieved with a change of 0.93 kW. Increasing the volume of th HP loop by 17.0% (from V0 to V1), the heating power at the lowest speed and low flow rat is 3.02 kW. Increasing the speed and the flow rate of the expansion valve, the heat output gradually increases to 4.21 kW. There has been an increase in power, but fluctuations have also been observed, potentially due to the different speeds and flow rates. It can be seen that the power range (compared to the V0 case) has widened by 27.66% (from 0.93 to 1.18 kW), but the overall limits of the achieved power control range have decreased, compared to the baseline, from 3.02 kW to 4.21 kW. The trend remained unchanged with further increases in system volume. For V2, the loop volume increased by 26.4% to 6.01 dm^3, and for V3 by 34.4% to 6.40 dm^3 compared to the baseline (case V0). For case V2, the control range of the heating power increased by 65.15% (from 0.93 to 1.53 kW), and for case V3 by 69.62% (from 0.93 to 1.57 kW) compared to the baseline case. On the other hand, the overall limits of the achieved power control range decreased by 1.74 to 3.27 kW and by 1.25 to 2.82 kW for V2 and V3, respectively.

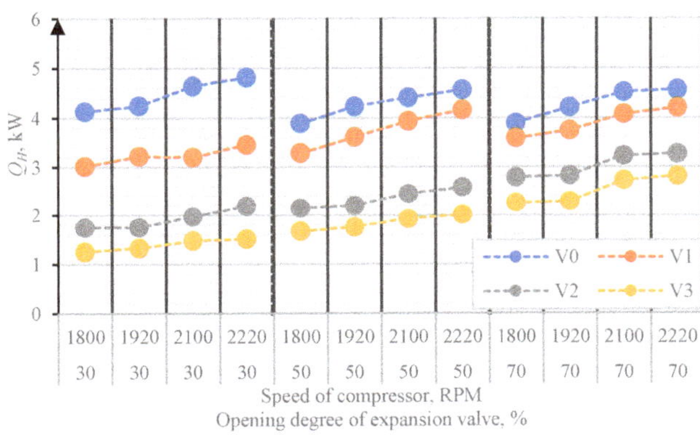

Figure 12. The heating power of the HP during the experiment.

It is noted that the use of additional controls by varying the volume of the heating circuit, in combination with conventional control measures, has allowed the heating power range to be extended, i.e., the difference between the maximum and the minimum power is achieved by varying the speed of rotation and the throttle capacity. This increase in the power range is as much as 0.7 times compared to the V0 case. It can be concluded that the volume expansion reduces the maximum heating power of the unit but allows it to be controlled over a wider range. A wider control range may be useful in cold climates during transition periods when a low HP output is required, which means that buffer capacity can be avoided. A buffer capacity is built in when the selected high-capacity HP cannot operate at low capacities, while the expansion of the volume reduces the power and increases the control range. This paper does not analyse the heat pump performance in the long term or in transient periods: this will be investigated in the next steps.

3.2.2. Efficiency

Adjusting the volume of the HP loop has allowed extending the range of the available heating power, but the general trend is that the COP decreases with the HP loop volume going up. The average COP for each volume increase dropped by 5.14% (V1), 18.03% (V2) and 26.53% (V3) compared to the base case V0 (Figure 13).

Figure 13 shows that in most cases, increasing the bandwidth and having a low speed slightly increases the efficiency for V0, V2 and V3. For V2 and V3, the highest efficiency is achieved with this style of operation.

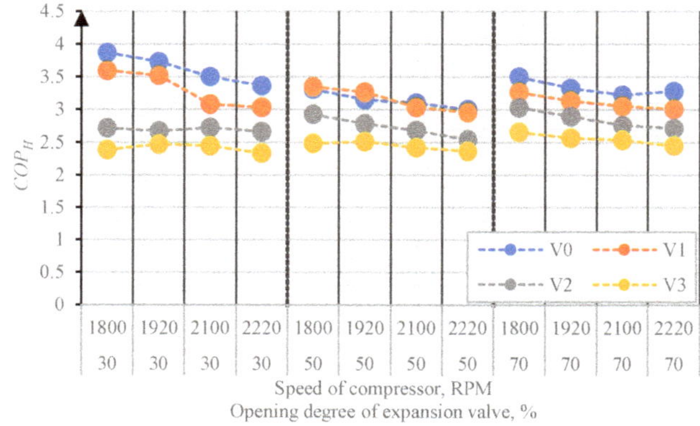

Figure 13. The COP during the experiment.

However, it has been observed that the variation in performance is also dependent o the outside air temperature, which shapes the fluctuations. The variation of the outdoc temperature has a negligible effect on the heating power achieved by the heat pump, with regression coefficient of 0.36 (Figure 14a), but the performance coefficient has been observe to be more sensitive to the outdoor air parameters, with a regression coefficient of 0.4 (Figure 14b).

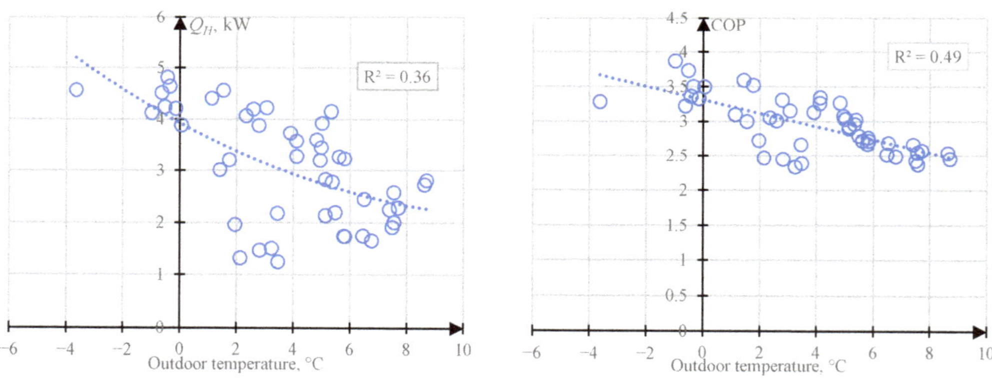

(**a**) (**b**)

Figure 14. The dependence of the heat pump parameters on the outside air temperature: (**a**) heatin power vs. outdoor air temperature; (**b**) COP vs. the outdoor air temperature.

Additionally, one general trend has been noted, where the efficiency of the heat pum increases with the outdoor air temperature decreasing. This is not a typical trend for hea pumps, and it may be due to the installed compressor power, which is higher to ensure higher efficiency even when the outdoor air temperature is very low.

From the above results, it can be seen that increasing the loop volume enables th heat pump to operate at a lower heating capacity than in the initial case (V0). This mean that the change of the heat pump circuit volume has resulted in a wider range of heatin power. This allows the unit to operate with a decent amount of stability over a wider rang of outdoor air temperatures. However, there is a downside to increasing the volume o the heat pump circuit, as it decreases the efficiency of the HP; however, the effect of th third stage on the seasonal efficiency has not been investigated. The evaluation of th seasonal performance factor will be carried out in future work where its performance in real system with real heat demands will be simulated.

With a strong increase in volume (cases V2 and V3), a phenomenon that has similarities with cases where there is partial leakage of the refrigerant from the heat pump system is observed [50,51]. It is therefore appropriate in future studies to add more refrigerant to the system than recommended and subsequently have a significantly higher performance potential.

3.3. Heat Pump Cycles in a p-h Diagram

The previous part of the results showed how increasing the system volume affects the output and the performance of the plant. This subsection shows the characteristic changes in the heat pump operation cycle in a *p-h* diagram when the plant is operated in the above-mentioned fashion. The graphs below are based on the experimental results.

Figure 15 shows a typical case of compressor control, with constant expansion valve capacity at 70%, the compressor speed varying between 1800 and 2220 (intermediate cycles of 1900 and 2100 are not shown), and the heat pump loop volume of V0. It can be seen that the variation of the compressor speed has a small effect on the differences between the superheating and sub-cooling temperatures of the cycle, which remain similar. In addition, increasing the compressor speed increased the condensation temperature and decreased the evaporation temperature. This trend comes into light because increasing the compressor speed results in a higher rate of refrigerant flow through the compressor, which also results in a higher flow rate through the expansion valve, and a lower pressure is generated behind the expansion valve than at speed. It can therefore be argued that the change in the compressor speed has a direct influence on the operating mode of the evaporator. Similar trends can be seen in other throttle cases as well.

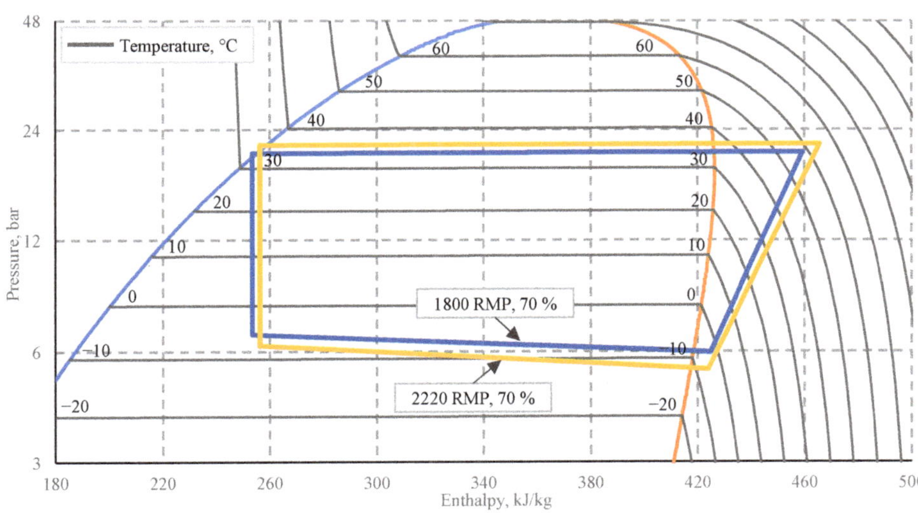

Figure 15. The operation cycles of the HP in a *p-h* diagram (compressor speeds: 1800 and 2220 rpm; expansion valve capacity: 70%; volume: V0).

Figure 16 shows a case where the heat pump is controlled by an expansion valve, its flow rate varying between 30% and 70%, with a fixed compressor speed and a fixed volume of the HP loop at V0. The change of the expansion valve flow rate had a negligible effect on the condensing temperature (the difference between the limiting cases is 2.75 °C), but a more significant effect on the evaporating temperature of 6.83 °C between the valve flow rate cases of 30% and 70%. In addition, controlling the heat pump with the expansion valve was able to affect the superheating and sub-cooling temperatures of the cycle. Reducing the valve permeability from 70% to 30% resulted in an increase in the differences between the superheating and sub-cooling temperatures of 19.33 °C and 6.87 °C, respectively. Therefore, the position of the expansion valve has a significantly greater impact on the operating mode of the evaporator than the case of changing the compressor mode.

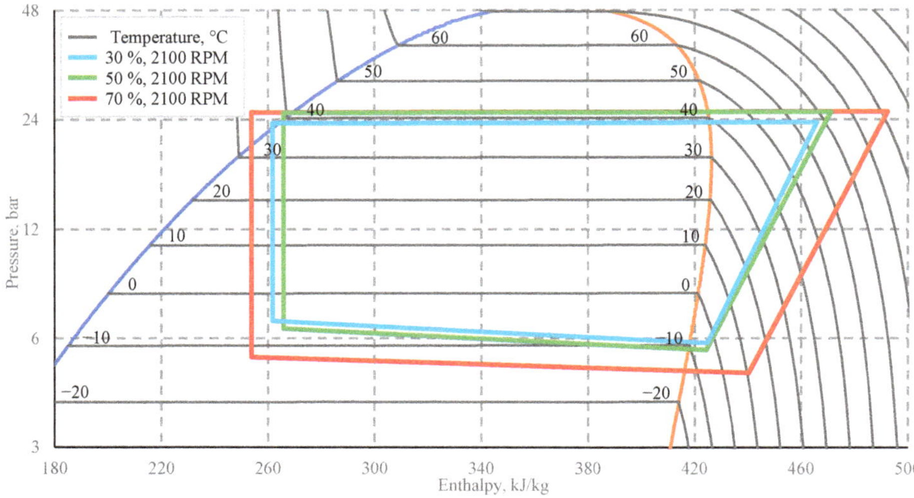

Figure 16. The operation cycles of the HP in a *p-h* diagram (compressor speed: 2100 rpm; expansion valve capacity: 30, 50, 70%; volume: V0).

Changing the volume of the heat pump circuit had the most significant effect on the heat pump operation cycle (Figure 17). Increasing the volume from V0 to V1 resulted in a virtually uniform "lowering" of the cycle in the graph. In this case, the condensation and evaporation isotherms were reduced by 8.30 °C and 8.22 °C, respectively, while the temperatures upstream and downstream of the compressor were similar. As the system volume was further increased to V2 and V3, the condenser and evaporator lines moved steadily in the direction of the pressure drop. However, the cycles did not reach/pass into the two-phase region, which means that no refrigerant condensation process occurred in the condenser and the whole cycle was in the superheated vapour region. The high temperature and pressure vapour leaving the compressor was cooled but not condensed in the condenser. In these cases (V2, V3), the enthalpy differences between the condenser and the evaporator were reduced by several times compared to the cases V0 and V1.

From the *p-h* diagrams, it can be seen that changing the volume of the system extends the operating range compared to standard measures. An increase in volume by 17% improves the controllability, but an increase in volume by more than 26% is detrimental to the performance as no evaporation or condensation are achieved. This benefit of volume variation can also be applied to cooling or refrigeration processes, since neither the compressor speed nor the throttle capacity is altered, and increasing the volume lowers the evaporating temperature with almost no change in the performance. An average reduction in the COP from V0 to V1 for a volume increase of 5.14% is observed.

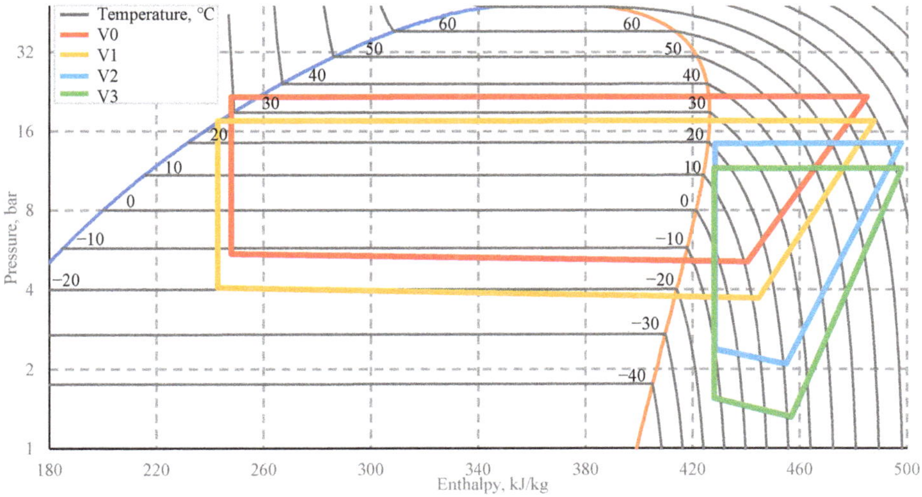

Figure 17. The operation cycles of the HP in a *p-h* diagram (compressor speed: 1800 rpm; expansion valve capacity: 30%; volumes: V0, V1, V2, V3).

4. Conclusions

This article presents experimental studies of a heat pump integrated into a ventilation unit. The experiments involved varying the operating modes of the unit by changing the compressor speed, the position of the expansion valve, and the system volume. The comparison of the performance characteristics showed that the performance trends of the variable-volume heat pump are similar to those of the typical constant-volume heat pumps controlled by standard means (compressor and expansion valve). The experimental results show that the trends are similar, indicating a sufficient degree of accuracy.

Additional control was achieved by changing the volume of the heat pump circuit, which, when combined with conventional controls, extended the heating power range. Increasing the volume of the system loop enables the heat pump to operate at a lower heating capacity than in the original case (V0), allowing the unit to operate with stability over a wider range of outdoor air temperatures.

Increasing the volume of the heat pump loop was found to decrease its efficiency. This phenomenon is similar to the case of partial refrigerant leakage from the heat pump system.

The analysis of the experimental heat pump operation cycles in a pressure-enthalpy diagram revealed that altering the compressor speed had a negligible impact on the temperature differences between the superheating and sub-cooling stages of the cycle. Increasing the compressor speed raises the condensation temperature and lowers the evaporation temperature. On the other hand, modifying the flow rate of the expansion valve has a more pronounced effect on the evaporation temperature than on the condensation temperature. The superheating and sub-cooling temperatures of the cycle were influenced by adjusting the heat pump with the expansion valve. An increase in the difference between these temperatures was achieved by reducing the valve's permeability. The operation cycle of the heat pump was significantly affected by increasing the volume of the heat pump circuit. Increasing the volume can lead to a more uniform "lowering" of the cycle in the diagram, resulting in lower temperatures for the condensation and evaporation isotherms.

However, it is important to note that controllability is positively affected when the increase in volume is 17% or less. Beyond this point, performance is negatively impacted as the evaporation and condensation process cannot be achieved. Volume variation can benefit the cooling or refrigeration processes without altering the compressor speed or throttle capacity. By increasing volume, the evaporating temperature can be lowered with minimal impact on the operating efficiency.

Experimental investigations have revealed a significant influence of control compo-nents on the performance of a variable-volume heat pump. When controlled by varying th volume, the heating capacity parameters of the heat pump respond with the highest degre of sensitivity compared to control by the compressor or expansion valve. This sensitivit indicates the potential for developing novel control methods and components through th implementation of variable circuit volume technology in heat pump regulation. Since ther are few new heat pump control components emerging in commercial devices (with only th existing components being improved instead), this technology could expand the operatin ranges of heat pumps in the future, thereby helping to reduce energy consumption durin their production and operation.

This study represents initial steps in exploring the potential of variable-volume hea pump control. Future research is recommended to investigate the volume change contro of the heat pump over a wider range of outdoor air temperatures, as well as in device of different capacities. Additionally, it is worth analysing the adaptability of this volum change method in refrigeration equipment, since it has been established that an increas in volume, with constant compressor speed and expansion valve position, reduces th evaporation temperature with almost no change in performance.

During the subsequent stages of the study, analysis will be conducted on the long-terr operation of a heat pump with a variable circuit volume in the ventilation system, conside ing its performance during transition periods and assessing seasonal efficiency indicator

Author Contributions: Conceptualization, A.F., J.B., R.D.-T. and V.M.; methodology, A.F. and J.B software, A.F.; validation, A.F. and J.B.; formal analysis, A.F.; investigation, A.F.; data curation, A.F writing—original draft preparation, A.F., J.B. and R.D.-T.; writing—review and editing, A.F., J.B R.D.-T. and V.M.; visualization, A.F. and J.B. All authors have read and agreed to the publishe version of the manuscript.

Funding: This research received no external funding.

Institutional Review Board Statement: Not applicable.

Informed Consent Statement: Not applicable.

Data Availability Statement: The data presented in this study are available on request from th corresponding author.

Acknowledgments: The authors are thankful to Vytautas Martinaitis for the inspiration, usefu discussion of theoretical frameworks, and valuable and constructive suggestions during the plannin and development of this study.

Conflicts of Interest: The authors declare no conflicts of interest.

References

1. International Energy Agency. *Market Report Series: Energy Efficiency 2O17*; IEA Publications: Paris, France, 2017; pp. 1–14 Available online: https://www.iea.org/reports/energy-efficiency-2017 (accessed on 1 May 2024).
2. European Commission. An EU strategy on heating and cooling 2016. *J. Chem. Inf. Model.* **2016**, *53*, 1689–1699. [CrossRef]
3. European Parliament and Council. Directive 2010/31/EU of the European Parliament and of the Council of 19 may 2010 due o the energy performance of buildings. *Off. J. Eur. Union* **2010**, *153*, 13–35.
4. Schlomann, B.; Kleeberger, H. *Report: Energieverbrauch des Sektors Gewerbe, Handel, Dienstleistungen (GHD) in Deutschland fü Die Jahre 2011 bis 2013*; Institute for Resource Efficiency and Energy Strategies: Karlsruhe, Germany, 2015; p. 30. Availabl online: https://irees.de/2020/06/15/energieverbrauch-des-sektors-gewerbe-handel-dienstleistungen-ghd-in-deutschlanc fuer-die-jahre-2011-bis-2013-2015 (accessed on 1 May 2024).
5. Jouhara, H.; Yang, J. Energy efficient HVAC systems. *Energy Build.* **2018**, *179*, 83–85. [CrossRef]
6. Zhu, Y.; Wu, S.; Li, J.; Jia, Q.; Zhang, T.; Zhang, X.; Han, D.; Tan, Y. Towards a carbon-neutral community: Integrated renewabl energy systems (IRES)–sources, storage, optimization, challenges, strategies and opportunities. *J. Energy Storage* **2024**, *83*, 11066 [CrossRef]
7. Olympios, A.V.; Song, J.; Ziolkowski, A.; Shanmugam, V.S.; Markides, C.N. Data-driven compressor performance maps and cos correlations for small-scale heat-pumping applications. *Energy* **2024**, *291*, 130171. [CrossRef]
8. Wen, Q.; Zhi, R.; Wu, Y.; Lei, B.; Ban, G. Experimental studies of semi hermetic single screw compressor with variable speed i heat pump. *Case Stud. Therm. Eng.* **2024**, *58*, 104399. [CrossRef]

Wu, C.; Li, X.; Wang, Z.; Xu, Z.; Xu, C.; Yang, Y. Optimizing the selection and combined operation of multiple air-source heat pumps for sustainable heating systems. *Energy Build.* **2024**, *310*, 114052. [CrossRef]

Koopman, T.; Zhu, T.; Rohlfs, W. Performance evaluation of air-source heat pump based on a pressure drop embed-ded model. *Heliyon* **2024**, *10*, e24634. [CrossRef]

Panteli, M.; Mancarella, P. Influence of extreme weather and climate change on the resilience of power systems: Impacts and possible mitigation strategies. *Electr. Power Syst. Res.* **2015**, *127*, 259–270. [CrossRef]

Pineda Quijano, D.; Fonseca Lima, B.; Infante Ferreira, C.; Brück, E. Seasonal COP of a residential magnetocaloric heat pump based on MnFePSi. *Int. J. Refrig.* 2024; *in press*. [CrossRef]

Wang, J.; Lu, X.; Adetola, V.; Louie, E. Modeling Variable Refrigerant Flow (VRF) systems in building applications: A comprehensive review. *Energy Build.* **2024**, *311*, 114128. [CrossRef]

Jain, N.; Alleyne, A.G. Thermodynamics-based optimization and control of vapor-compression cycle operation: Optimization criteria. In Proceedings of the 2011 American Control Conference, San Francisco, CA, USA, 29 June–1 July 2011; pp. 1352–1357. [CrossRef]

Jensen, J.B.; Skogestad, S. Optimal operation of simple refrigeration cycles. Part I: Degrees of freedom and optimality of sub-cooling. *Comput. Chem. Eng.* **2007**, *31*, 712–721. [CrossRef]

Aprea, C.; Mastrullo, R.; Renno, C. Experimental analysis of the scroll compressor performances varying its speed. *Appl. Therm. Eng.* **2006**, *26*, 983–992. [CrossRef]

Dongellini, M.; Abbenante, M.; Morini, G.L. A strategy for the optimal control logic of heat pump systems: Impact on the energy consumption of a residential building. In Proceedings of the 12th IEA Heat Pump Conference, Rotterdam, The Netherlands, 15–18 May 2017.

Xu, Z.; Li, H.; Xu, W.; Shao, S.; Wang, Z.; Gou, X.; Zhao, M.; Li, J. Investigation on the efficiency degradation characterization of low ambient temperature air source heat pump under partial load operation. *Int. J. Refrig.* **2022**, *133*, 99–110. [CrossRef]

Chae, S.; Bae, S.; Nam, Y. Performance improvement of air-source heat pump via optimum control based on artificial neural network. *Energy Rep.* **2023**, *10*, 460–472. [CrossRef]

Wang, W.; Hu, B.; Wang, R.Z.; Luo, M.; Zhang, G.; Xiang, B. Model predictive control for the performance improvement of air source heat pump heating system via variable water temperature difference. *Int. J. Refrig.* **2022**, *138*, 169–179. [CrossRef]

Hlanze, P.; Jiang, Z.; Cai, J.; Shen, B. Model-based predictive control of multi-stage air-source heat pumps integrated with phase change material-embedded ceilings. *Appl. Energy* **2023**, *336*, 120796. [CrossRef]

Putrayudha, S.A.; Kang, E.C.; Evgueniy, E.; Libing, Y.; Lee, E.J. A study of photovoltaic/thermal (PVT)-ground source heat pump hybrid system by using fuzzy logic control. *Appl. Therm. Eng.* **2015**, *89*, 578–586. [CrossRef]

Yang, T.; Ge, T. Performance study of a heat pump fresh air unit based on desiccant coated heat exchangers under different operation strategies. *Energy* **2024**, *296*, 131182. [CrossRef]

Rohrer, T.; Frison, L.; Kaupenjohann, L.; Scharf, K.; Hergenr, E. Deep Reinforcement Learning for Heat Pump Control. *arXiv* **2022**, arXiv:2212.12716. [CrossRef]

Park, C.; Lee, H.; Hwang, Y.; Radermacher, R. Recent advances in vapor compression cycle technologies. *Int. J. Refrig.* **2015**, *60*, 118–134. [CrossRef]

Al-Obaidi, A.S.M.; Naif, A.; Al-Harthi, T.K. Optimization of the Performance of Vapour Compression Cycle using Liquid Suction Line Heat Exchanger. *J. Therm. Eng.* **2020**, *6*, 201–210. [CrossRef]

Pottker, G.; Hrnjak, P. Experimental investigation of the effect of condenser subcooling in R134a and R1234yf air-conditioning systems with and without internal heat exchanger. *Int. J. Refrig.* **2015**, *50*, 104–113. [CrossRef]

Kwan, T.H.; Shen, Y.; Wu, Z.; Yao, Q. Performance analysis of the thermoelectric device as the internal heat exchanger of the trans-critical carbon dioxide cycle. *Energy Convers. Manag.* **2020**, *208*, 112585. [CrossRef]

Liu, X.; Fu, R.; Wang, Z.; Lin, L.; Sun, Z.; Li, X. Thermodynamic analysis of transcritical CO_2 refrigeration cycle integrated with thermoelectric subcooler and ejector. *Energy Convers. Manag.* **2019**, *188*, 354–365. [CrossRef]

Qureshi, B.A.; Zubair, S.M. The effect of refrigerant combinations on performance of a vapor compression refrigeration system with dedicated mechanical sub-cooling. *Int. J. Refrig.* **2012**, *35*, 47–57. [CrossRef]

Khan, J.U.R.; Zubair, S.M. Design and rating of an integrated mechanical-subcooling vapor-compression refrigeration system. *Energy Convers. Manag.* **2000**, *41*, 1201–1222. [CrossRef]

Huff, H.; Lindsay, D.; Radermacher, R. Positive displacement compressor and expander simulation. In Proceedings of the International Compressor Engineering Conference: 8 (Paper 1527), West Lafayette, IN, USA, 16–19 July 2002; Available online: https://docs.lib.purdue.edu/icec/1527/ (accessed on 26 February 2024).

Wang, M.; Zhao, Y.; Cao, F.; Bu, G.; Wang, Z. Simulation study on a novel vane-type expander with internal two-stage expansion process for R-410A refrigeration system. *Int. J. Refrig.* **2012**, *35*, 757–771. [CrossRef]

Boumaraf, L.; Haberschill, P.; Lallemand, A. Investigation of a novel ejector expansion refrigeration system using the working fluid R134a and its potential substitute R1234yf. *Energy Econ.* **2014**, *45*, 148–159. [CrossRef]

Cao, X.; Liang, X.; Shao, L.; Zhang, C. Performance analysis of an ejector-assisted two-stage evaporation single-stage vapor-compression cycle. *Appl. Therm. Eng.* **2022**, *205*, 118005. [CrossRef]

Cho, H.; Chung, J.T.; Kim, Y. Influence of liquid refrigerant injection on the performance of an inverter-driven scroll compressor. *Int. J. Refrig.* **2003**, *26*, 87–94. [CrossRef]

37. Pathak, A.; Binder, M.; Ongel, A.; Ng, H.W. Investigation of a multi stage vapour-injection cycle to improve air-conditionin system performance of electric buses. In Proceedings of the 14th International Conference on Ecological Vehicles and Renewabl Energies, EVER, Monte-Carlo, Monaco, 8–10 May 2019; pp. 1–7. [CrossRef]
38. Lee, H.; Hwang, Y.; Radermacher, R.; Chun, H.H. Performance investigation of multi-stage saturation cycle with natural workin fluids and low GWP working fluids. *Int. J. Refrig.* **2015**, *51*, 103–111. [CrossRef]
39. Yang, M.; Wang, B.; Li, X.; Shi, W.; Zhang, L. Evaluation of two-phase suction, liquid injection and two-phase injection fc decreasing the discharge temperature of the R32 scroll compressor. *Int. J. Refrig.* **2015**, *59*, 269–280. [CrossRef]
40. Jacob, T.A.; Shah, N.; Park, W.Y. Evaluation of hybrid evaporative-vapor compression air conditioners for different global climate *Energy Convers. Manag.* **2021**, *249*, 114841. [CrossRef]
41. Bagarella, G.; Lazzarin, R.; Noro, M. Sizing strategy of on-off and modulating heat pump systems based on annual energ analysis. *Int. J. Refrig.* **2016**, *65*, 183–193. [CrossRef]
42. Huang, B.; Jian, Q.; Luo, L.; Zhao, J. Experimental study of enhancing heating performance of the air-source heat pump by usin a novel heat recovery device designed for reusing the energy of the compressor shell. *Energy Convers. Manag.* **2017**, *138*, 38–4 [CrossRef]
43. Wang, L.; Ma, G.; Ma, A.; Zhou, F.; Li, F. Experimental study on the characteristics of triplex loop heat pump for exhaust air hea recovery in winter. *Energy Convers. Manag.* **2018**, *176*, 384–392. [CrossRef]
44. Guoyuan, M.; Qinhu, C.; Yi, J. Experimental investigation of air-source heat pump for cold regions. *Int. J. Refrig.* **2003**, *26*, 12–1 [CrossRef]
45. Lee, Z.; Gupta, K.; Kircher, K.J.; Zhang, K.M. Mixed-integer model predictive control of variable-speed heat pumps. *Energy Buil* **2019**, *198*, 75–83. [CrossRef]
46. Clauß, J.; Georges, L. Model complexity of heat pump systems to investigate the building energy flexibility and guidelines fc model implementation. *Appl. Energy* **2019**, *255*, 113847. [CrossRef]
47. Lee, S.H.; Jeon, Y.; Chung, H.J.; Cho, W.; Kim, Y. Simulation-based optimization of heating and cooling seasonal performance of an air-to-air heat pump considering operating and design parameters using genetic algorithm. *Appl. Therm. Eng.* **2018**, *14* 362–370. [CrossRef]
48. Jensen, J.B.; Skogestad, S. Control and optimal operation of simple heat pump cycles. *Comput. Aided Chem. Eng.* **2005**, *2* 1429–1434. [CrossRef]
49. Frik, A.; Bielskus, J.; Dzikevics, M. Experimental Study on the Control of the Positions of the Cycle Isotherms of the Heat Pump i the Air Handling Unit. *Environ. Clim. Technol.* **2023**, *27*, 889–899. [CrossRef]
50. Zhang, Y.; Li, M.; Dong, J.; Zhang, C.; Li, X.; Han, Z. Study on the impacts of refrigerant leakage on the performance and reliabilit of datacenter composite air conditioning system. *Energy* **2023**, *284*, 129336. [CrossRef]
51. Pelella, F.; Viscito, L.; Mauro, A.W. Combined effects of refrigerant leakages and fouling on air-source heat pump performances i cooling mode. *Appl. Therm. Eng.* **2021**, *204*, 117965. [CrossRef]

applied sciences

MDPI

rticle

Scheme Design and Energy-Saving Optimization of Cold and Heat Energy Supply System for Substation Main Control Building in Cold Area

ng Wang [1,2], Xu Jin [3,*], Jiapeng Zhang [3], Cong Zeng [2], Xiuyun Gao [1], Lei Zhao [1] and Shuai Sha [3]

1 Economic and Technological Research Institute of State Grid Heilongjiang Electric Power Co., Ltd., Harbin 150036, China; wyj_1129@hrbeu.edu.cn (Y.W.); zc_nedu@126.com (X.G.); zhaolei19870212@126.com (L.Z.)
2 Key Laboratory of Electric Power Infrastructure Safety Assessment and Disaster Prevention of Jilin Province, Northeast Electric Power University, Jilin 132012, China; zc_1113@126.com
3 School of Energy and Power Engineering, Northeast Electric Power University, Jilin 132012, China; 17360822859@163.com (J.Z.); shashuai11@163.com (S.S.)
* Correspondence: jinxu7708@sina.com

Abstract: In the context of global climate change, the implementation of building energy conservation and carbon reduction, as well as the realization of zero-energy buildings, is a key measure to cope with climate change and resource depletion. A substation is an indispensable building in the process of urbanization construction. However, in existing cold areas, the heating form of substations generally adopts electric heating, which consumes a large amount of energy. This paper optimizes the existing HVAC form of substations through the rational utilization of surrounding environmental resources and puts forward reasonable building energy-saving and carbon-reduction methods. It demonstrates the feasibility of combining solar photovoltaic power generation systems, air source heat pumps, and natural ventilation to optimize energy savings and carbon reduction in the main control building of a substation in a cold area. The computational fluid dynamics (CFD) method is used to demonstrate the feasibility of natural ventilation during the summer and transition seasons. The data indicate that the installation of a solar photovoltaic power generation system results in an average annual power generation of 18.75 MWh. Additionally, using an air source heat pump can save 44.5% of electricity compared to electric heating. When both a solar photovoltaic power generation system and an air source heat pump are used to provide a building with cold and heat sources, the annual emissions of CO_2 can be reduced by 4.90 tons compared to a traditional electric heating system.

Keywords: solar energy; near-zero-energy buildings; natural ventilation

tation: Wang, Y.; Jin, X.; Zhang, J.; ng, C.; Gao, X.; Zhao, L.; Sha, S. heme Design and Energy-Saving otimization of Cold and Heat ergy Supply System for Substation ain Control Building in Cold Area. pl. Sci. 2024, 14, 1562. https:// i.org/10.3390/app14041562

cademic Editors: Daniel nchez-García and David envenido Huertas

ceived: 16 January 2024
vised: 10 February 2024
cepted: 12 February 2024
blished: 15 February 2024

1. Introduction

In the present social context, with people's increasing demand for a better quality of life, energy consumption is on the rise. Building energy consumption constitutes a significant portion of the total social energy consumption. Among the various components of building energy consumption, energy consumed by daily operations such as air conditioning, heating, lighting, and hot water accounts for more than 80%. Based on statistics from the International Energy Agency, terminal energy consumption in buildings accounts for 36% of social carbon emissions, with carbon emissions reaching 39%. Furthermore, during a building's operational phase, carbon emissions account for 70% to 80% of its entire life cycle [1]. Although building energy efficiency has received increasing attention, many countries have introduced the concepts of low-energy buildings, ultra-low-energy buildings, near-zero-energy buildings, and zero-energy buildings [2,3]. It is crucial to control carbon emissions during the building operation stage to achieve the goal of carbon peaking and carbon neutrality. Achieving zero-energy and zero-carbon buildings is a general trend [4]. Although building energy consumption is increasing, it is important for buildin design to propose a reasonable carbon emission mechanism and reduce building energ consumption. A building energy conservation development model should be establishe as soon as possible, specifically targeting the existing domestic environment.

Zeyghami et al. [5] investigated the realization of zero-energy buildings by studyin buildings in a specific area of Egypt. The authors aimed to utilize natural cold heat t provide the necessary temperature control for the buildings under examination. They co cluded that in certain areas with favorable natural conditions, the use of solar photovolta systems can significantly reduce the costs of transforming buildings into near-zero-energ consumption buildings by collecting and storing renewable energy through solar pane on the roof. Lee et al. [6] conducted a field study on a building photovoltaic system. Th study shows that the building's annual energy consumption was 104,602.4 kWh, and th power generation of the building's photovoltaic system was 105,266.6 kWh. This prove that the building meets the requirements of zero-energy buildings. Cao et al. [7] analyze the relationship between the body type coefficient and the energy consumption of buildin cooling and heating. The data indicate a positive linear correlation between the buildin size coefficient and total energy consumption. In other words, as the building size coeff cient increases, so does the total energy consumption. However, this correlation is not a pronounced in mild and sunny areas during the heating period. Li et al. [8] investigate the use of photovoltaic glass in cold regions of China. They evaluated the energy-savin performance of photovoltaic glass by altering its orientation, visible light transmission rati and comprehensive heat transfer coefficient. When the photovoltaic system is arrange facing south, the energy-saving effect is as high as 60%. However, if the system is facin east, the energy-saving is reduced to about 13%. It is important to note that the orientatio of the photovoltaic glass has a significant impact on the energy-saving effect.

Lebied et al. [9] analyzed the impact of the perimeter structure and system equipmer operation on the energy consumption of an office building in Denmark. They sorte the degree of impact of different variables through sensitivity analysis. Balali et al. [1(conducted a study on a single-sided lighting office. They used computer software t evaluate various influencing factors, such as building size and the ratio of windows to wal By analyzing the impact of different design variables on the total energy consumption of th building, they obtained the most energy-efficient combination design scheme of influencin factors. Sivaram et al. [11] conducted a study on office buildings in Tokyo, Japan. The investigated the impact of four design variables of the outer envelope structure on the col and heat load of buildings. These variables were the solar heat gain coefficient of the oute window, the heat transfer coefficient of the outer window, the heat transfer coefficient c the opaque envelope, and the solar reflectivity of the opaque envelope. Tariq [12] provide the comprehensive mechanism of using the digital twin and the impact assessment c interventions at the local as well as global scales. The measures of passive energy storag based on phase-change energy storage materials are studied, and the energy efficiency ca be increased by 40% by adding relevant interventions.

Natural ventilation refers to the use of natural airflow to achieve indoor air circulatio and renewal in a ventilation mode. The principle of natural ventilation mainly include three aspects: the buoyancy principle, the wind pressure difference principle, and the win force principle. Natural ventilation is a fundamental and sustainable solution to reduc the energy needs of buildings for ventilation and cooling [13]. Research in this field ha demonstrated that natural ventilation has the potential to save energy. Tong et al. [14 conducted a study on the energy-saving potential of natural ventilation in various Chines cities. The results indicated that natural ventilation could save between 8% and 78% o cooling energy. Similarly, in warm weather, the use of natural ventilation can avoid th need for 54.4% of the power required for cooling [15]. Stasi et al. [16] investigated th impact of mixed ventilation on zero-energy residential buildings. The results indicate tha the combination of night ventilation and mechanical ventilation can reduce the deman for cooling capacity by 14.4%, while natural cooling can reduce power consumption b

7.7%. Domjan et al. [17] analyzed the importance of natural cooling and mixed ventilation systems for near-zero-energy buildings. Studies on all-glass buildings have shown that energy consumption can be reduced by up to 26% through night ventilation and mixed ventilation cooling.

In summary, numerous scholars have conducted extensive research on the influence of various parameters, including window-to-wall ratio, building size coefficient, and thermal resistance of outer protective structures, on building energy conservation. It is worth noting that China's construction industry accounts for over 20% of the country's total energy consumption, with winter heating alone accounting for 20% of the total building energy consumption [13]. The energy-saving potential of HVAC systems in substation sites located in cold areas requires further investigation. Substations are typically situated in remote areas, far from residential areas. As a result, there is no urban heating network available, and traditional urban central heating is not feasible. Additionally, heating pipes cannot pass through the transformer room, distribution unit room, and other electrical equipment rooms in most substations. Therefore, given the extreme outdoor conditions of −30 °C in winter in high-cold areas, improving heating efficiency and reducing heating energy loss in substations are crucial to enhancing the energy consumption structure and efficiency of substation buildings. This is necessary to ensure the normal operation of zero-carbon building design concepts and typical room equipment in substations under low-temperature conditions. Based on theoretical calculations and software simulations, this paper provides recommendations for selecting ventilation and heating equipment for typical rooms in high-cold area substations.

2. Architectural Models and Research Methods

The research subject is the main control building of a substation located in Yichun City, Heilongjiang Province, China. It consists of a 10 kV equipment distribution room, a secondary equipment room, a secondary equipment operation monitoring room, and other rooms. The main transformer is located outside. The building has a construction area of 293.02 m^2 and a total height of 5.00 m. Please refer to Figure 1 for the building plan. The red portion in Figure 1 represents the distribution of electrical equipment within the main control building of the substation. Currently, the building is equipped with a mechanical exhaust system for summer ventilation. The main HVAC systems comprise split air conditioning units for cooling in the summer and electric heating for warming in the winter.

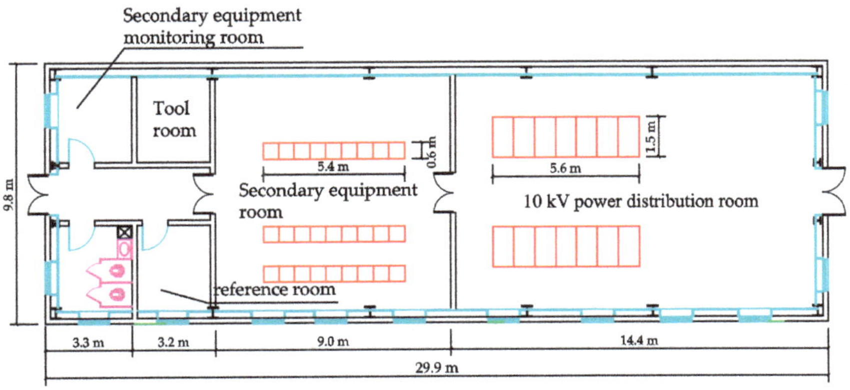

Figure 1. Substation main control building plan.

The diagram in Figure 2 illustrates the actual exterior wall structure of the building. Based on the physical properties and thickness of the wall materials, the exterior wall heat transfer coefficient was calculated to be 0.22 W/(m^2·K) using the GBSWARE2024 software. The thermal performance of the reference building in Table 1 was determined based on parameters such as wall and window materials. The reference building's therma performance was obtained according to the regulations for thermal transfer coefficient for doors and windows outlined in the "Unified Standard for Energy-saving Design c Industrial Buildings" GB51245-2017 [18]. The building's thermal performance complie with energy-saving standards.

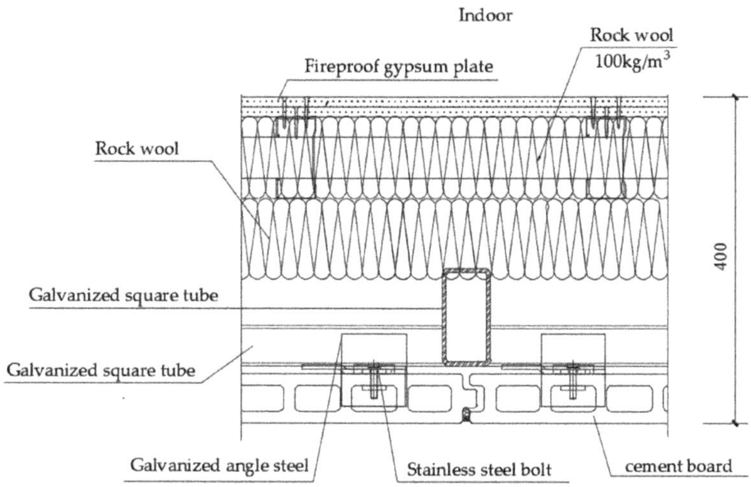

Figure 2. Actual building exterior wall construction.

Table 1. Comparison of heat transfer coefficients.

Name	Actual Building	Reference Building	Unit
Shape coefficient	0.49	0.49	
Roof heat transfer coefficient	0.34	0.35	W/(m^2·K)
Heat transfer coefficient of the outer wall	0.22	0.35	W/(m^2·K)
Thermal resistance of the surrounding ground	3.76	1.10	(m^2·K)/W
Heat transfer coefficient of the window	0.9	0.9	W/(m^2·K)

The heat load is calculated based on the principle of steady-state heat transfer, conside ing both heat loss and gain in the room. The focus of the heat load calculation is on heat los To achieve heat balance, the calculation includes heat lost through the enclosure structur heat lost through the heating of cold air, and heat produced by equipment operation.

According to Section 4.5.5 of the "35 kV~110 kV Substation Design Code" GB5005 2012 [19]: "Distribution unit room, reactor room, and other electrical equipment room should have a mechanical ventilation system set up, and the indoor temperature should b maintained at no more than 40 °C in summer." According to Section 8.3.2 of the "Technica Regulations for the Design of 220 kV~750 kV Substation" DL/T5218-2012 [20]: "The mai control room, computer room, relay room, communication room, and other rooms require by process equipment of the substation should be equipped with air conditioning." Th indoor temperature and humidity of the air-conditioned room should meet the proces requirements, and the process has no special requirements. The design temperature i summer is 26 °C to 28 °C, the design temperature in winter is 18 °C to 20 °C, and th relative humidity should not exceed 70%. Generally, there is no backup for air conditioner The working time of the substation is set by BESI, the energy consumption calculatio software of GBWARE2024, and the indoor temperature is set at 20 °C in the heating seaso and 28 °C in the cooling season. The hourly temperature throughout the year is obtaine by invoking meteorological parameters, with the highest temperature being 33.3 °C an the lowest temperature being −26.7 °C, as shown in Figure 3.

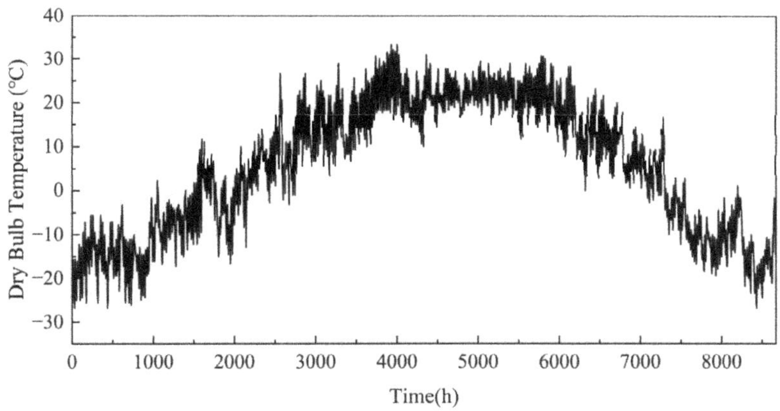

Figure 3. Hourly dry bulb temperature and load chart throughout the year.

Meteorological parameters of outdoor design are given in Table 2.

Table 2. Outdoor design meteorological parameters.

Outdoor Parameter	Design Value	Unit
Dry bulb temperature calculation for summer air conditioning	30.7	°C
Summer ventilation outdoor temperature calculation	26.7	°C
Outdoor relative humidity calculation for summer air conditioning	62%	
Calculation of outside temperature for winter air conditioning	−27.1	°C
Calculation of relative humidity for winter outdoor air conditioning	73%	

Formula (1) shows the heat consumption Q_j (W)of the enclosure structure.

$$Q_j = F \cdot K \cdot (t_n - t_w) \cdot a \quad (1)$$

where F(m^2) represents the area of the enclosure structure, K is the heat transfer coefficient of the enclosure structure (W/m^2·K), t_n is the design temperature of the heating room (K), t_w is the calculated temperature of the heating outside (K), and a is the temperature difference correction coefficient of the maintenance structure.

There are gaps and openings in the building envelope, including doors and windows. These gaps and openings allow outdoor air to penetrate into the room due to the pressure difference between indoor and outdoor environments caused by factors such as wind pressure and temperature differences. It is important to seal these gaps and openings to prevent energy loss and maintain indoor comfort. This results in the infiltration of cold or hot air, depending on the outdoor temperature. When calculating the heat load, the usually considered factor is the permeable air load. The calculation of the air infiltration load Q_x (W) is as follows:

$$Q_x = \frac{1}{3.6} \cdot \rho_w \cdot L \cdot c_p \cdot (t_w - t_n) \quad (2)$$

where ρ_W represents the air density (kg/m^3) outside the heating room at the calculated temperature, c_p is the specific heat coefficient of air (J/(kg K)), and L is the amount of permeated cold air (m^3/h).

The ventilation and air conditioning system are designed to dissipate the heat generated by the electrical equipment, ensuring efficient operation and preventing equipment failure due to excessive temperature. The calculation of the equipment's heat load includes the heat loss of the transformer, high-voltage switchgear, and low-voltage switchgear, as well as the power loss of the cable, heat gain of the enclosure structure, and heat dissipatio of the room lighting. Formula (3) is used to calculate the equipment heat load Q_e (W).

$$Q_e = n_1 \cdot n_2 \cdot n_3 \cdot N \cdot C_{cl} \quad (3)$$

where n_1 is the coefficient of simultaneous use, n_2 is the installation coefficient, n_3 is th load coefficient, N is the installation power of the electric heating equipment, and C_{cl} is th cooling load coefficient of the heat dissipation of electric equipment and appliances.

Based on the formula and relevant software modeling analysis described abov specific parameters are depicted in Figure 4. The conclusive calculation results reveal th the peak heat load of the substation's main control building is 9.842 kW, with a peak coolin load of 5.052 kW. Additionally, the annual heating heat consumption is 19,425.519 kW while the cooling consumption is 2524.104 kWh.

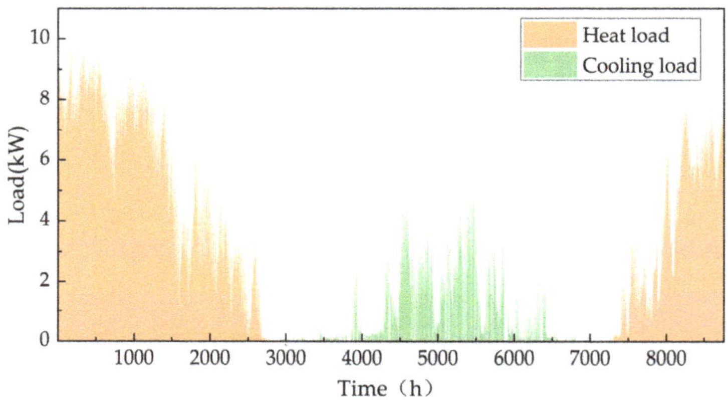

Figure 4. Hourly load of substation main control building throughout the year.

This paper uses CFD to calculate the distribution of indoor flow velocity and airflo direction, presenting a comprehensive view of indoor wind speed and airflow organizatio The turbulence model reflects the fluid flow state. When conducting numerical simulatior of fluid mechanics, it is important to select an appropriate turbulence model for differer fluid flows to simulate the real flow field value as accurately as possible. In this paper, th indoor flow field is calculated using the standard k-ε turbulence model.

The CFD method is used to solve the wind field by establishing the conservatio of mass, momentum, and energy of the fluid flow in the computational domain. This i carried out by establishing a mathematical governing equation, which has the followin general form:

$$\frac{\partial(\rho\phi)}{\partial t} + div(\rho U\phi) = div(\Gamma_\phi grad\phi) + S_\phi \quad (4)$$

ϕ in this formula can be physical quantities such as velocity (m/s), turbulent kinetic energ (m^2/s^2), turbulent dissipation rate (m^2/s^3), and temperature (T), as shown in Table 3.

μ is the dynamic viscosity of air (Pa·s), μ_t is the component of the fluid along th t direction, ρ is the density of air (kg/m^3), α is the thermal diffusivity (m^2/s), G_k is th turbulent momentum caused by the average velocity gradient,$C_\mu = 0.0845$, $C_{1\varepsilon} = 1.$ $C_{2\varepsilon} = 1.42$, and σ is the turbulence parameter.

The equations are discretized using the second-order upwind scheme, which meet the accuracy requirements for general fluid simulation calculations.

Table 3. Governing equations of computational fluid mechanics.

Name	Variable	Γ_ϕ	S_ϕ
Equation of continuity	1	0	0
Velocity in x direction	u	$\mu_{eff} = \mu + \mu_t$	$-\frac{\partial P}{\partial x} + \frac{\partial}{\partial x}(\mu_{eff}\frac{\partial u}{\partial x}) + \frac{\partial}{\partial y}(\mu_{eff}\frac{\partial v}{\partial x}) + \frac{\partial}{\partial z}(\mu_{eff}\frac{\partial w}{\partial x})$
Velocity in y direction	v	$\mu_{eff} = \mu + \mu_t$	$-\frac{\partial P}{\partial y} + \frac{\partial}{\partial x}(\mu_{eff}\frac{\partial u}{\partial y}) + \frac{\partial}{\partial y}(\mu_{eff}\frac{\partial v}{\partial y}) + \frac{\partial}{\partial z}(\mu_{eff}\frac{\partial w}{\partial y})$
Velocity in z direction	w	$\mu_{eff} = \mu + \mu_t$	$-\frac{\partial P}{\partial z} + \frac{\partial}{\partial x}(\mu_{eff}\frac{\partial u}{\partial z}) + \frac{\partial}{\partial y}(\mu_{eff}\frac{\partial v}{\partial z}) + \frac{\partial}{\partial z}(\mu_{eff}\frac{\partial w}{\partial z}) - \rho g$
Turbulent kinetic energy	k	$\alpha_k \mu_{eff}$	$G_k + G_B - \rho\varepsilon$
Turbulent dissipation	ε	$\alpha_z \mu_{eff}$	$C_{1\varepsilon}\frac{\varepsilon}{k}(G_k + C_{3\varepsilon}G_B) - C_{2\varepsilon}\rho\frac{\varepsilon^2}{k} - R_\varepsilon$
Temperature	T	$\frac{\mu}{\mathbf{Pr}} + \frac{\mu_t}{\sigma_T}$	S_T

The inlet boundary conditions for the calculation model mainly consist of wind speed and direction data under different working conditions. The inlet wind speed is calculated using Formula (5), while the free flow is used as the exit boundary condition. The wind field's two side boundaries and top boundary are set as sliding wall surfaces. This ensures that the air flow is not affected by wall surface friction, allowing for a simulation of real outdoor wind flow. The ground boundary of the wind field is set as a non-slip wall, which affects the air flow due to ground friction.

$$v_{in} = v_R \cdot \left(\frac{Z}{Z_R}\right)^{\alpha_1} \tag{5}$$

where v is the average wind speed of any point, Z represents the height of the point, and subscript R represents the parameter under the standard height. Additionally, $\alpha 1$ represents the ground roughness coefficient.

The multi-area network method is used to simulate and calculate the natural ventilation volume of multi-room for natural ventilation, in order to obtain the ventilation times. The software extracts the wind pressure of single building doors and windows through outdoor ventilation calculation and then calculates the air exchange times of the whole building through the multi-area network method.

The heating and air conditioning systems used in the above-mentioned system are electric heating and split air conditioning, respectively. However, according to Article 9.1.2 of the "Design Code for Heating, Ventilation and Air Conditioning of Industrial Buildings" [21] GB50019-2015, industrial plants and auxiliary buildings should not use electric direct heating equipment as a heat source for heating and air conditioning, unless they meet one of the following conditions and cannot use heat pumps. Isolated buildings without access to central heating and lacking any other heat source can be challenging to heat. In such cases, the use of fuel oil and coal-fired equipment is strictly restricted due to environmental and fire safety concerns. In areas with sufficient electricity supply and peak and valley electricity prices, heat can be stored during the low power period at night and used during peak hours or periods of non-use. Additionally, important electrical premises that cannot be heated by hot water or steam may require alternative heating solutions. Renewable energy can be used to generate electricity, which can then be used for electric heating. Therefore, the original heating system does not meet current standards and requires improvement. Considering local meteorological conditions, solar energy can be converted into electricity through solar photovoltaic power generation to supply power to equipment on site, reducing energy loss.

The main control building system of the substation has the following issues:

(1) The consideration of renewable energy or heat pump forms to meet the heating load requirements was not considered.
(2) The building's ventilation system design is imperfect.
(3) There is potential for more energy-efficient heating methods that need to be further optimized.

3. Analyzing Systems Improvement Programs

The ideal energy-saving building should aim to minimize energy consumption whil also ensuring the maintenance of thermal and humidity conditions indoors throughou different seasons, ensuring necessary ventilation and air exchange indoors, and rationa utilization of solar radiation in various seasons and regions. Energy-saving method in building primarily involve several aspects: minimizing reliance on non-renewabl energy sources, enhancing energy utilization efficiency, reducing energy loss in buildin maintenance structures, and maximizing the use of renewable energy sources. Achievin these goals is closely related to cutting-edge technology in today's society. By combinin existing scientific technology with traditional building structures, a new passive low-carbo building structural system has been proposed for optimizing and renovating existin substation building systems, focusing on the rational use of renewable energy and reducin equipment operating energy consumption.

3.1. Efficient Use of Solar Energy Resources

China possesses a vast territory and abundant solar energy resources. The sola energy resources on Earth are generally expressed in terms of the total annual irradiatio (J/(m^2·a)) and the annual sunshine duration (h). The solar radiation energy received b the land surface of China is about 5.0 $\times$ 10^{19} kJ. More than two-thirds of the countr experiences annual sunshine hours exceeding 2200 h, and the total radiation amount i higher than 5.86 $\times$ 10^6 kJ/(m^2·a) [22]. Figure 5 illustrates the monthly distribution of tota solar radiation in the Yichun area.

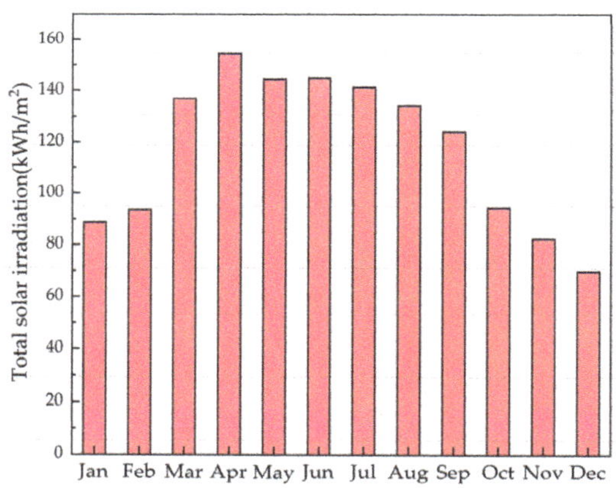

Figure 5. Monthly distribution of total solar radiation in the Yichun area.

According to the Assessment Method for Solar Energy Resource classification metho GB/T37526-2019 [23], global horizontal irradiance (GHR) is divided into four levels: mos abundant (A), very abundant (B), abundant (C), and general (D), as shown in Table 4.

Table 4. Classification and regional distribution of total solar radiation in China.

Name	GHR		Grade Symbol
	(MJ/m^2)	(kWh/m^2)	
Most abundant	GHR $\geq$ 6300	GHR $\geq$ 1750	A
Very abundant	5040 $\leq$ GHR < 6300	1400 $\leq$ GHR < 1750	B
Abundant	3780 $\leq$ GHR < 5040	1050 $\leq$ GHR < 1400	C
General	GHR < 3780	GHR < 1050	D

The total annual horizontal irradiance of the area where the substation is located i 4118.1 MJ/m^2, which belongs to class C, indicating an area rich in solar energy resources. I can be considered to use solar power generation to provide electricity for the HVAC system on the site to achieve a zero-energy consumption building.

To simulate the building's solar irradiance and other parameters, the Building Photovoltaic Software (BPV2024) in GBWARE2024 is used. By setting the project location and constructing the building model, the solar resource data can be reasonably analyzed, and the best tilt angle and installation position can be calculated, along with the power generation of the photovoltaic panels. As shown in Figure 6, the substation unit is modeled.

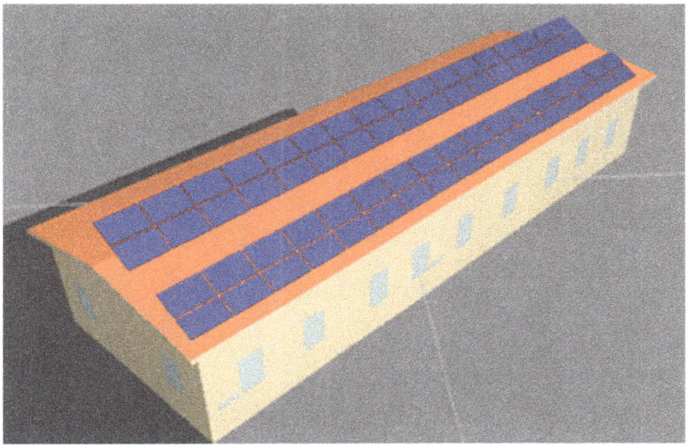

Figure 6. Photovoltaic power generation simulation modeling diagram.

The calculation of a building's photovoltaic system's power generation should consider the solar energy resources of the location, the design of the photovoltaic system, the arrangement of the photovoltaic array, and environmental conditions. To obtain the power generation value of a photovoltaic system, one should refer to the "Design Code for Photovoltaic Power Station" GB 50797 [24] and other relevant standards. The formula for calculating power generation (E_P) is as follows:

$$E_p = \frac{H_A}{E_S} \cdot P \cdot K_t \tag{6}$$

where H_A is the total solar irradiance of the horizontal plane (kWh/m^2), E_S is the irradiance (constant) under standard conditions (1 kW/m^2), P is the installed capacity (kWp), and K_t is the comprehensive efficiency coefficient. The latter is affected by various parameters, including inverter efficiency, collector line loss coefficient, photovoltaic module surface pollution coefficient, and correction coefficient. The specific parameters are given in Table 5.

Table 5. Calculation parameters for photovoltaic system.

Information about Photovoltaic Systems			
Component type	Monocrystalline silicon	Number of components	68
Total installed capacity	17.68 kW	Component installation method	Fixed integration
Component area	111 m^2	Inverter efficiency	1%
Power of inverter	6.75 kW	Collector line loss coefficient	1%
Photovoltaic module surface pollution coefficient	1%	Correction coefficient	1%
Comprehensive efficiency coefficient	85%		

A total of 68 photovoltaic modules with a surface area of 111 m^2 and a total installe capacity of 17.68 kW can be installed through the rational use of the roof of the main contro building. The photovoltaic system has a life cycle of 25 years and is estimated to generate total of 468.7 MWh during this period. Figure 7 illustrates the monthly power generatio during the first year, with April recording the highest generation of 2.33 MWh. Figure shows the annual power generation of a building's PV system over its 25-year life cycl with the first year generating 21.25 MWh.

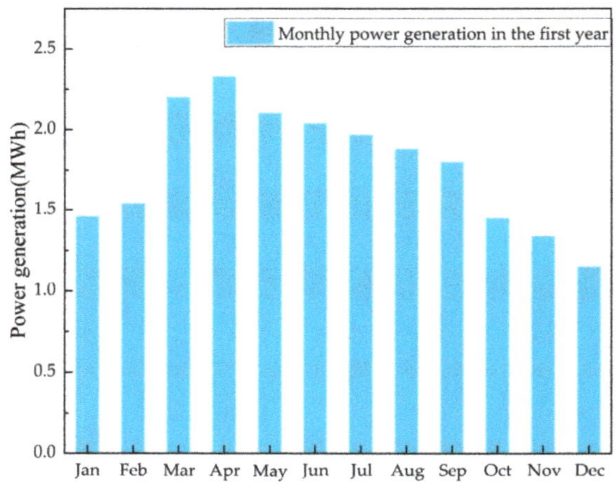

Figure 7. Monthly electricity generation during the first year.

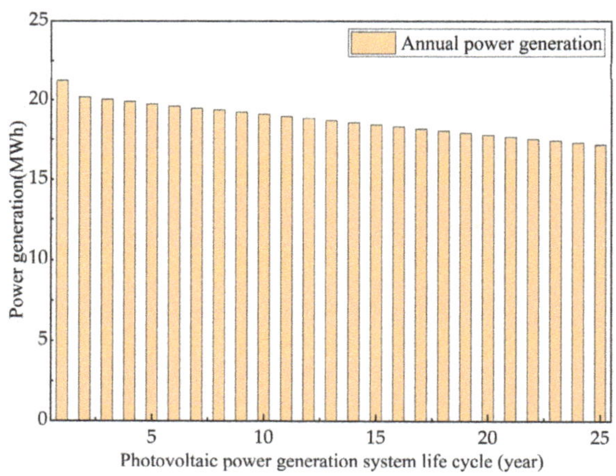

Figure 8. Annual production of electricity.

The economic analysis of photovoltaic power generation involves evaluating th costs and benefits of solar photovoltaic power generation systems, both quantitatively an qualitatively. This paper selects photovoltaic modules with a peak power of 260 Wp, costin 5 RMB/W, and a total installed capacity of 17.68 kW. The total investment, with componen investment accounting for 50% of the total investment ratio, is 176,800 yuan. Table provides specific data. Figure 9 illustrates the cost benefits of investing in photovoltai power generation systems. The data show that the investment can be recovered afte 13 years of operation.

Table 6. Economic benefit analysis of solar power generation systems.

Cost		Income	
Total installed capacity	17.68 kW	Power generation in the first year	21.25 MWh
Cost	5 RMB/W	Total power generation	468.7 MWh
Components as a percentage of total investment	50%	Gross income	0.7 RMB/kWh
Total investment	176,800 RMB	Collector line loss coefficient	328,100 RMB

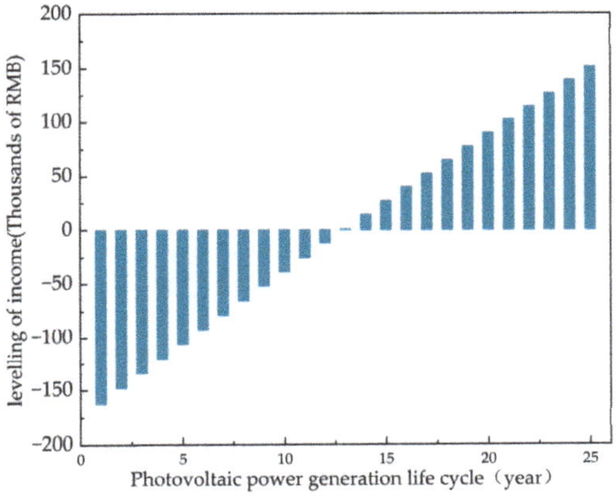

Figure 9. Profitability of photovoltaic power generation.

The photovoltaic power generation system offers several advantages, including no depletion risk, no fuel consumption, and no pollution emissions. These advantages play a crucial role in achieving China's carbon-neutral goal. According to the Annual Development Report of China Power Industry 2022, Table 7 presents the calculated environmental impact of solar photovoltaic power generation systems and the effect of energy conservation based on the statistics of CO_2, SO_2, and nitrogen oxide emissions per unit of thermal power generation.

Table 7. Analysis of the benefits of emission reduction from photovoltaic systems.

Name	Converted Value	Conventional Unit	Annual Average Value	Gross	Unit
Power generation	-	-	18.75	468.72	MWh
Standard coal	0.3015	kg/kWh	5650	141,330	kg
Electric dust and fume	0.0022		40	1030	
CO_2	0.5703		10,692	267,300	
SO_2	0.0101		190	4730	
NO_X	0.0152		290	7130	

In summary, this project has a PV module installation area of 111 m^2, a total installed capacity of 17.68 kW, and a system efficiency of 85.3%. The power generation in the first year is expected to be 21.3 MWh, with a total power generation of 468.7 MWh expected over 25 years. The investment required is 176,800 yuan, with an income of 328,100 yuan. Additionally, the project is expected to emit approximately 267.3 tCO_2.

3.2. Minimize the Power Consumption of the Device

Based on the structural parameters of the main control building of the substation, a hexahedral unstructured mesh, which is simple to partition and shape, was used for modeling. Since the quality of the mesh partition directly affects the accuracy of calculations, the "Normal" method was employed for partitioning. Furthermore, areas with significar temperature and velocity gradients, such as air supply and exhaust outlets, underwer local refinement after coarse partitioning. The maximum size of the mesh did not excee 0.2 m, and the mesh quality was above 0.8. This simulation involved a total of 166,685 mes elements and 181,139 nodes. The simulation results and analysis are discussed below.

Electric heating is a natural convection heat transfer terminal device, which ha the advantages of low investment cost, but it can lead to uneven heating and coolin high surface temperature, and high maintenance costs. Aiming to address the issue c uneven heating and cooling, Airpak3.0.16 is used to simulate the temperature field c each room in the main control building during electric heating and analyze the possibilit of condensation in the room. Figure 10 displays the temperature field distribution c the main control building of the substation at four different heights (0.2 m, 0.5 m, 1.5 n and 2.1 m) with an outdoor ambient temperature of −30 °C. It can be observed that th temperature of the 10 kV equipment distribution room, secondary equipment room, an secondary equipment room monitoring room falls within the range of 17 to 20 °C, meetin the design requirements. No low-temperature phenomenon is observed in any part of th room. Figure 11 displays the distribution diagram of indoor temperature when the hea supply is at its peak and the outdoor ambient temperature is −20 °C. It is evident from th figure that an increase in outdoor temperature leads to a significant rise in temperatur (by 4 to 6 °C) of the 10 kV equipment distribution room, secondary equipment room, an secondary equipment room monitoring room, if the heat supply of the radiator is nc reduced. The reason for this is that the electric heater in the substation cannot be adjuste for heat production, and the equipment's heat supply is designed based on the maximur heat load. As a result, when the outdoor temperature increases, the required heat loa decreases, but the heat supply remains the same, causing the indoor temperature to ris This results in unnecessary energy waste and makes it difficult to achieve the goal of energ conservation and emission reduction.

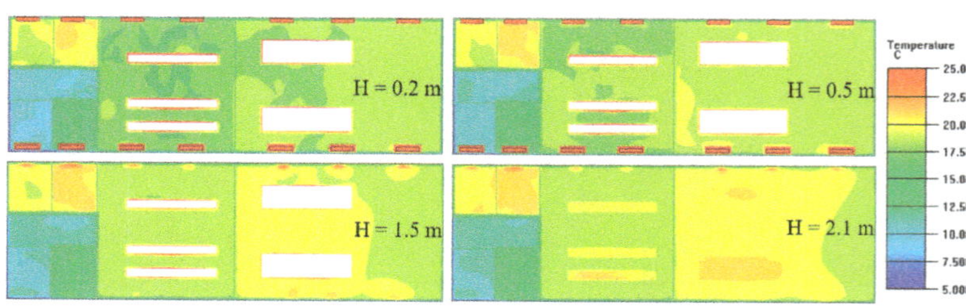

Figure 10. Outdoor ambient temperature −30 °C and temperature distribution in the room whe electric heating is used.

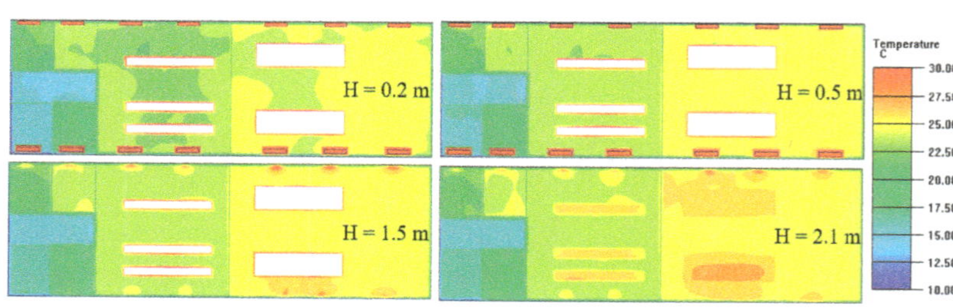

Figure 11. Outdoor ambient temperature −20 °C and temperature distribution in the room whe electric heating is used.

The air source heat pump utilizes a compressor to convert low-grade heat energ from the air into high-grade heat energy that can be used for industrial production wit the aid of electric energy. This makes it possible to save non-renewable energy. As ai is an inexhaustible renewable energy source, air source heat pumps can significantly reduce carbon dioxide emissions during winter heating or even achieve zero emissions. In comparison to electric heaters that consume non-renewable energy, such as electricity and gas, to produce heat, air source heat pumps can generate several times more heat under the same electrical energy input. They are energy-efficient, safe, and have adjustable heating capabilities at a lower cost [25]. Therefore, the air source heat pump combined with a fan coil is being considered as a replacement for electric heating in the main control building.

According to the "Fire Protection Standard for Design of Thermal Power Plants and Substations" GB50229-2019 [26], heating pipes must not pass through rooms containing electrical equipment, such as transformer and distribution device rooms. As a result, separate heating must be provided to the 10 kV equipment distribution room and secondary equipment room. The terminal heating is achieved through fan coil units. Regardless of the type of heating used, it is crucial to maintain a stable indoor temperature and humidity environment. To achieve this, an air source heat pump can be installed to simulate the temperature field inside the substation. Figure 12 illustrates the indoor temperature distribution when a fan coil is used at an outdoor ambient temperature of −20 °C. The figure illustrates that the indoor temperature distribution is more uniform with this heating method compared to direct electric heating. This method can adjust indoor temperature and humidity changes by detecting indoor and outdoor temperature changes and adjusting the indoor supply wind speed and heating temperature of the air source heat pump.

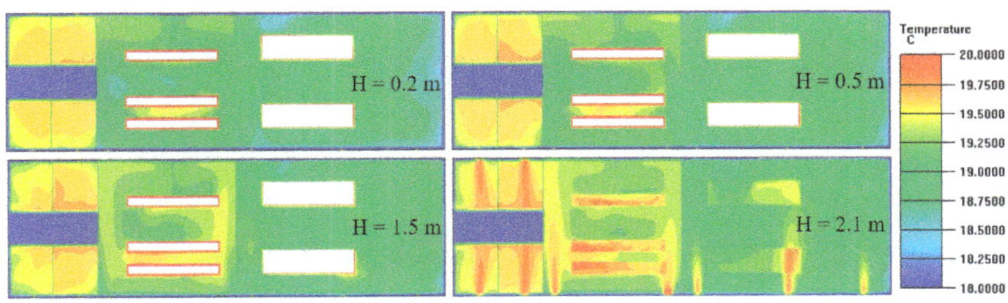

Figure 12. When the outdoor temperature drops to −20 °C, an air source heat pump is used to distribute heat throughout the room.

In accordance with the standard GB/T25127.1 [27] for low ambient temperature air source heat pump (water chilling) packages—Part 1: Heat pump (water chilling) packages for industrial and commercial and similar applications, the energy efficiency ratio of the system at −25 °C is determined to be 1.8. This means that for every 1 kW of electricity consumed, the system is capable of producing 1.8 kW of heat. It is important to note that the performance of the air source heat pump system is significantly influenced by the ambient temperature. As the ambient temperature rises, the heating energy efficiency ratio of the system also increases. The use of an air source heat pump system can provide real-time adjustment of indoor temperature and other parameters, reduce personnel requirements, and save at least 55% of annual electricity consumption.

As stated in Section 3.1, solar radiant panels generate an average of 18.748 MWh of power annually, while the annual heat consumption for heating and cooling, calculated in Section 2, is 19.425 MWh and 2.524 MWh, respectively. When using electric heating, the solar power generated annually can meet 96.51% of the annual heating heat consumption. If the air source heat pump requires no more than 65.04% of the annual power generated by the photovoltaic system, it can satisfy the annual cooling and heating needs. However, Figure 13 shows that the monthly power generation of the solar photovoltaic system does not align with the cooling and heating demands. Therefore, while the building has achieved near-zero-energy consumption in terms of total energy usage, it is necessary to fully achieve this goal by increasing the installed capacity of the photovoltaic system and reducing the heat load required for winter heating.

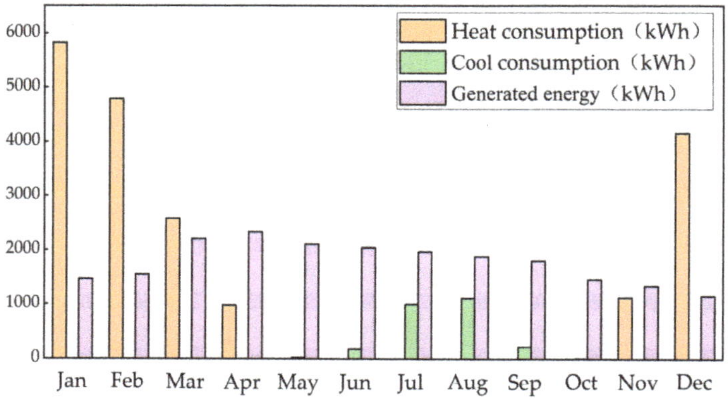

Figure 13. Comparison chart of monthly heat consumption, cooling consumption, and electricity generatio

The carbon emissions during the operational phase of the building should be dete mined based on the energy consumption of different types of systems and the carbo emission factors of different types of energy. The total carbon emissions per unit buildin area (C_M) during the operational phase of the building should be calculated according t the following formula:

$$C_M = \frac{\left[\sum_{i=1}^{n} (E_i EF_i) - Cp\right] y}{A} \quad (7$$

$$E_i = \sum_{j=1}^{n} (E_{i,j} - ER_{i,j}) \quad (8$$

where C_M represents the carbon emissions per unit area during the operational phase c the building ($kgCO_2/m_2$), E_i denotes the annual consumption of the ith energy type in th building (units/year), EF_i stands for the carbon emission factor of the ith type of energ $E_{i,j}$ represents the consumption of the ith energy type by the jth type of system (units/year $ER_{i,j}$ is the amount of the ith energy type provided by renewable energy systems for th jth type of system (in units per year), i refers to the type of terminal energy consumptio in the building (including electricity, gas, and oil), j denotes the type of energy system i the building (including HVAC and lighting), Cp represents the annual carbon reductio amount of the building's green space carbon sink system ($kgCO_2$/year), y is the design lif of the building (in years), and A is the building area (m^2).

Table 8 presents the total CO_2 emissions during the operational phase of the buildin over its lifecycle (50 years).

Table 8. Building operational carbon emissions.

Type	Cooling/Heating Load (kWh/Year)	Electricity Consumption (kWh/Year)	Carbon Emission Factor ($kgCO_2$/kWh)	Carbon Emission (tCO_2/Year)
Air source heat pump heating system	19,426	10,792		6.2
Heating system for electric heating	19,426	19,426	0.5703	11.1
Cold supply system	2524	841		0.5
Illuminating system	-	2746		1.6
Equipment operation system	-	21,900		12.5
		Total		20.3

The energy consumption during the operational phase of the research object mainly comes from electricity, and the emission factor for the electricity grid is selected based on the data released in February 2023 by the Ministry of Ecology and Environment in the notice "Regarding the Management of Greenhouse Gas Emission Reports for Power Generation Enterprises for the Period of 2023–2025." The average emission factor for the national electricity grid in 2022 was 0.5703 kg CO_2/kWh. From Table 8, it can be observed that the heating system using an air source emits 4.9 fewer tCO_2 per year compared to the electric heating system. Over the 50-year lifecycle of the building, the carbon emissions per unit area during the operational phase of the building are 1638 CO_2/m^2.

3.3. Efficient Use of Solar Energy Resources

Table 9 presents the primary wind directions and speeds for the winter and summer seasons in the region where the main control building is located. The data were obtained by consulting the relevant literature. Simulations of the outdoor airflow around the substation's main control building were conducted for both winter and summer seasons using the Ventilation Analysis module in the GBWARE2024, based on the main control building model and related parameters. Figures 14 and 15 show the wind speed distribution at a height of 1.5 m outdoors during winter and summer, respectively. The wind speed on the windward side of the building ranges from 0.4 to 0.7 m/s under typical summer outdoor conditions.

Table 9. Main wind direction and wind speed in each season.

Season	Wind Speed (m/s)	Wind Direction
Winter	3.20	WNW
Summer	2.00	ENE
Transition season	2.00	ENE

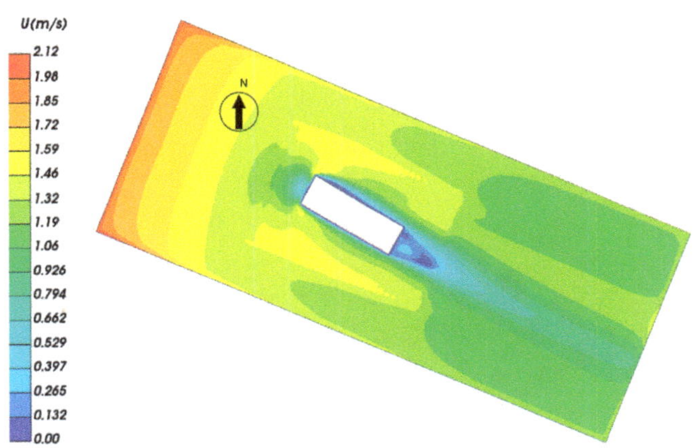

Figure 14. Diagram showing wind speed at 1.5 m in the outdoor wind field during winter.

Due to the indoor design temperature not exceeding 28 °C in the summer and the calculated outdoor ventilation temperature in the Yichun area being 26.7 °C during the summer, natural ventilation can be used to remove the heat entering the room and the heat generated by the room's equipment. Figure 16 shows the distribution of indoor wind speed at a height of 1.5 m in the 10 kV distribution room under summer natural ventilation conditions. Figure 17 shows the cloud map of indoor temperature distribution at a height of 1.5 m in the 10 kV distribution room under natural ventilation conditions in summer. Figure 18 presents the streamline map of indoor conditions in the same room under the same conditions. The maximum indoor wind speed, as shown in the figures, is 0.384 m/s. However, Figure 17 shows a temperature field distribution of around 27.8 °C, which is relatively uniform. The central area between two rows of equipment has a slightly highe temperature of 28.1–28.4 °C due to poor ventilation conditions caused by the obstructiv effect of equipment heat load and obstacles, which prevent effective dissipation of equip ment heat through natural ventilation. Despite its effectiveness in regulating temperatur in most summer conditions, natural ventilation has limitations when it comes to indoo environmental control. Specifically, there is an upper limit on outdoor meteorologica temperature beyond which natural ventilation alone cannot effectively reduce indoor hea load and create a comfortable indoor environment. Therefore, the use of air conditionin and fresh air systems is necessary for maintenance. Research indicates that when the indoc temperature reaches 28 °C, the outdoor ambient temperature for the study object is 26.4 °C due to changes in outdoor temperature. Therefore, activating the air source heat pump i necessary to maintain optimal indoor temperature conditions when the outdoor ambien temperature exceeds 26.4 °C.

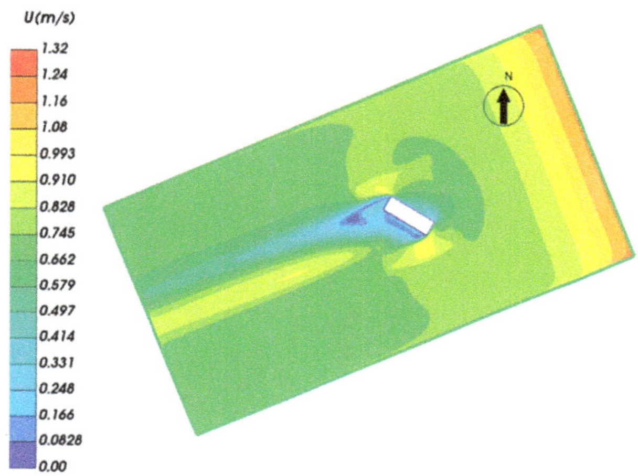

Figure 15. Diagram showing wind speed at 1.5 m in the outdoor wind field during summer.

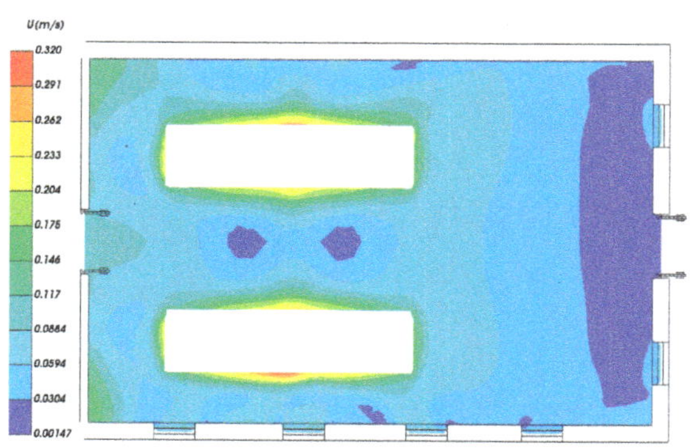

Figure 16. Wind speed distribution cloud map of the 10 kV distribution room during summer.

During the heating season, the indoor design temperature should not be set lowe than 20 °C. In transitional seasons, outdoor ventilation temperatures can fluctuate widel By adjusting the opening range of windows judiciously, excess indoor heat load can b effectively dissipated. This study simulates conditions with an outdoor temperature o 20 °C as the benchmark for transitional seasons. Figure 19 analyzes the distributior of indoor wind speeds at a height of 1.5 m in the 10 kV distribution room during the transitional season under natural ventilation conditions. Figure 20 illustrates the clouc map of indoor temperature distribution at the same height in the 10 kV distribution roorr under natural ventilation conditions during the transitional season. Figure 21 presents the streamline map of indoor conditions in the same room under the same conditions. During the transitional seasons, the maximum indoor wind speed is 0.467 m/s, and the temperature ranges from 20.3 °C to 21 °C. Natural ventilation is an effective method for removing indoor heat load, resulting in significant energy savings compared to air conditioning or mechanical ventilation.

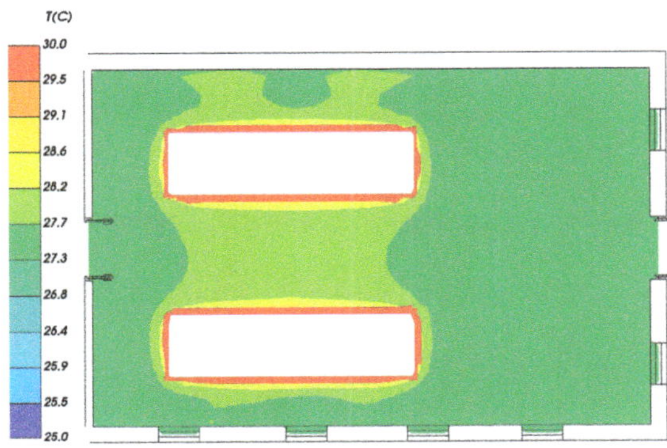

Figure 17. Temperature distribution cloud map of the 10 kV distribution room during summer.

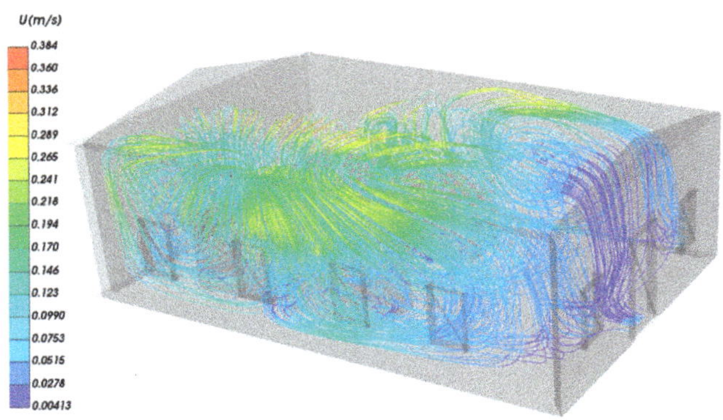

Figure 18. Indoor air streamline diagram of 10 kV distribution room during summer.

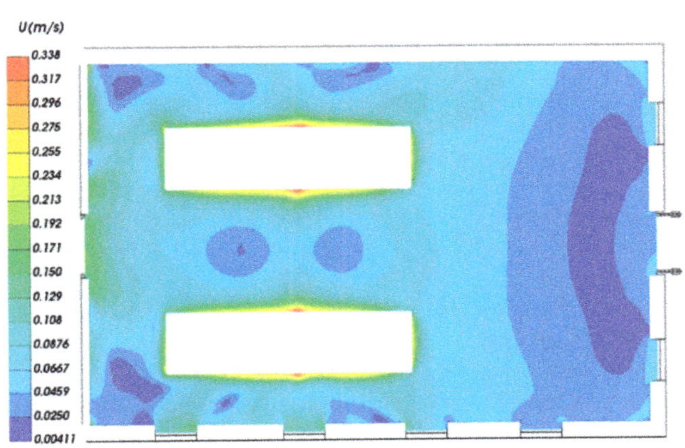

Figure 19. Wind speed distribution cloud map of the 10 kV distribution room during transition season.

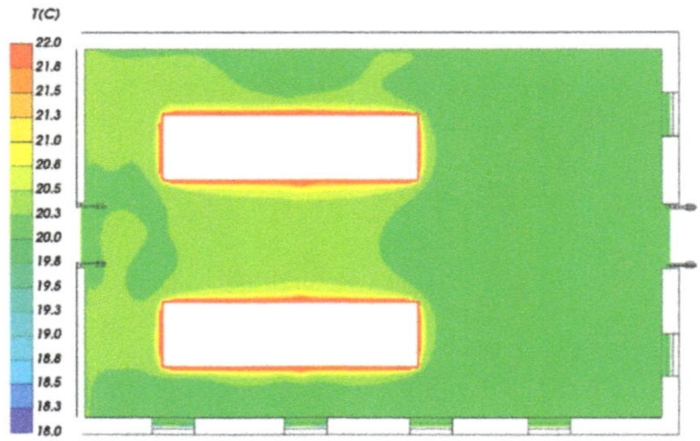

Figure 20. Temperature distribution cloud map of the 10 kV distribution room during transition season.

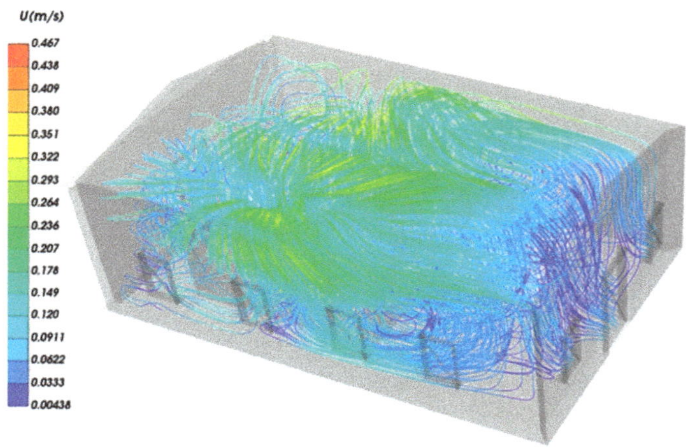

Figure 21. Indoor air streamline diagram of 10 kV distribution room during transition season.

In addition to studying the temperature and velocity fields during the summer and transition seasons, the substation must also ensure a sufficient number of indoor air changes. Based on the above calculation parameters, it has been determined that the maximum number of indoor air exchanges that can be guaranteed for the 10 kV distribution room is 16 times in the summer and 17 times in the excessive season.

4. Conclusions

In the context of actively achieving the goals of carbon peaking and carbon neutrality in China, building energy consumption is influenced by various factors such as geographical conditions, climate, and architectural form. Therefore, energy-saving and carbon reduction in substation buildings should be tailored to local conditions and comprehensively considered. This paper focuses on substation buildings in cold regions as the research object. It employs two methods: photovoltaic power generation systems and computational fluid dynamics (CFD) simulations. By integrating the climatic characteristics of cold regions and practical engineering considerations, suitable energy-saving technology routes are proposed, emphasizing two aspects: the effective utilization of surrounding renewable energy sources and the reduction in equipment operational energy consumption. To expand new avenues of innovation in the technological advancement of energy conservation and carbon reduction in substation buildings, the relevant results are as follows:

(1) The solar photovoltaic power generation system in this study has an annual average electricity generation of 18.75 MWh. Solar power generation can meet the annual

electricity supply for HVAC load. Using a solar photovoltaic power generation system in the building energy supply system can save 10.7 tCO_2.

(2) Additionally, using an air source heat pump can save 44.5% of electricity when compared to electric heating. When both a solar photovoltaic power generation system and an air source heat pump are used to provide a building with a cold and heat source, the annual emissions of CO_2 can be reduced by 4.90 t compared to a traditional electric heating system. It provides some design ideas for realizing zero-energy consumption building and building energy savings and carbon reduction.

(3) To ensure the efficient operation of indoor equipment in the main control building of the substation, natural ventilation can be used during summer and transitional seasons when the outdoor air temperature is between 20 °C and 26.4 °C. Automated control can be implemented to maximize energy-saving and carbon-reduction measures.

Author Contributions: Conceptualization, X.J. and Y.W.; methodology, J.Z and Y.W.; software, J.Z and L.Z.; validation, S.S., Y.W. and C.Z.; formal analysis, J.Z.; investigation, X.G.; writing—original draft preparation, J.Z.; writing—review and editing, X.J., Y.W., X.G. and C.Z.; visualization, S.S.; funding acquisition, C.Z. All authors have read and agreed to the published version of the manuscript.

Funding: This research was funded by the Science and Technology Project of Heilongjiang Power Grid Company, grant number 522448220005, "Research on the key technology of energy-saving systems for power transmission and transformation projects in high-cold areas".

Institutional Review Board Statement: Not applicable.

Informed Consent Statement: Not applicable.

Data Availability Statement: The data presented in this study are available on request from the corresponding author. The data are not publicly available due to privacy.

Conflicts of Interest: Author Ying Wang, Xiuyun Gao and Lei Zhao was employed by Economic and Technological Research Institute of State Grid Heilongjiang Electric Power Co. Ltd. The remaining authors declare that the research was conducted in the absence of any commercial or financial relationships that could be construed as a potential conflict of interest. The authors declare that this study received funding from Heilongjiang Power Grid Company. The funder was not involved in the study design, collection, analysis, interpretation of data, the writing of this article or the decision to submit it for publication.

eferences

Lin, Y.; Shan, P.; Huang, W.; Lin, D.; Huang, H.; Zhang, C.; Zhuo, Z. Evaluation method of building carbon emission in Fujian Province based on CIE-AHP. *Fujian Constr. Sci. Technol.* **2021**, *02*, 77–80.

Voss, K.; Goetzberger, A.; Bopp, G.; Häberle, A.; Heinzel, A.; Lehmberg, H. The self-sufficient solar house in Freiburg-Results of 3 years of operation. *Sol. Energy* **1996**, *58*, 17–23. [CrossRef]

Zhang, S. *Research on Energy Saving Potential and Technical Path of Ultra-Low Energy Consumption Buildings*; Harbin Institute of Technology: Harbin, China, 2016.

Wang, L.; Zheng, L.; Liu, C.; Zheng, N. There are technical routes for zero energy consumption transformation of public buildings in cold areas. *Build. Energy Effic.* **2023**, *51*, 98–104.

Zeyghami, M.; Goswami, D.Y.; Stefanakos, E. A review of clear sky radiative cooling developments and applications in renewable power systems and passive building cooling. *Sol. Energy Mater. Sol. Cells* **2018**, *178*, 115–128. [CrossRef]

Lee, J.B.; Park, J.W.; Yoon, J.H.; Baek, N.C.; Shin, U.C. An empirical study of performance characteristics of BIPV (Building Integrated Photovoltaic) system for the realization of zero energy building. *Energy* **2014**, *66*, 25–34. [CrossRef]

Cao, V.D.; Pilehvar, S.; Salas-Bringas, C.; Szczotok, A.M.; Bui, T.Q.; Carmona, M.; Rodriguez, J.F.; Kjøniksen, A.-L. Thermal performance and numerical simulation of geopolymer concrete containing different types of thermoregulating materials for passive building applications. *Energy Build.* **2018**, *173*, 678–688. [CrossRef]

Li, Z.; Peng, W.; Yujiao, H.; Wei, T.; Yong, S. Relationships between design parameters of see-through thin film photovoltaic facade and energy performance of office building in China cold zone. *Energy Procedia* **2018**, *152*, 401–406. [CrossRef]

Lebied, M.; Sick, F.; Choulli, Z.; El Bouardi, A. Improving the passive building energy efficiency through numerical simulation–A case study for Tetouan climate in northern of Morocco. *Case Stud. Therm. Eng.* **2018**, *11*, 125–134. [CrossRef]

). Balali, A.; Hakimelahi, A.; Valipour, A. Identification and prioritization of passive energy consumption optimization measures in the building industry: An Iranian case study. *J. Build. Eng.* **2020**, *30*, 101239. [CrossRef]

11. Sivaram, P.; Mande, A.B.; Premalatha, M.; Arunagiri, A. Investigation on a building-integrated passive solar energy technolog for air ventilation, clean water and power. *Energy Convers. Manag.* **2020**, *211*, 112739. [CrossRef]
12. Tariq, R.; Torres-Aguilar, C.E.; Sheikh, N.A.; Ahmad, T.; Xamán, J.; Bassam, A. Data engineering for digital twining an optimization of naturally ventilated solar façade with phase changing material under global projection scenarios. *Renew. Energ* **2022**, *187*, 1184–1203. [CrossRef]
13. Liu, X.; Ge, Q.; Jiang, L.; Cui, L.; Li, B.; Du, X. Thinking on the path of total coal consumption control in China. China Populatio *Resour. Environ.* **2019**, *29*, 160–166.
14. Tong, Z.; Chen, Y.; Malkawi, A.; Liu, Z.; Freeman, R.B. Energy saving potential of natural ventilation in China: The impact c ambient air pollution. *Appl. Energy* **2016**, *179*, 660–668. [CrossRef]
15. Oropeza-Perez, I.; Østergaard, P.A. Energy saving potential of utilizing natural ventilation under warm conditions—A case stud of Mexico. *Appl. Energy* **2014**, *130*, 20–32. [CrossRef]
16. Stasi, R.; Ruggiero, F.; Berardi, U. The efficiency of hybrid ventilation on cooling energy savings in NZEBs. *J. Build. Eng.* **202** *53*, 104401. [CrossRef]
17. Gilani, S.; O'brien, W. Natural ventilation usability under climate change in Canada and the United States. *Build. Res. Inf.* **202** *49*, 367–386. [CrossRef]
18. *GB51245-2017*; Unified Standard for Energy Saving Design of Industrial Buildings. China Planning Press: Beijing, China, 2017.
19. *GB50059-2012*; 35~110 kV Substation Design Code. China Planning Press: Beijing, China, 2012.
20. *DL/T5218-2012*; Technical Regulations for the Design of 220~750 kV Substation. China Planning Press: Beijing, China, 2012.
21. *GB50019-2015*; Design Code for Heating, Ventilation and Air Conditioning of Industrial Buildings. China Planning Press: Beijin China, 2012.
22. An, Y. *Thermal Performance Research and Optimization Design of U-Shaped Glass Vacuum Tube Solar Collector*; Beijing Institute of Civi Engineering and Architecture: Beijing, China, 2011.
23. *GB/T 37526-2019*; Assessment Method for Solar Energy Resource. State Administration for Maket Regulation. Standardizatio Administration: Beijing, China, 2016.
24. *GB 50797-2012*; Design Code for Photovoltaic Power Station. China Planning Press: Beijing, China, 2012.
25. Zhang, Q.; Li, Y.; Liu, B.; Di, H. Study on suitability of low temperature air source heat pump heating method in northern heatin area. *Build. Sci.* **2015**, *31*, 140–145.
26. *GB50229-2019*; Fire Protection Standard for Design of Thermal Power Plants and Substations. China Planning Press: Beijin China, 2019.
27. *GB/T25127.1*; Low Ambient Temperature Air Source Heat Pump (Water Chilling) Packages—Part 1: Heat Pump (Water Chilling Packages for Industrial & Commercial and Similar Application. State Administration for Market Regulation. Standardizatio Administration: Beijing, China, 2020.

applied sciences

MDPI

rticle

Active Strategies Based on Parametric Design for Applying Shading Structures

o-Soon Choi

Department of Architecture, Gachon University, 1342 Seongnamdaero, Sujeong-gu, Seongnam-si 13120, Gyeonggi-do, Republic of Korea; hosoon@gachon.ac.kr; Tel.: +82-31-750-5519

Abstract: This study aimed to increase the energy independence of buildings by utilizing solar energy to produce renewable energy. The subject of this study was a shading structure installed in an outdoor space to provide solar energy. Solar panels were applied to the shaded structures to generate energy actively. The solar panels were designed to be moved according to the optimal tilt angle each month to produce the optimal amount of renewable energy. The architectural design of the shading structure and the energy simulation of the solar panels were conducted using a parametric design. The results of the energy simulation showed the generation of 31,570 kWh· year^{-1} of renewable energy. This amount of energy is 10% higher than that produced by fixed solar panels. Thus, the moving solar panel system developed in this study not only increases the energy independence of buildings, but also has the advantage of higher renewable energy production compared with fixed solar panels. Additionally, various types of shading structures can be designed depending on the combination of solar panel modules; in particular, moving solar panels have the potential to facilitate ecofriendly designs when applied to the exterior of buildings.

Keywords: active design strategies; energy-efficient architecture; solar panel system; parametric design; shading structure

tation: Choi, H.-S. Active rategies Based on Parametric esign for Applying Shading ructures. *Appl. Sci.* **2024**, *14*, 974. tps://doi.org/ 3390/app14030974

:ademic Editors: David Bienvenido uertas and Daniel Sánchez-García

ceived: 3 January 2024
vised: 15 January 2024
cepted: 22 January 2024
blished: 23 January 2024

1. Introduction

Climate change caused by environmental pollution is progressing rapidly. Industries need to be reorganized to reduce environmental pollution to preserve the global environment for the survival of mankind. The construction sector directly contributes to environmental pollution through energy consumption. Changes in the energy consumption in the construction field are an important factor affecting the global environment, given that the energy consumption of the construction sector accounted for 37% of the global energy consumption in 2022 [1]. Consequently, technological development related to efficient energy consumption and renewable energy production in buildings is essential. Accordingly, the International Energy Agency announces energy reduction goals for the construction sector every year. To achieve these goals, each country specifically reduces building energy consumption and CO_2 emissions through Building Energy Sufficiency policies [2,3]. In addition, energy-related studies in the field of architecture are actively being conducted to generate renewable energy in buildings.

Clean and efficient building energy systems can be divided into active and passive energy-saving systems. Active energy-saving systems generate energy within a building using solar thermal panels, photovoltaic systems, wind power, and geothermal energy. Passive energy-saving systems conserve building energy and minimize building energy consumption through the insulation of the exterior and roof of the building and the installation of highly energy-efficient windows [4–7]. Recently, studies have been conducted on complex, clean, and efficient building energy systems that combine active and passive energy-saving systems. Studies have been conducted to develop an energy system optimization model and policies for model operation under the premise that a method combining active and passive energy-saving systems has the potential to reduce buildin energy costs [8,9]. An empirical study was conducted to apply the developed energy-savin system model to specific buildings. To increase the energy use efficiency of public instit tions and schools, studies have been conducted on the development of energy combinatio models and algorithms [10,11] and on active and passive energy-saving systems to increas the energy independence of residential buildings [12,13].

This study conducted architectural research based on active design strategies. Th active design strategy pursued in this study involved increasing the energy independenc of buildings by utilizing solar energy from various natural energy sources. Second, thi study targeted buildings that were easy to construct. In this study, buildings that coul be constructed within a short period without applying professional construction method were selected. From an active design perspective, buildings that are easy to constru were selected as the research objects to maximize the effect of renewable energy generate by the building compared with construction costs. The solar panel used to utilize sola energy is not a fixed type, but a mobile model based on a parametric design for optima energy harvesting. The experimental process was designed to present a model of creativ architectural design.

The building selected for this study was a shaded structure. The modularization c the elements that constitute the structure and the ease of the construction process satisf the attributes of active design. This empirical study serves as an experimental model for a energy-efficient architectural design that can be used as a shelter for outdoor spaces neede for outdoor activities.

2. Literature Review

2.1. Active Design Based on Solar Panel System for Energy-Efficient Architecture

The photovoltaic effect, which generates energy from sunlight, was first introduced b the French physicist Alexandre Edmond Becquerel in 1839. Solar cells were first used as a energy source when they were deployed as a power source for communication networks i Americus, Georgia in 1955 [14,15]. Solar panels were first applied in the architectural fiel by American architects and engineers just before World War II up to the late 1950s. Afte the war, as concerns about energy supply and global environmental problems emerge interest in the development of ecofriendly architectures that utilize solar energy increase However, as oil energy was used as the main energy source at the time, interest in sola energy decreased, whereas interest in renewable energy increased again following th oil crisis of the 1970s [16]. Today, solar-energy-based active designs are being applied t various buildings because of the global interest in low carbon emissions and energy saving in all industrial fields.

The methods for utilizing solar energy in the construction sector can be broadl divided into two types. First, some buildings produce electrical energy from solar energ This method involves installing a solar power generation system in a building. Solar cell are installed on the roofs of buildings to convert the natural energy of sunlight into electrica energy. Second, some buildings produce heat energy from solar energy. Solar collectors ar installed on the roofs of buildings to produce energy for heating water [17,18].

Solar panels installed in buildings are placed on the rooftops of the buildings to expos the solar panels fully to sunlight and facilitate their maintenance. Building-integrate photovoltaics (BIPVs) are widely used to install solar panels on sloped roofs and buildin façades rather than on flat roofs [19]. Before applying BIPVs, the selection of the opt mal solar panel location and various climatic conditions are considered to increase th energy production efficiency [20]. Once the location of the solar panel is selected, th panel is attached to the exterior wall of the building, and such an integration change the building's envelope from an energy consumer to an energy producer. Solid-base active building envelopes are systems in which solar panels are integrated into the buildin roof and exterior walls. Currently, owing to the development of solar panel material solid-based active building envelopes are applied not only to opaque building shells bu also to transparent windows [21]. Semitransparent solar panels have been inserted into double-glazed windows to maximize the amount of renewable energy produced by building envelopes [22–24]. In particular, the active energy façade system installed on the southern vertical surface of a building façade maximizes the energy production efficiency through the attachment of solar panels to moveable louvers installed to control the solar influx [25,26]. Solar-panel-integrated louvers with fluid properties have the potential to be applied to the exterior of a building and used as a design element for creative façades and to produce renewable energy. Research related to the kinetic façade system installed on the exterior wall of a building aims to use technology to increase the energy efficiency of buildings through architectural design. Investigations have been conducted on kinetic façades based on geometric patterns inspired by traditional architectural designs in the Middle East [27]. In addition, researchers have developed a kinetic biomimetic system in which plants respond to light as a kinetic façade [28]. Glass façades of various colors with different geometric shapes and a kinetic façade abstracting a biomimetic system were developed on a building façade. These studies aimed to improve daylight performance within the building space using the developed kinetic façade. Daylight performance was measured in a virtual space via parametric simulations.

2.2. Parametric Design in Architecture

"Parameter", the core of parametric systems applied to various fields today, has a broad meaning. The etymology of "parameter" is a combination of the ancient Greek words "para" and "metron," where "para" means "besides, before, or instead of," and "metron" means "measure." According to ancient Greek etymology, "parameter" refers to a term that represents or determines another measure. Parametric design in the field of architecture was first introduced in 1971 as the concept of "relationships between measures" [29]. These parametric concepts were implemented as buildings in a virtual space using computer software.

Architectural parameters were used to draw conclusions based on the diverse variables necessary for the architectural design. In other words, the parameters in the field of architecture are characterized by a range of possible values rather than by obtaining a constant result, such as a mathematical formula. The advantage of parametric design is that it can produce various architectural design results by controlling several parameters [30]. Proposals of various architectural designs using this parameter system have the advantage of not only deriving creative designs but also judging the effectiveness of the design according to each parameter.

Parametric tools originated in SketchPad [31], a computer program developed by Ivan Sutherland in 1963. SketchPad, the first computer-aided system (CAD) program, is a pen equipped with a laser pointer [32]. It allows not only the creation of simple basic shapes, but also design changes through parametric equations. Additionally, as the parameters change, the shapes are recalculated, and design changes become possible. While early CAD systems simply digitized design concepts expressed on paper using basic logic, parametric design is currently evolving to present architectural concepts and new design perspectives, owing to the development of parametric software programs.

Parametric design has been applied in the architectural field since the early 1990s. Parametric modeling has significantly influenced the design process for architectural development and the establishment of new design concepts. Parametric design was established as a new architectural theory called "Parametricism" and began to be used as a concept to signify a new global architectural style beyond a simple design tool [33]. In particular, the digital revolution of the 20th century accelerated the application of parametric design in the architectural field.

Parametric design software have developed rapidly since the early 2000s. The most commercialized software in the architectural field today is Grasshopper, a plug-in program for Rhinoceros, and Dynamo, a plug-in program for Revit developed by Autodesk under the concept of building information modeling [34,35]. Designers use parametric design software, and design is performed through the definition of an algorithm, which is a integrated system that combines diverse variables for architectural design. An algorithm i a system that solves a given problem using well-defined processes and basic unambiguou instructions [36]. When limited data are input into the algorithm, special results ar obtained. Various algorithms have been built for optimal design, and the best method tha satisfies the criteria for design decisions is determined by generating positive or negativ results based on the input data.

2.3. Shading Structure

Shading structures are minimal architectural structures installed to avoid exposure t excessive sunlight. As an architectural structure, the shading structure has its origins in th velarium of the ancient Roman era. The velarium is a canvas-type folding system compose of ropes and masts, and a representative example is the one installed on the roof of th Colosseum in ancient Rome. The velarium installed on the roof of the Colosseum has th advantage of low installation costs because it uses minimal construction technology, pr vides shade to spectators, and simultaneously facilitates ventilation inside the Colosseu through the sloping roof [37]. Shading structures, which were developed for long-ter outdoor activities (starting with the velarium), are still in good use. Shaded structures ar installed on public streets in countries with hot climates. The awning installed at the top c a public street not only protects pedestrians from excessive solar radiation but also protect public spaces from dust and glare. Additionally, through material development, shadin structures are responsible for the architectural research field of membrane architecture [38 The flexibility of the materials that constitute the shading structure and the lightweigh architectural structure that supports the flexibility of the materials not only play a functiona role in blocking sunlight but are also used as an aesthetic element in architectural desig In particular, sunsails that make up sunshades are an important architectural elemen in that they can variably construct and control spaces and provide users with variou spatial experiences depending on light, shadow, and internal temperature [39]. Shadin structures, which have architectural functions and aesthetic potential, are widely use today in external spaces, stadium roofs, and shading designs on the exteriors of building

From the perspective of building energy utilization, the sunsail-type shading structur is based on a passive energy system. However, current buildings must have active energ systems that produce a certain amount of renewable energy within the building itself. Fro this perspective, shading structures installed outdoors have the potential to utilize passiv and active energy systems, because of the direct exposure of the surrounding environme to sunlight, the area of the structure, and the flexible nature of the structure.

3. Methodology

This study consisted of five stages. First, an architectural design was develope for the shading structure. The shading structure was designed to be variable and eas to install. The main architectural elements that constituted the shading structure wer integrated solar panels with a solar-energy-based active design. It was assumed that th shaded structure was installed in an external space without shadow interference fro surrounding structures. The purpose of installing a shading structure in an external spac was to build a multipurpose architectural space to facilitate various outdoor activities an simultaneously utilize solar energy to increase the renewable energy production efficienc of the building. Second, the architectural design was implemented in a virtual space usin a computer program. In this study, a shaded structure was designed using the Autodes Revit 2020 software. The Revit software is a representative architectural design progra that integrates various data from the architectural design process and digitally realizes a the stages from architectural planning and design to construction and operation, whic are the final stages of construction [40]. Third, a parametric design was developed. Th purpose of the parametric design was to build a solar-based active architectural desig Dynamo, Revit's plug-in software, was used to build algorithms for the parametric desig and perform the design according to the variables. Fourth, the active energy design of the shaded structure was simulated. The design simulation aimed to measure the amount of renewable energy generated by the shading structure. This study used Revit's plug-in software, the Insight program, to measure the amount of solar energy generated by the solar panels. Finally, the parametric design results were derived. The amount of renewable energy generated from the shading structure over one year and the overall appearance of the shading structure were evaluated (Figure 1).

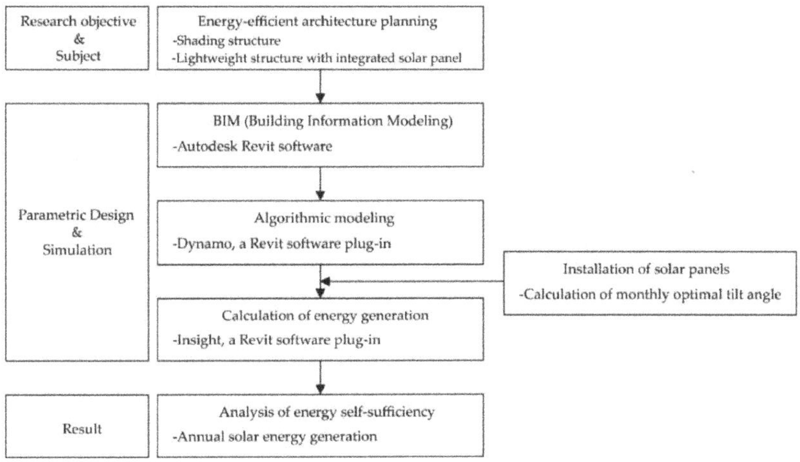

Figure 1. Flowchart of methodology.

4. Design

4.1. Arrangement of Solar-Panel-Integrated Modules That Form the Shading Structure

The shading structure was implemented using the Autodesk Revit 2020 program [41]. The module comprising the shading structure was an integrated solar panel and was 0.5 m wide and 0.7 m long. All the panels were placed facing south to ensure maximum sunlight exposure (Figure 2A). The module size was set based on the specifications of solar panels that are currently being produced. The solar panel selected for this study (Q. PEAK DUO XL-G10, 1 m wide and 2.1 m long) is used on building roofs [42]. To apply a solar panel to a shaded structure, the panel must be in the form of a small module. As the shading structure is easy to install and has a variable structure, the solar panels used on the roofs of buildings must be lightweight. By dividing the solar panels installed on the roof of the building selected for this study into six parts, a minimum unit shading structure module with a width of 0.5 m and a length of 0.7 m was obtained. The size of this single module was determined to ensure no panel loss after segmenting the existing large solar panels. The modules on the upper floor were placed at 1 m intervals up to a maximum height of 7 m. The modules were not placed up to 1 m above the ground to prevent damage to them from pedestrians, and they were placed up to 7 m, the height of two floors of a regular building, to create sufficient shade space on the ground floor. Additionally, the modules were arranged inclinedly at intervals of 0.75 m on the sides and 1 m on the top to prevent shadow interference between the modules (Figure 2B). In the plan, each module was arranged at intervals of 1 m on the side and 0.75 m at the top and bottom. A total of 57 unit modules were placed at a site with a width of 11 m and a length of 5 m (Figure 2C). An area of 55 m^2, including the empty space between the 57 solar modules, was created using the shading structure. The 55 m^2 area was planned to satisfy Korea's minimum installation standard for public open spaces of 45 m^2 [43]. Excluding the space between the 57 solar panels in the 55 m^2 area, a 45 m^2 external space was created for people to occupy under the shading structure. First, through the arrangement of this module, the energy production efficiency was maximized because there was no shadow interference between the modules. Second, a small amount of sunlight entered the shade curtain through th gap between the modules; therefore, the inside of the shade cloth was not completely dar The number of modules can be varied according to the size of the external space in whic the shading structure is installed and the purpose of the shading structure.

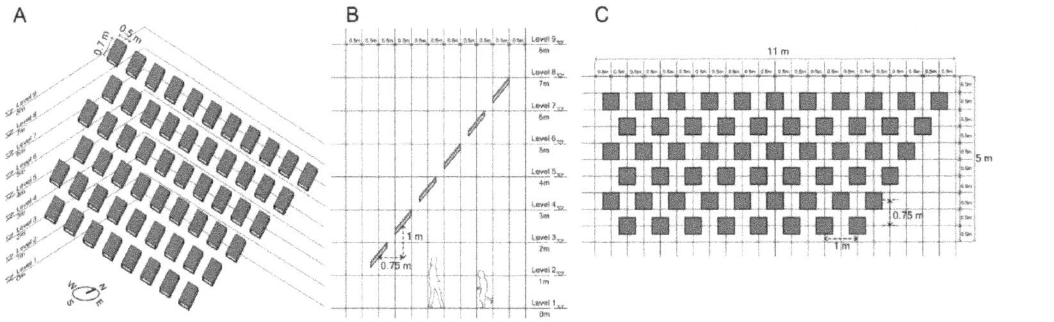

Figure 2. Layout of solar-panel-integrated modules. (**A**) Bird's-eye view of shading structur (**B**) Cross-section of shading structure, (**C**) Layout of shading structure.

4.2. Flexible Shading Structure Plan for Energy Optimization

This study utilized solar panels for an active design that produces energy within th building itself and employed flexible solar modules to maximize the energy generatio efficiency. In a flexible solar module, the installation angle of the solar panel changes t increase the energy generation efficiency according to the movement trajectory of the su which changes monthly during the year. These flexible solar panels can maximize energ generation compared with existing fixed solar panels. They have the potential to be applie to the exterior of buildings as a "kinetic façade" and be used as an important element c architectural design. The solar modules were installed according to the optimal tilt ang for each month to achieve the optimal energy generation. The monthly optimal tilt ang value is calculated using the Calabrò function (1) [44]. In Equation (1), β is the optimal ti angle value, and a_1 and a_2 are coefficients derived from the solar declination value.

$$\beta = a_1 + a_2 \times \varphi \quad (1$$

The monthly optimal tilt angle value (β) is derived from the coefficients a_1 and a from January to December. The device supporting the solar module was designed to mov 90° relative to the south-facing central axis of the panel. The solar modules were place according to the optimal tilt angle for each month using a lightweight metal device tha could move in response to sunlight. In each solar panel, PV cells, each with a thicknes of 6 mm, were placed on top of a lightweight steel structure with a thickness of 1 mn The solar panel had a rotating device installed at the bottom such that it could be move according to the optimal tilt angle. The rotating device integrated with the solar panel wa placed on thin aluminum tubes (d = 1 cm) (Figure 3). The total weight of a unit solar pane including the lightweight steel structure, was approximately 0.5 kg. The total weight of th 57 solar panels that constituted the shading structure was expected to be 28.5 kg, and thi weight was sufficiently supported by the tensile-supporting structure. In particular, th tensile-supporting structure of the thin aluminum tube was a flexible structural form tha moved in response to wind action. This flexible structure is a structurally safe model i which the structure deforms without collapsing owing to external shock [45].

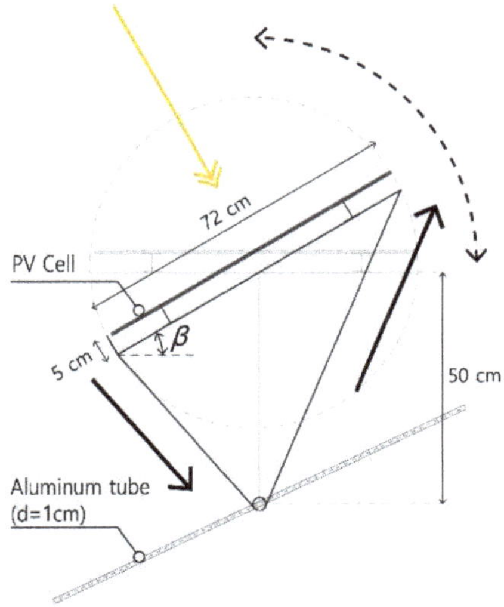

Figure 3. Principle of flexible solar-panel-integrated module.

The study was conducted assuming that the shading structure was installed in Seoul, the capital of Korea (37.5° N, 126.9° E). By substituting the latitude of Seoul (φ = 37.5) and the coefficients a_1 and a_2 from January to December into Equation (1), the optimal tilt angle values are derived (Table 1).

Table 1. Optimal tilt angle from January to December for Seoul (37.5° N, 126.9° E).

Month	Coefficients		Optimal Tilt Angle (Deg)
	a_1	a_2	β
January	31.33	0.68	56.8°
February	16.25	0.86	48.5°
March	6.80	0.84	38.3°
April	−6.07	0.87	26.6°
May	−14.95	0.87	17.7°
June	−19.27	0.87	13.4°
July	−15.65	0.83	15.5°
August	−4.23	0.75	23.9°
September	6.42	0.77	35.3°
October	15.84	0.83	47.0°
November	23.61	0.84	55.1°
December	30.56	0.76	59.1°

5. Parametric Modeling and Simulation

5.1. Building Algorithms for Parametric Modeling

The aforementioned shading structure implemented in a virtual space using the Revit software was composed of solar-panel-integrated modules of a certain size. In addition, the modules moved according to the optimal tilt angle every month to maximize the renewable energy generation. An energy simulation for the proposed shading structure was conducted using the size of the solar-panel-integrated module and the optimal tilt angle at which the module was installed as variables. Parametric modeling is the process of designing through the definition of an algorithm that integrates these variables. This study implemented parametric modeling using Dynamo [46], a plug-in program for the Revit software used for the implementation of the shading structure.

Dynamo is a programming language that allows the development of Revit programs. This tool can develop functions based on Revit to suit the user's environment [47]. Through the language system built with Dynamo, simple repetitive tasks owing to various environmental changes are shortened, and large amounts of input data are easily processed. In addition, Dynamo is a visual graphic programming tool that makes it easy to understand the flow of parametric modeling, because it is a programming language that connects nodes for executing commands and nodes for inputting variables. This parametric modeling process involves building an algorithm. The schematic of the algorithm for this study is described as follows. First, the solar panel is selected from among the various architectural elements that constitute the shading structure. A specific solar panel module can be selected from several solar panels. Second, the size of the solar panel module is determined. Finally, the solar panel module is placed at the desired installation angle (Figure 4).

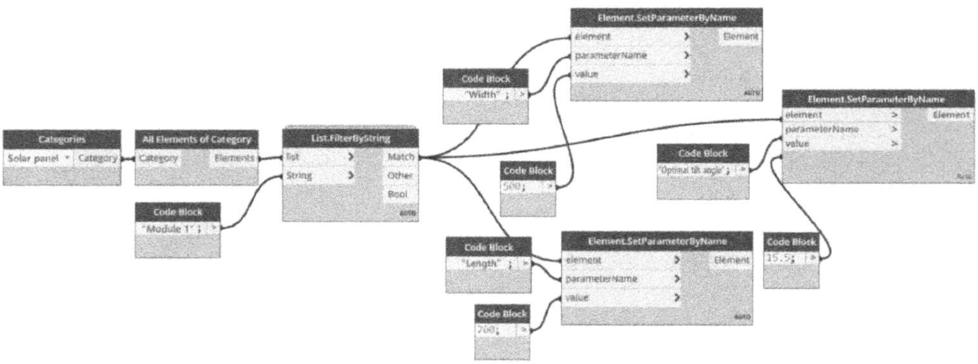

Figure 4. Schematic of the algorithm, with the width set to 500 mm and the length set to 700 mm for solar panel "Module 1" and an optimal tilt angle of 15.5° given as input.

5.2. Simulation for Solar Energy Generation

The simulation for renewable energy generation was conducted using the Revit plug-in, the Insight program [48]. The Insight program was developed to simulate solar energy generation in buildings using Revit. The input variables for running the Insight program include information about the site where the building will be constructed. When the name of a city is entered, the program automatically recognizes its latitude and longitude. Second, a solar panel attached to a building was selected for the simulation. We could select the type of solar panel to be installed on the exterior wall, and we could also select all or a few solar panels. Third, the simulation period was set. Solar panel modules were placed according to the monthly optimal tilt angle from January to December 2023 derived using the Calabrò function in Section 4.2. Subsequently, a solar energy simulation was conducted by setting sunrise and sunset times from 6 a.m. to 8 p.m. (Figure 5).

Table 2 presents the amount of energy generated from the 57 solar panel modules (total panel area, 20 m^{-2}) that constitute the shading structure. Based on the energy simulations, the months with the highest and lowest solar energy production were June and December, respectively. In June, a total of 3474 kWh·$month^{-1}$ was generated from solar panels installed at an optimal tilt angle of 13.4°. On the other hand, in December, a total of 1963 kWh·$month^{-1}$ was generated from solar panels installed at an optimal tilt angle of 59.1°. The total amount of energy generated from the solar panels during the year from January to December 2023 was 31,570 kWh·$year^{-1}$.

Figure 5. Solar panel module simulation generated by entering the optimal tilt angle of 35.3° corresponding to September 2023 into the Insight software.

Table 2. Solar energy generated according to optimal tilt angle and fixed optimal tilt angle (50°) from January to December.

Month	Total Area of Solar Modules	Optimal Tilt Angle	Energy Generation	Fixed Optimal Tilt Angle	Energy Generation
	m^{-2}	β	$kWh{\cdot}Month^{-1}$	β	$kWh{\cdot}Month^{-1}$
January	20	56.8°	2112	50°	2023
February	20	48.5°	2442	50°	2434
March	20	38.3°	3169	50°	3162
April	20	26.6°	2654	50°	2505
May	20	17.7°	3173	50°	2750
June	20	13.4°	3474	50°	2780
July	20	15.5°	1979	50°	1669
August	20	23.9°	2877	50°	2622
September	20	35.3°	2899	50°	2846
October	20	47.0°	2713	50°	2702
November	20	55.1°	2115	50°	2090
December	20	59.1°	1963	50°	1907
	Total energy generation		31,570 $kWh{\cdot}year^{-1}$		29,490 $kWh{\cdot}year^{-1}$

6. Results and Discussion

This study aimed to conduct empirical research to increase energy independence in the field of architecture. A shading structure composed of solar panel modules was selected as the research subject because, first, the shading structure itself is light and can be easily installed in an external space. As the structure is easy to install, various shading structure designs that utilize natural environments, such as trees, can be proposed. In addition, solar panels of various colors created through advancements in solar panel materials have the potential to be used as creative design elements for shaded structures (Figure 6). Second, the shading structure is directly exposed to sunlight; therefore, it is the optimal architecture to apply an active energy design that produces renewable energy from sunlight.

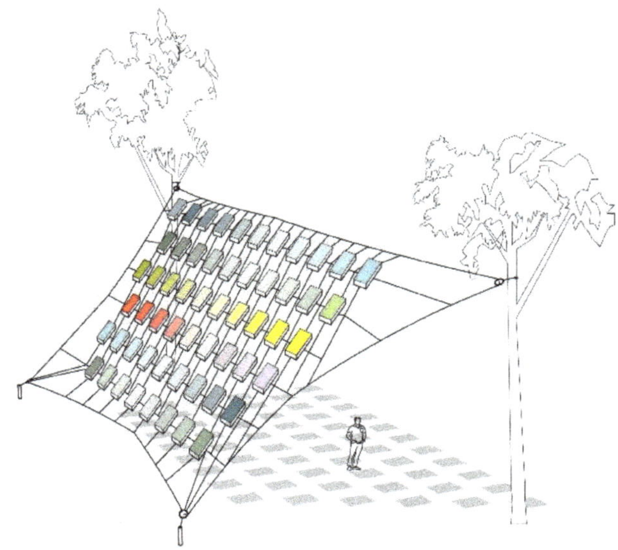

Figure 6. Virtual image of a shading structure with colored solar panels.

The shaded structure was constructed using a parametric design process. Parametri design is a process wherein a design is completed by inputting limited data into an alg rithm built by a designer. Parametric design is evolving with the development of variou software programs. In this study, a shading structure was implemented in a virtual spac using Autodesk Revit. The Revit program is a preliminary step in applying parametri design and ensures that all the elements that constitute a building are implemented in three-dimensional space. After the shading structure was built using the Revit progran an algorithm for the parametric design was built using Dynamo, Revit's plug-in softwar The Dynamo algorithm was designed with the inputs of the size and installation angle c the solar panel module to be applied to the shading structure. A parametric design wa conducted by inputting the optimal tilt angle information from January to December fc 57 solar panel modules with a width of 0.5 m and a length of 0.7 m. After the physical forr of the shading structure was determined using the Dynamo algorithm, an energy simu lation was performed using the Revit plug-in software, the Insight program. The Insigh program produces solar-based renewable energy upon inputting the location informatio of the shading structure and the period for energy simulation. In this study, an energ simulation was conducted for one year, from January to December 2023, assuming Seou to be the location where the shading structure will be installed. Consequently, the energ generation was the highest in June (3474 kWh·month^{-1}), and the lowest in Decemb (1963 kWh·month^{-1}). The total energy produced through the solar panels in 2023 wa 31,570 kWh·year^{-1}. Assuming that LED street lights (100 W·h^{-1}) installed in outdo spaces operate for 12 h every day, the amount of generated energy (31,570 kWh·year^{-} derived from this study is sufficient to supply power to 72 LED street lights. These result demonstrate that parametric design in the architectural field is highly efficient in terms c construction and renewable energy generation.

As an empirical study related to kinetic façades, a building with a parametric desig applied to the façade was completed in Germany in 2020. The optimal tilt angle for opt mizing solar power generation was derived through a parametric simulation, and sola modules were placed at different angles on the south, southwest, and west sides of th building. As of 2022, the total renewable energy generated from building façades wa equivalent to 14% of the total building energy consumed [49]. Solar panels at different a gles were installed on the three building façades using the kinetic façade concept, increasing the energy independence of the building.

The shading structure based on kinetic solar panels, whose positions change every month, as proposed in this study, can increase energy generation compared to solar panels fixed at optimal tilt angles. The generated renewable energy not only supplies power inside the building but can also power a number of external streetlights around the shading structure.

7. Conclusions

Since the onset of the COVID-19 pandemic, various external activities have been undertaken. This study was conducted with the objective of planning architectural spaces that serve as rest areas for various outdoor activities and increase the energy independence of architecture. A shading structure was used as the research object to provide the resting space necessary for outdoor activities, and an integrated solar panel module was used as the main structure of the shading structure. The novelty of the structure proposed herein is that, rather than being passive, where existing solar panels are fixed and installed on the roof or façade of a building, this structure was designed to move actively according to the optimal tilt angle for optimal solar energy generation every month. The advantages of these "kinetic solar panels" are, first, that renewable energy generation increases compared with existing fixed solar panels. When installing fixed solar panels in Seoul, the subject of this study, the optimal solar panel angle was 50° [50]. When the shading structure proposed herein was installed at a fixed angle of 50° and energy simulation was performed using the Insight software for one year in 2023, a total of 29,490 kWh·year^{-1} was derived (Table 2). The energy generation of 29,490 kWh·year^{-1} from solar panels fixed at 50° is equivalent to 93% of the energy (31,570 kWh·year^{-1}) generated by the kinetic solar panels proposed in this study. Second, by using moving solar panels, the scale and shape of the shading structure can be varied by varying the size and combination of various panel modules. Ultimately, the kinetic solar-panel-based shading structure proposed herein can increase energy generation compared with fixed solar panels and supply power to several external streetlights around the shading structure. Additionally, in terms of design, kinetic solar panels can be applied not only to simple lightweight structures, such as shading structures, but also to the exteriors of various buildings, and can be applied as an important element of architectural design.

The proposed kinetic solar panel was designed in a virtual space. Assuming that a shaded structure is built in a real space, this study has the limitation of not having a specific plan regarding the mechanism of the device for installing solar panels on the structure. Future research will aim to develop a solar panel system that moves independently according to the monthly optimal tilt angle. In particular, analyzing the power consumption required to move solar panels is an important research topic that determines the energy efficiency of kinetic solar panels. This study has the potential to initiate new convergence research through a collaboration between computer and electrical energy engineering for the design of kinetic solar panels, focusing on the field of architectural design.

Funding: This study was supported by a National Research Foundation of Korea (NRF) grant funded by the Korean Government, Ministry of Science, and ICT (MSIT) (No. 2022R1F1A1065442).

Institutional Review Board Statement: Not applicable.

Informed Consent Statement: Not applicable.

Data Availability Statement: The data presented in this study are available in the article. Further inquiries for the data presented in this study are available on request from the corresponding author.

Conflicts of Interest: The author declares no conflict of interest.

References

1. International Energy Agency. Final Energy Consumption of Buildings Relative to Other Sectors. 2022. Available online: https://www.iea.org/data-and-statistics/charts/final-energy-consumption-of-buildings-relative-to-other-sectors-2022 (accessed on 11 August 2023).
2. Hu, S.; Zhou, X.; Yan, D.; Guo, F.; Hong, T.; Jiang, Y. A systematic review of building energy sufficiency towards energy and climate targets. *Renew. Sustain. Energy Rev.* **2023**, *181*, 113316. [CrossRef]
3. Park, D.J.; Yu, K.H.; Yoon, Y.S.; Kim, K.H.; Kim, S.S. Analysis of a building energy efficiency certification system in Korea. *Sustainability* **2015**, *7*, 16086–16107. [CrossRef]
4. Ma, H.; Zhou, W.; Lu, X.; Ding, Z.; Cao, Y. Application of low cost active and passive energy saving technologies in an ultra-low energy consumption building. *Energy Procedia* **2016**, *88*, 807–813. [CrossRef]
5. de Gracia, A.; Navarro, L.; Coma, J.; Serrano, S.; Romaní, J.; Pérez, G.; Cabeza, L.F. Experimental set-up for testing active and passive systems for energy savings in buildings–lessons learnt. *Renew. Sustain. Energy Rev.* **2018**, *82*, 1014–1026. [CrossRef]
6. Omrany, H.; Ghaffarianhoseini, A.; Ghaffarianhoseini, A.; Raahemifar, K.; Tookey, J. Application of passive wall systems for improving the energy efficiency in buildings: A comprehensive review. *Renew. Sustain. Energy Rev.* **2016**, *62*, 1252–1269. [CrossRef]
7. Benaddi, F.Z.; Boukhattem, L.; Ait Nouh, F.; Cesar Tabares-Velasco, P.; Benhamou, B. Energy-saving potential assessment of a classroom building envelope through sensitivity analysis and multi-objective optimization under different climate types. *Build. Serv. Eng. Res. Technol.* **2023**, *44*, 309–332. [CrossRef]
8. Wu, Z.; Shi, X.; Fang, F.; Wen, G.; Mi, Y. Co-optimization of building energy systems with renewable generations combining active and passive energy-saving. *Appl. Energy* **2023**, *351*, 121514. [CrossRef]
9. Shukla, A.K.; Yadav, A.K.; Prakash, R. Active and passive methods for cooling load reduction in a tropical building: A case study. *Energy Conv. Manag.* **2023**, *293*, 117490. [CrossRef]
10. Yu, S.; Liu, X.; Yang, J.; Han, F.; Wei, J. The optimization research on the coupled of active and passive energy supplying in public institutions in China. *Energy Built Environ.* **2024**, *5*, 288–299. [CrossRef]
11. Díaz-López, C.; Serrano-Jiménez, A.; Lizana, J.; Lopez-Garcia, E.; Molina-Huelva, M.; Barrios-Padura, A. Passive action strategies in schools: A scientific mapping towards eco-efficiency in educational buildings. *J. Build. Eng.* **2022**, *45*, 103598. [CrossRef]
12. Taherian, H.; Peters, R.W. Advanced Active and Passive Methods in Residential Energy Efficiency. *Energies* **2023**, *16*, 3905. [CrossRef]
13. Almehmadi, F.A.; Alqaed, S.; Mustafa, J.; Jamil, B.; Sharifpur, M.; Cheraghian, G. Combining an active method and a passive method in cooling lithium-ion batteries and using the generated heat in heating a residential unit. *J. Energy Storage* **2022**, *49*, 104181. [CrossRef]
14. McEvoy, A.; Markvart, T.; Castaner, L. *Practical Handbook of Photovoltaics*, 2nd ed.; Academic Press: Cambridge, MA, USA, 2011; p. 15.
15. Marques Lameirinhas, R.A.; Torres, J.P.N.; de Melo Cunha, J.P. A photovoltaic technology review: History, fundamentals and applications. *Energies* **2022**, *15*, 1823. [CrossRef]
16. Barber, D.A. *A House in the Sun: Modern Architecture and Solar Energy in the Cold War*; Oxford University Press: Oxford, UK, 2016; p. 53.
17. Kasaeian, A.; Nouri, G.; Ranjbaran, P.; Wen, D. Solar collectors and photovoltaics as combined heat and power systems: A critical review. *Energy Conv. Manag.* **2018**, *156*, 688–705. [CrossRef]
18. Nayak, P.K.; Mahesh, S.; Snaith, H.J.; Cahen, D. Photovoltaic solar cell technologies: Analysing the state of the art. *Nat. Rev. Mater.* **2019**, *4*, 269–285. [CrossRef]
19. Al Dakheel, J.; Tabet Aoul, K. Building Applications, opportunities and challenges of active shading systems: A state-of-the-art review. *Energies* **2017**, *10*, 1672. [CrossRef]
20. Unguresan, P.V.; Porumb, R.A.; Petreus, D.; Pocola, A.G.; Pop, O.G.; Balan, M.C. Orientation of facades for active solar energy applications in different climatic conditions. *J. Energy Eng.* **2017**, *143*, 04017059. [CrossRef]
21. Luo, Y.; Zhang, L.; Bozlar, M.; Liu, Z.; Guo, H.; Meggers, F. Active building envelope systems toward renewable and sustainable energy. *Renew. Sustain. Energy Rev.* **2019**, *104*, 470–491. [CrossRef]
22. Musameh, H.; Alrashidi, H.; Al-Neami, F.; Issa, W. Energy performance analytical review of semi-transparent photovoltaics glazing in the United Kingdom. *J. Build. Eng.* **2022**, *54*, 104686. [CrossRef]
23. Li, D.H.; Aghimien, E.I.; Alshaibani, K. An Analysis of Real-Time Measured Solar Radiation and Daylight and Its Energy Implications for Semi-Transparent Building-Integrated Photovoltaic Façades. *Buildings* **2023**, *13*, 386. [CrossRef]
24. Ioannidis, Z.; Rounis, E.-D.; Athienitis, A.; Stathopoulos, T. Double skin façade integrating semi-transparent photovoltaics: Experimental study on forced convection and heat recovery. *Appl. Energy* **2020**, *278*, 115647. [CrossRef]
25. Haselsteiner, E. *Robust Architecture. Low Tech Design*; DETAIL: Munich, Germany, 2023; pp. 68–71.
26. Vassiliades, C.; Agathokleous, R.; Barone, G.; Forzano, C.; Giuzio, G.; Palombo, A.; Buonomano, A.; Kalogirou, S. Building integration of active solar energy systems: A review of geometrical and architectural characteristics. *Renew. Sustain. Energy Rev.* **2022**, *164*, 112482. [CrossRef]
27. Hosseini, S.M.; Mohammadi, M.; Schröder, T.; Guerra-Santin, O. Integrating interactive kinetic façade design with colored glass to improve daylight performance based on occupants' position. *J. Build. Eng.* **2020**, *31*, 101404. [CrossRef]

28. Sommese, F.; Hosseini, S.M.; Badarnah, L.; Capozzi, F.; Giordano, S.; Ambrogi, V.; Ausiello, G. Light-responsive kinetic façade system inspired by the Gazania flower: A biomimetic approach in parametric design for daylighting. *Build. Environ.* **2024**, *247*, 111052. [CrossRef]
29. Rostagni, C. *Luigi Moretti 1907–1973*; Mondadori Electa: Milano, Italy, 2008; p. 18.
30. Jabi, W. *Parametric Design for Architecture*; Hachette: London, UK, 2013; pp. 9–11.
31. Sketchpad Software. Available online: https://sketch.io/sketchpad/ (accessed on 22 January 2024).
32. Sutherland, I.E. Sketchpad: A man-machine graphical communication system. In Proceedings of the Spring Joint Computer Conference, Cambridge, UK, 21–23 May 1963.
33. Schumacher, P. Parametricism: A new global style for architecture and urban design. *Archit. Des.* **2009**, *79*, 14–23. [CrossRef]
34. Assai, R. Parametric Design, A Historical and Theoretical Overview. In Proceedings of the International Conference on Emerging Technologies in Architectural Design (ICETAD2019), Toronto, Canada, 17–18 October 2019.
35. Eltaweel, A.; Yuehong, S. Parametric design and daylighting: A literature review. *Renew. Sustain. Energy Rev.* **2017**, *73*, 1086–1103. [CrossRef]
36. Di Paola, F.; Mercurio, A. *Parametric Experiments in Architecture*; Springer: Cham, Switzerland, 2023; p. 3.
37. Ramzy, N. Sustainable spaces with psychological values: Historical architecture as reference book for biomimetic models with biophilic qualities. *ArchNet-IJAR* **2015**, *9*, 248–267. [CrossRef]
38. Sackmann, E. Biological membranes architecture and function. *Struct. Dyn. Membr.* **1995**, *1*, 1–63.
39. Carl, T.; Stepper, F.; Schein, M. Solar Spline. *Comput. A Better Tomorrow* **2018**, *1*, 149.
40. Smith, P. BIM implementation–global strategies. *Procedia Eng.* **2014**, *85*, 482–492. [CrossRef]
41. Revit Software. Available online: https://www.autodesk.com/products/revit/overview?plc=RVT&term=1-YEAR&support=ADVANCED&quantity=1 (accessed on 14 April 2023).
42. Hanwha Q CELLS'. Available online: https://qcells.com/kr/get-started/complete-energy-solution/solar-panel-detail?slrPnlId=SRPL211201071943005 (accessed on 28 October 2023).
43. Seoul Metropolitan Government. Available online: https://news.seoul.go.kr/citybuild/archives/511445 (accessed on 9 January 2024).
44. Calabrò, E. An algorithm to determine the optimum tilt angle of a solar panel from global horizontal solar radiation. *J. Renew. Energy* **2013**, *2013*, 307547. [CrossRef]
45. Georgantzia, E.; Gkantou, M.; Kamaris, G.S. Aluminium alloys as structural material: A review of research. *Eng. Struct.* **2021**, *227*, 111372. [CrossRef]
46. Dynamo Software. Available online: https://primer.dynamobim.org/en/02_Hello-Dynamo/2-1_launching_dynamo.html (accessed on 20 April 2023).
47. Thabet, W.; Lucas, J.; Srinivasan, S. Linking life cycle BIM data to a facility management system using Revit Dynamo. *Organ. Technol. Manag. Constr.* **2022**, *14*, 2539–2558. [CrossRef]
48. Insight Software. Available online: https://www.autodesk.com/products/insight/overview (accessed on 2 May 2023).
49. Hofmeister, S. *Detail, Circular Economy*; Detail Business Information GmbH: München, Germany, 2022; p. 22.
50. Yoo, S.H. Simulation for an optimal application of BIPV through parameter variation. *Sol. Energy* **2011**, *85*, 1291–1301. [CrossRef]

applied sciences

MDPI

Article

Natural Ventilation for Cooling Energy Saving: Typical Case of Public Building Design Optimization in Guangzhou, China

Menglong Zhang, Wenyang Han, Yufei He, Jianwu Xiong and Yin Zhang *

School of Architecture, Southwest Minzu University, Chengdu 610225, China
* Correspondence: cdzhangyin@163.com; Tel.: +86-134-8891-8589

Featured Application: The architectural design, modeling, mechanical dynamic simulation, and corresponding analysis have been applied to a practical new construction program, a science museum located in Guangzhou, China.

Abstract: Heating ventilation and air conditioning systems account for over one-third of building energy usage, especially for public buildings, due to large indoor heat sources and high ventilation and thermal comfort requirements compared to residential buildings. Natural ventilation shows high application potential in public buildings because of its highly efficient ventilation effect and energy-saving potential for indoor heat dissipation. In this paper, a building design is proposed for a science museum with atrium-centered natural ventilation consideration. The floor layout, building orientation, and internal structure are optimized to make full use of natural ventilation for space cooling under local climatic conditions. The natural ventilation model is established through computational fluid dynamics (CFD) for airflow evaluation under indoor and outdoor pressure differences. The preliminary results show that such an atrium-centered architectural design could facilitate an average air exchange rate over 2 h^{-1} via the natural ventilation effect. Moreover, indoor thermal environment simulation results indicate that the exhaust air temperature can be about 5 °C higher than the indoor air mean temperature during the daytime, resulting in about 41.2% air conditioning energy saving ratio due to the free cooling effect of natural ventilation. This work can provide guidance and references for natural ventilation optimization design in public buildings.

Keywords: public building; natural ventilation; wind pressure; energy saving; indoor environment

Citation: Zhang, M.; Han, W.; He, Y.; Xiong, J.; Zhang, Y. Natural Ventilation for Cooling Energy Saving: Typical Case of Public Building Design Optimization in Guangzhou, China. *Appl. Sci.* **2024**, *14*, 610. https://doi.org/10.3390/app14020610

Academic Editors: David Bienvenido Huertas and Daniel Sánchez-García

Received: 25 November 2023
Revised: 4 January 2024
Accepted: 8 January 2024
Published: 10 January 2024

1. Introduction

The building construction sector is considered one of the largest consumers of natural resources and energy. Buildings consume 30–40% of all primary energy and natural resources over their lifespan (construction, operation, maintenance, and demolition) and account for 30% of the global emissions of greenhouse gases [1]. A significant proportion of the increase in energy use was due to the spread of HVAC installations in response to the growing demand for better thermal comfort within the built environment. In general, in developed countries, HVAC is the largest energy end-use, accounting for about half of the total energy consumption in buildings, especially non-domestic buildings [2]. Public building energy consumption constitutes a large proportion of total building energy use. The overall energy consumption per m^2 in public buildings is over two times higher than that in residential buildings in China [3]. Meanwhile, China faces constant growth in the amount of public building area and energy consumption [4]. The energy consumption intensity of public buildings has also increased rapidly. Heating ventilation and air conditioning (HVAC) systems account for nearly 50% of the energy consumption of public buildings [5]. In recent years, China has been undergoing fast urbanization, with the public building areas reaching 11.6 billion m^2 by 2020, and urbanization causes thermal elevation which increases household energy consumption through air conditioning to reduce human heat stress [6,7]. In fact, in the building sector, HVAC systems represent between 40 and 60% of energy consumption in Europe and more than 50% in the United States [8]. Therefore, to reduce building energy, it is essential to reduce heating and cooling energy consumption. Public buildings mainly adopt concentrated air conditioning systems. For public buildings with certain thermal physical properties of the building envelope, there are two main factors affecting the cooling and heating load requirements [9]. One main impact factor is the outdoor climatic conditions (such as outdoor air temperature and humidity, and solar radiation intensity), which determine the basic amount of the cooling and heating load [10]. In particular, large-scale structures such as science and technology museums typically rely on air conditioning systems for indoor temperature control, a practice associated with increasing energy consumption. In alignment with national carbon reduction policies, there is a growing emphasis on harnessing the natural ventilation potential in regions characterized by hot summers and warm winters [11]. Studies have shown that occupants of naturally ventilated buildings are comfortable over a wider range of temperatures than occupants of buildings with centrally controlled HVAC systems [12]. Hence, many researchers have focused on natural ventilation system modeling, testing, and performance simulation of natural ventilation systems. Li and Chen [13] conducted a study on when and where natural ventilation cooling can be considered in the pre-design phase in China, and concluded that Kunming is the top priority for one-year mixed-ventilation air-conditioned buildings through a study of 100 cities. Zhang et al. [14] and González-Cruz et al. [15] combined the natural ventilation system with advanced air conditioning terminal devices such as floor radiation heating for an integrated system configuration investigation. Krusaa et al. [16] utilized building performance simulation software to analyze the effectiveness of natural ventilation in a local public building. The authors proposed a strategy that combines natural ventilation with air conditioning and verified that this strategy could reduce energy consumption by 13.1%. Ji et al. [17] conducted an analysis of the impact of air pollution and climate changes on the potential for natural ventilation in 74 Chinese cities from 2014 to 2019, identifying key factors limiting the utilization of natural ventilation. Maite Gil-Baez et al. [18] analyzed and compared the effectiveness of air renewal in two school buildings with a mechanical ventilation system compared to a natural ventilation system, and showed that by using a natural ventilation system, energy savings of 18–33% could be achieved while maintaining classroom comfort. Furthermore, Rosato et al. [19] and Prieto et al. [20] conducted a comprehensive summary of various natural ventilation methods and the current status of natural cooling methods, pointing out the research direction of achieving natural cooling through natural ventilation.

Considering the energy-saving potential of natural ventilation in the early stages of building design, especially in China, is currently the subject of more limited research. Kyosuke Hiyama et al. proposed a preliminary design methodology for naturally ventilated buildings utilizing a target air change rate [13].

The heavy dependence of sizable public structures on air conditioning systems for maintaining indoor thermal comfort leads to substantial energy consumption within these buildings. Consequently, mitigating the energy use of air conditioning systems in public structures has emerged as a pressing concern demanding immediate attention. However, although some scholars have considered the impact of natural ventilation potential on building energy efficiency in the early stages of building design, research based on a typical case of considering the cooling potential of natural ventilation in the early stages of building design is still insufficient.

Even though existing studies offer various concepts, methods, and ideas for building ventilation and air conditioning systems, with comprehensive consideration from experimental, technical, economic, and thermal perspectives, attention has been paid mainly to natural ventilation performance and operation from the viewpoint of the system designer. Designers of air conditioning systems pay relatively limited attention to air conditioning energy consumption, and reducing it is considered a challenging task. Therefore, the authors of this paper suggest that it is possible to approach this issue from the perspective of architects. By incorporating considerations of building energy consumption in the early stages of architectural design, it becomes feasible to effectively reduce overall building energy consumption. This approach offers an effective reference and design strategy for achieving energy efficiency in buildings at the architectural design level. Consequently, this paper addresses several key questions: How can internal land use and space planning of buildings be determined from the perspective of building design as a way to reduce energy consumption? How can building design be optimized to leverage natural ventilation in local wind environments? What could be the pre-estimated energy-saving potentials of such naturally ventilated public buildings, compared to traditional reference design modes? This study presents tentative answers to these questions using the new construction project of a science museum in Guangzhou, China, as a typical illustrative example. The main body of the paper is structured as follows: (1) The natural ventilation dynamic models are established based on the local wind environment and climatic conditions. (2) The detailed architectural design and space planning are demonstrated and optimized with an atrium-centered floor layout to make full use of the free cooling effect through natural ventilation. (3) The energy saving potential for such a design is evaluated with dynamic building load simulation and annual energy consumption comparison. This work can provide a typical design reference and application prototype for new construction with natural ventilation considerations, especially for big public buildings with air conditioning energy efficiency improvement expectations.

2. Methods

2.1. Natural Ventilation Mechanism

Building ventilation is a key passive strategy for designing energy-efficient buildings and improving indoor air quality [21]. In the natural environment, the ventilation phenomenon in which a building relies on the airflow between the indoor and outdoor spaces to introduce fresh air and remove dirty air is usually called natural ventilation. Natural ventilation is usually referred to as the behavior of indoor personnel consciously opening building vents during the use of the building (especially during the transition season) to make full use of natural air to cool and dehumidify the building, thus reducing the indoor cooling and humidity loads of the building; air infiltration is usually defined as the physical phenomenon of air flowing from high-pressure to low-pressure areas through cracks, crevices, or non-deliberate openings in the enclosure structure during the operation phase of the building (e.g., air conditioning and heating seasons) when there is a difference in pressure between the indoor and outdoor spaces. This is the physical phenomenon of high-pressure to low-pressure flow [22]. Natural ventilation can be defined as a type of anticipatory ventilation, whereas air infiltration is a disorganized and unintended exchange of air. Based on their physics, both intended ventilation and unintended air infiltration are, in essence, natural ventilation. The driving force for natural ventilation is usually the wind pressure created by atmospheric movement on the building surface or the thermal pressure (which is essentially a difference in air density) caused by the difference in air temperature between the interior and exterior of the building [23]. Therefore, heat pressure and wind pressure are the two main factors affecting natural ventilation. In addition, another necessary condition for natural ventilation is the existence of cracks or openings in the building envelope for airflow. Ventilation caused by wind-driven forces is called wind-pressure ventilation, and ventilation caused by buoyancy-driven forces is called thermal-pressure ventilation.

Thermal pressures are caused by density differences due to differences in air temperatures between the interior and exterior of the building, and their effect on the building is known as the chimney effect. The root cause of the chimney effect in high-rise buildings is the temperature difference between indoor and outdoor air, which further leads to a density difference between indoor and outdoor air. This density difference results in a difference in the gravity of the indoor and outdoor air columns, which in turn results in a difference in the distribution of the static pressure gradient between the indoor and outdoor air, resulting in a difference in air pressure. The greater the temperature difference between indoor and outdoor air when the building height is certain, or the higher the building is when the temperature difference between indoor and outdoor air is certain, the greater the heat pressure generated and the more significant the chimney effect caused.

When natural wind blows towards the building, due to the shielding of the building, the outdoor airflow will change direction around the building to produce a winding flow. The windward side of the building causes the airflow to be obstructed, resulting in the formation of a positive pressure zone, which is characterized by a decrease in dynamic pressure and an increase in static pressure. The top and sides of the windward side of the building force the air to produce a flow boundary layer on the surface of the building, which develops into a localized vortex, resulting in a reduction in static pressure on the top and back of the building, or even a negative pressure [23]. As a result of the atmospheric movement, the static pressure of the air around the building changes, and this pressure is often referred to as wind pressure. Wind pressure forces a differential pressure on both sides of an opening or cracks in the envelope, which manifests itself as a flow of air from a high-pressure area to a low-pressure area (Figure 1).

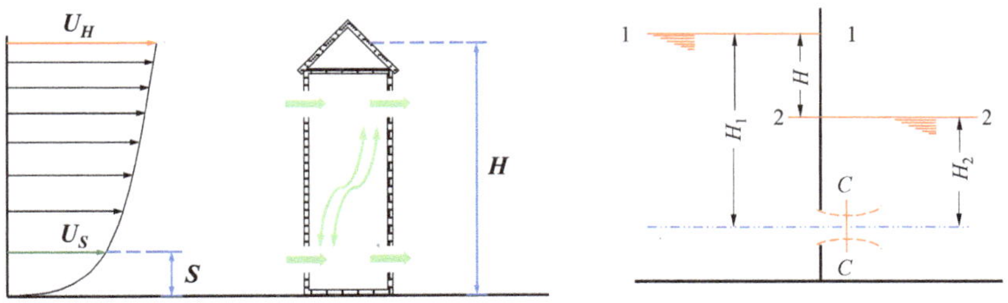

Figure 1. Schematic diagram of building natural ventilation driven by pressure difference.

2.2. Mathematical Modeling

The strength of wind pressure on a building surface is closely related to outdoor wind speed, wind direction, temperature, air density, and other parameters. However, in the actual environment, wind conditions are characterized by rapid changes and unpredictability, and the wind pressure on a building surface is strongly fluctuating and transient. Due to the complexity and variability of wind pressure, the accurate measurement of wind pressure on a building surface is a challenging task, especially when accurate transient wind pressure data are required. The airflow field around a building is related to the influencing factors such as the shape of the building, the angle between the building orientation and the wind direction, and the shading of the surrounding environment (e.g., buildings, structures, and vegetation), which will result in the wind pressure coefficient on the surface of the building being difficult to obtain directly from theoretical calculations. From the above analysis, it can be seen that both expected natural ventilation and unintended air infiltration are natural ventilation in nature [23]. According to the findings of fluid mechanics, the physical model of natural ventilation is the gas orifice flooding outflow model. As Figure 1 shows, according to the Bernoulli equation,

$$H_1 + \frac{P_1}{\rho g} + \frac{\alpha_1 v_1^2}{2g} = H_2 + \frac{P_2}{\rho g} + \frac{\alpha_2 v_2^2}{2g} + \zeta_1 \frac{v_c^2}{2g} + \zeta_2 \frac{v_c^2}{2g}, \tag{1}$$

Then, the airflow velocity caused by natural ventilation can be expressed by

$$v_c = \frac{1}{\sqrt{\zeta_1 + \zeta_2}} \cdot \sqrt{2gH_0}, \tag{2}$$

where ζ_1 is the local resistance coefficient of the liquid through the orifice and ζ_2 is the local resistance coefficient of the liquid that suddenly expands after the contraction of the section (ζ_2 = 1 because the 2–2 section is much larger than the C–C section). Air outflow is generally submerged outflow, due to the small gas capacity, which can be ignored in the total head difference before and after the orifice of the location of the head term (bit pressure). Furthermore, when there is a pressure difference between the two sides of the geometrically large openings in the building envelope, the natural ventilation airflow pattern is turbulent flow. Thus, after taking the building opening area into consideration, the ventilation airflow rate is

$$Q_V = \mu A \sqrt{\frac{2\Delta P_0}{\rho}}, \tag{3}$$

where ρ and μ are density (kg/m^3) and opening airflow coefficient, respectively. Currently, the commonly used methods for determining the wind pressure coefficient mainly include field measurements, wind tunnel experiments, computational fluid dynamics numerical simulation (CFD), or semi-empirical models [24].

According to the Navier–Stokes fluid mechanic theory, partial differential equations can be deduced for incompressible fluid and flows. Generally speaking, natural ventilation generates turbulent airflow in big public buildings, which makes it difficult to solve Navier–Stokes equations because of the greatly varying mixing-length scales for any turbulent airflow [25]. In this paper, the k–ε turbulence model is used to govern the Reynolds-averaged Navier–Stokes (RANS) formulas, combined with practical computational fluid dynamics (CFD) applications. The mass continuity equation of incompressible fluid is:

$$\frac{d\rho}{dt} + \nabla \cdot (\rho V) = 0, \tag{4}$$

where V is the flow velocity vector (m/s). The airflow density remained unchanged:

$$\nabla \cdot (\rho V) = \sum_i \frac{\partial \rho V_i)}{\partial x_i} = -\frac{d\rho}{dt} = 0. \tag{5}$$

Then, according to Newton's second law of motion,

$$\frac{d(\rho V)}{dt} = F + \nabla \cdot P, \tag{6}$$

where F and P represent body force and mass force, respectively. Then, integrated with viscous stress, the momentum equation can be expressed by

$$\frac{\partial(\rho V)}{\partial t} + V_j \frac{\partial(\rho V_i)}{\partial x_j} = F - \frac{\partial P}{\partial x_i} + \mu \nabla^2 V, \tag{7}$$

where μ is the kinematic viscosity (kg/m^{2}s) and P is the static pressure (Pa). For the air energy transfer and conversion,

$$\frac{\partial \rho H}{\partial t} + \frac{\partial}{\partial x_i}(V_i \rho H) = \frac{\partial}{\partial x_j}(\frac{\lambda}{c_p}\frac{\partial H}{\partial x_j}) + S, \tag{8}$$

where H, λ, and c_p represent the enthalpy (kJ/kg), thermal conductivity (W/mK), and specific heat (kJ/kgK), respectively, and S represents the heat source (kW). According to the Boussinesq eddy viscosity hypothesis, the turbulent model using the k–ε model changes into

$$\rho \frac{\partial K}{\partial t} + \rho V_i \frac{\partial K}{\partial x_i} = \frac{\partial}{\partial x_i}[(\mu + \frac{\mu_t}{\sigma_k})\frac{\partial K}{\partial x_i}] + \mu_t \frac{\partial V_i}{\partial x_j} \nabla \cdot V - C_a \rho \frac{k^{\frac{3}{2}}}{l}, \tag{9}$$

where C_a is a constant, and μ_t and σ_k represent the kinematic viscosity for turbulent flow and Prandtl number of turbulence kinetic energy, respectively. The heat dissipation in the turbulent airflow can be depicted by

$$\rho\frac{\partial\varepsilon}{\partial t} = \frac{\partial}{\partial x_i}[(\mu + \frac{\mu_t}{\sigma_k})\frac{\partial\varepsilon}{\partial x_i}] + C_b\frac{\varepsilon}{K}(G_k + C_c) - C_d\rho\frac{\varepsilon^2}{K}. \tag{10}$$

The above governing equations can be recast in a conservative form, and then solved through the finite volume method, combined with the semi-implicit method for pressure-linked equation-consistent algorithm [26].

2.3. Air Exchange Rate

This project adopts the multi-area network method to calculate the number of indoor air changes in the building. In the multi-area network method, the indoor rooms are divided into different ventilation and air changes in different areas, with the wind pressure of windows and doors as the boundary conditions; the different areas are connected through the connected windows and doors for the transmission of data, and ultimately to obtain the number of air changes in each room.

The calculation of the number of air changes in the room is derived from the calculation of the air quality flow in the ventilation path, and the air quality flow based on the multi-area network method is calculated as follows:

$$Q = C_d A\sqrt{\frac{2\Delta P}{\rho}} \tag{11}$$

where Q represents the room volume flow (m^2/s); P represents the difference in wind pressure between the doors and windows of adjacent rooms; and C represents the flow coefficient, which, for large building openings, is 0.5, for narrow openings, 0.65, and for the project calculations, 0.6; A represents the area of the opening (m^2), ρ represents the density of air (kg/m^2).

After obtaining the volume flow rate Q of a room by the above method, the calculation of the number of air changes in the room can be carried out:

$$Acr = \frac{Q \times 3600}{V} \tag{12}$$

where Q denotes the room volume flow rate (m^3/s); Acr denotes the number of air changes (times/h); and V denotes the room volume (m^3).

In the following section, an atrium-based natural ventilation system in a newly designed science museum in Guangzhou is taken as the illustrative example to investigate the free cooling potential of pressure-difference-driven ventilation under local climatic conditions for building thermal design optimization.

2.4. Simulation and Solution

CFD analysis was performed using a polyhedral mesh grid generated through ANSYS Meshing (Ansys-Fluent v22.2). Considering the change in outdoor wind direction, the inlet and outlet of wind speed in the geometric model will be changed, and the power shadow area has a large impact on the building surroundings, so the size of the calculation area has a significant impact on the accuracy of the numerical calculation results. If the computational domain is too small it will cause problems of flow field distortion and insufficient flow development, and if the computational domain is too large it will cause unnecessary waste of computational resources [23]. Regarding the setting of the computational domain, the COST guideline stipulates that for the simulation of a single building, the distance between the two sides of the building and the two sides of the computational domain should be more than 5 times the height of the building, the distance between the top of the building and the top boundary of the computational domain should be more than 5 times the height of the building, and the distance between the boundary of the air outlets and the building should be at least 15 times the height of the building. Based on the above considerations, the dimensions of the computational domain for the simulation in this study are set as shown in Figure 2. Taking the base model as an example, the distance of the wind speed inlet from the building is usually 5 H (H is the building height), the distance of the outlet from the building is 10 H, and the boundaries of the left and right calculation domains are usually 5 H from the building. Before conducting outdoor wind field calculations, it is essential to determine the size of the computational domain involved, termed the computational domain in fluid mechanics, which typically constitutes a rectangular or cubic space encompassing a cluster of buildings.

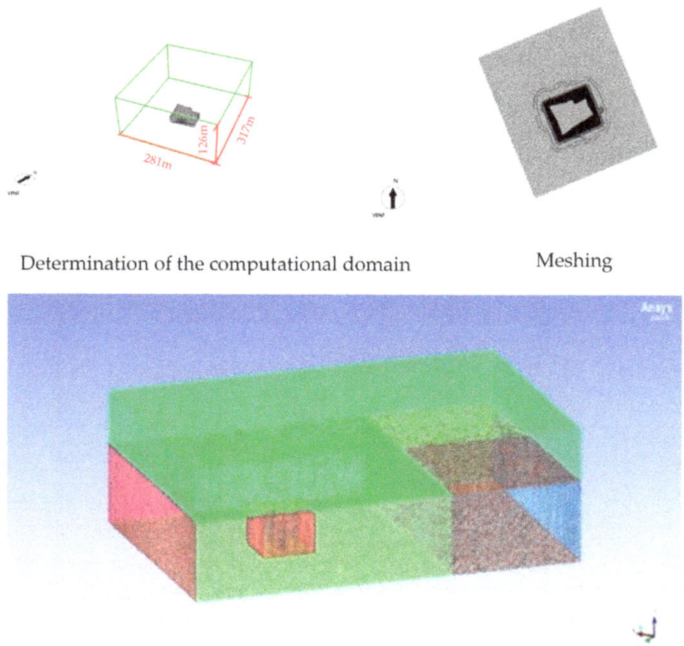

Figure 2. Schematic diagram of CFD simulation and meshing (colours difference denoting computational domains for building surrounding zones and city wind environment respectively).

Figure 2 displays a cross-sectional diagram of the wind field computational domain, the grid generation, and the allocation of boundary conditions for each region. The dimensions of the wind field are 317 m in length, 281 m in width, and 126 m in height. The total grid count amounts to 1,687,448.

Figure 3 illustrates the research workflow of this study. Initially, the objective was to explore the potential of natural ventilation in reducing temperatures and conserving energy, coupled with geographical and climatic data for preliminary architectural design. Subsequently, from an architectural perspective, energy-saving considerations were incorporated, encompassing the preliminary design of ventilation corridors, architectural self-shading, and the augmentation of ventilation. Thirdly, an architectural model was established using simulation software, followed by the configuration of various parameters for simulation. Finally, the simulation results validated the substantial energy-saving effects of the architectural energy-saving methods considered during the initial design phase.

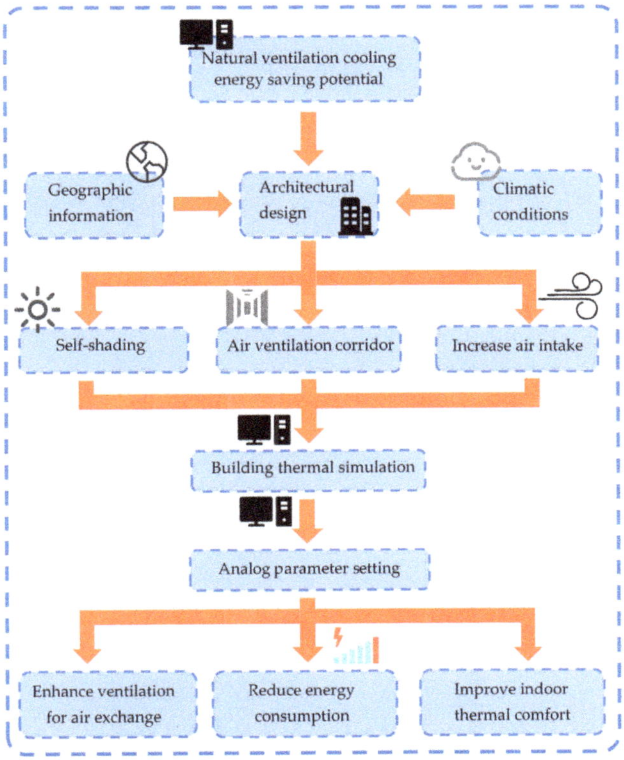

Figure 3. Research flow chart.

3. Geographical Information

Guangzhou City is situated in the southern region of mainland China, in the central and southern parts of Guangdong Province, at the northern edge of the Pearl River Delta. It is the confluence point for the Beijiang (North River), Xijiang (West River), and Dongjiang (East River) as they flow into the South China Sea. Geographically, the city is positioned between approximately 112°57′ to 114°3′ E longitude and 22°26′ to 23°56′ N latitude (Figure 4).

This paper focuses on the hottest region with hot summers and warm winters among China's five climate zones, where cooling consumption is an important component of building operation consumption [27,28]. So, Guangzhou was selected as the illustrative example with the following main climate parameters: the annual average temperature is 22.5 °C, the average temperature in the hottest month is 32.9 °C, the annual maximum temperature is 36.1 °C, and the annual total solar radiation is 1072.1 kWh/year. Figure 5 gives the key climatic parameter variations throughout a typical year. China divides building climate zones into five main climate zones and 20 climate subzones. The main division index of building climate zones is the average monthly temperature in January and July. The names of thermal zones are cold region, hot summer and cold winter region, hot summer and warm winter region, and mild region. The case study building is located in Guangzhou City and belongs to the hot summer and warm winter region, that is, the average temperature in January is greater than 10 °C, and the average temperature in July is 25–29 °C. Guangzhou falls within a subtropical climate zone, characterized by distinct seasons and abundant precipitation. Summer, the rainy season, is hot and humid, influenced by tropical monsoons. Winter is warm, dry, relatively mild, and experiences minimal rainfall, shaped by the influence of northern monsoon winds. Spring and autumn boast moderate temperatures and lower humidity levels.

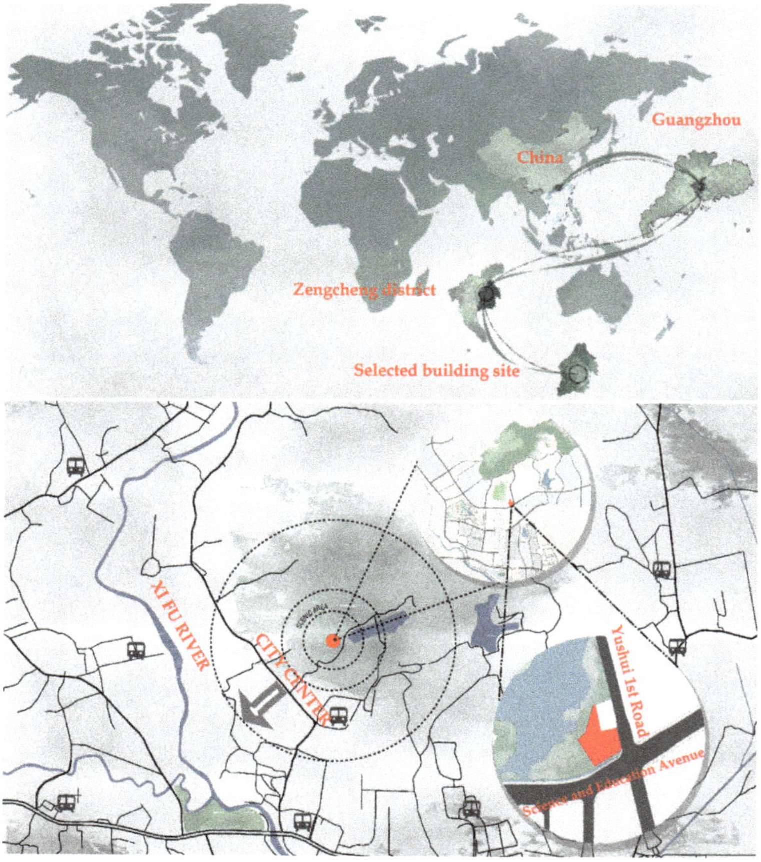

Figure 4. Location map of the case study city, Guangzhou, in southern China (drawn by authors).

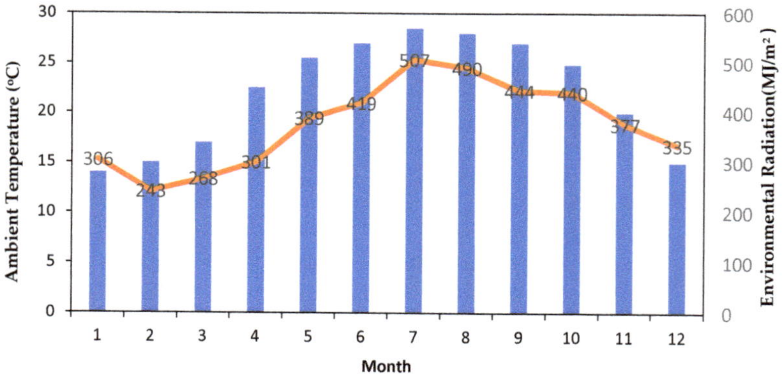

(**a**)

Figure 5. *Cont.*

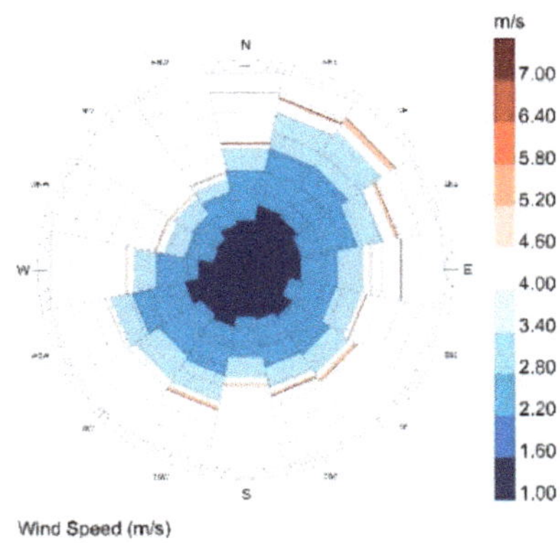

(**b**)

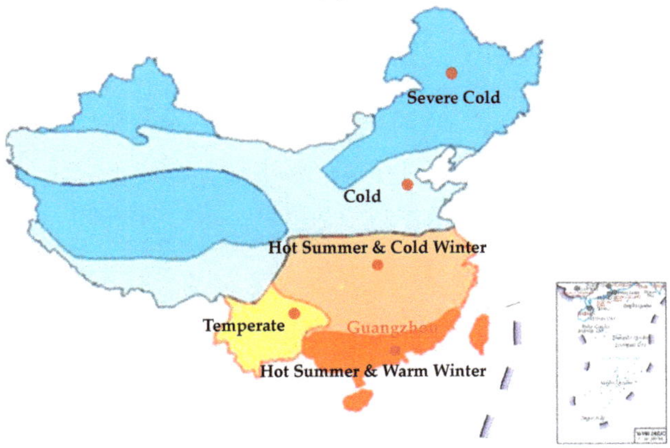

(**c**)

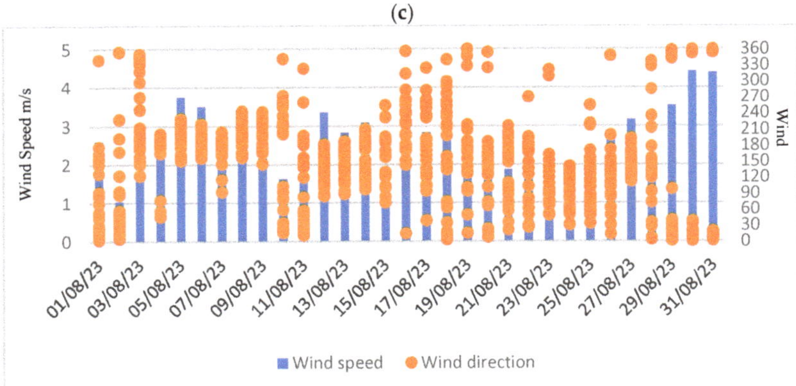

(**d**)

Figure 5. Key climatic parameters in Guangzhou (hot summer and warm winter climatic zone). (**a**) Average monthly temperature and solar radiation (Monthly average temperature in blue, monthly ambient radiation in orange). (**b**) Wind rose diagram (The color difference represents the wind speed in each wind direction zone). (**c**) Map of China's 5 climate zones (The color difference represents different climate zones in China) [29]. (**d**) Information on the wind environment at the site during a typical summer month (Blue for wind speed, orange for wind direction).

Guangzhou experiences its highest annual temperatures in August, reaching 37 °C. During the summer, temperature fluctuations are relatively minor. The lowest temperatures occur in January and February, dropping to 6 °C, and during the winter, temperature variations are more significant. The monthly normal direct radiation remains around 250 MJ/m^2 with relatively minor annual fluctuations. The monthly total horizontal radiation exhibits a trend of transitioning from low to high and then decreasing from January to December, reaching its peak in August at approximately 530 MJ/m^2, and reaching its lowest point in February at approximately 260 MJ/m^2. The wind rose diagram indicates the predominant wind directions are from the northeast and southwest in Guangzhou. Specifically, the average wind speed for the northeast wind can reach a maximum of 6.40 m per second, while the southwest wind reaches a maximum average speed of 5.80 m per second. It is evident that the average wind speed for the northeast wind is significantly higher than that for the southwest wind.

In Figure 5d, the wind speed and direction around the building site during typical summer months are depicted. In Figure 5d, wind speed recordings were taken at one-hour intervals each day, and the maximum daily wind speed was used as the value for Figure 5d. And according to the statistics, the lowest recorded wind speed for the month was 0.09 m/s, while the highest was 4.4 m/s. The scatter plot in the figure indicates that the predominant wind direction during this period is concentrated between 90° and 180°, primarily coming from the southeast.

4. Results

4.1. Building Design Optimization

The chosen study case here is a typical science museum, with a total building area of 11,447 m^2, a volume of 55,726.81 m^3, and shape coefficient of 0.30 (Table 1). The technology museum implemented several strategies to enhance the cooling and energy-saving potential through natural ventilation. These included the use of ventilation corridors, structural shading elements integrated into the façade from top to bottom, and an expanded main entrance to augment natural airflow. Serving as a representative case study, it demonstrates the application of these tactics to encourage natural ventilation in buildings and effectively reduce energy consumption.

Table 1. Building thermal physical properties.

Parameters	Value	Unit
total area	19,012.41	m^2
external envelope area	7771.2	m^2
building volume	55,726.81	m^3
shape coefficient	0.30	
ratio of window to wall	0.34	
window opening area (south)	265.83	m^2
heat transfer coefficient of external wall	1.07	W/m^2·K
heat transfer coefficient of external window	3.5	W/m^2·K
solar radiation absorption coefficient	0.08	
penetration coefficient of solar radiation through the glazing	0.8	
shading coefficient	1	

As shown in Figure 6 the building's south façade gradually extends outward from the bottom to the top, achieving self-shading for the structure. This design embodies the concept of integrating gray spaces, shading, and epidermal ventilation. The outward expansion of the building's exterior structure serves the following key functions: Firstly, it provides a gray space for the building's main entrance, serving not only as a shading canopy structurally but also as a brief stopover and gathering space for people, further enhancing the functionality and pleasantness of the building's entrance. Secondly, the extension of the structure outward increases the inflow volume of the ventilation corridor, facilitating the harnessing of the cooling potential of natural ventilation. This plays a vital role in improving the interior comfort of the science museum and reducing its energy consumption.

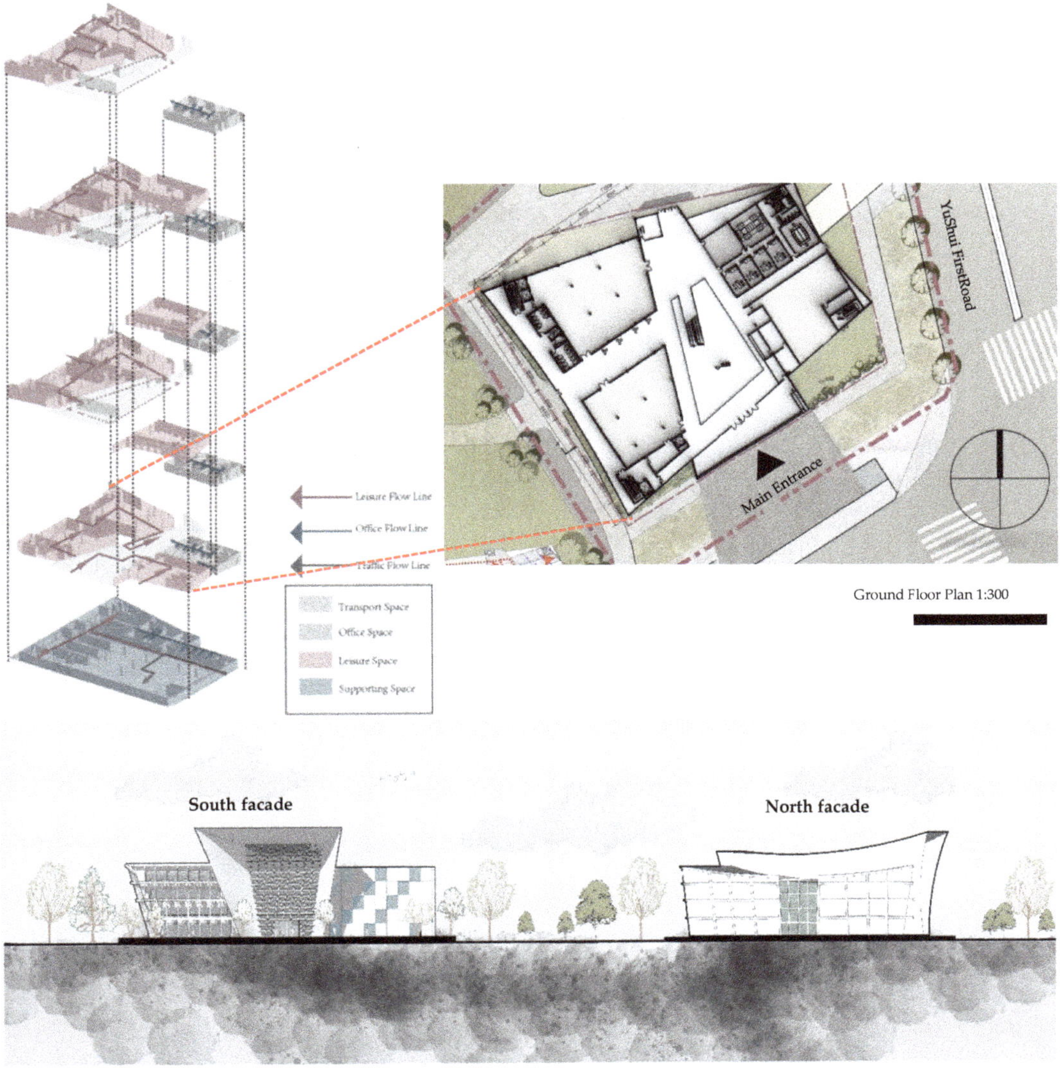

Figure 6. Architectural design details with floor layouts and external façades (drawn by authors).

Figure 6 depicts the ground floor plan of the science museum, where the primary functions of this level are categorized into exhibition areas and office spaces. Notably, the ground floor incorporates a gradually tapering north–south-oriented ventilation corridor. This ventilation corridor serves a dual purpose: firstly, it functions as a spatial divider within the architectural layout; secondly, it effectively enhances the natural ventilation capacity of the building, thereby facilitating the exploitation of the cooling potential inherent in natural ventilation. This architectural feature is strategically designed to improve the overall thermal performance of the building by optimizing natural airflow and cooling mechanisms.

Figure 7 illustrates the main passive building thermal design optimization approaches including natural ventilation, solar radiation absorption, smart shading and rain collection, etc. This pathway initiates from the main entrance on the southern façade of the building, and traverses through a ventilation corridor, driven by wind pressure, to fulfill the lateral ventilation needs of the structure. Simultaneously, natural air enters the interior through the central atrium of the building and influenced by thermal pressure, meets the vertical ventilation requirements of the indoor space. Notably, the glass surface at the top of the building's atrium serves as a rainwater collection system, contributing to the supplementary cooling of the interior. Additionally, the architectural extensions extending outward enhance the spatial capacity of the main entrance area, thereby increasing the effectiveness of natural ventilation.

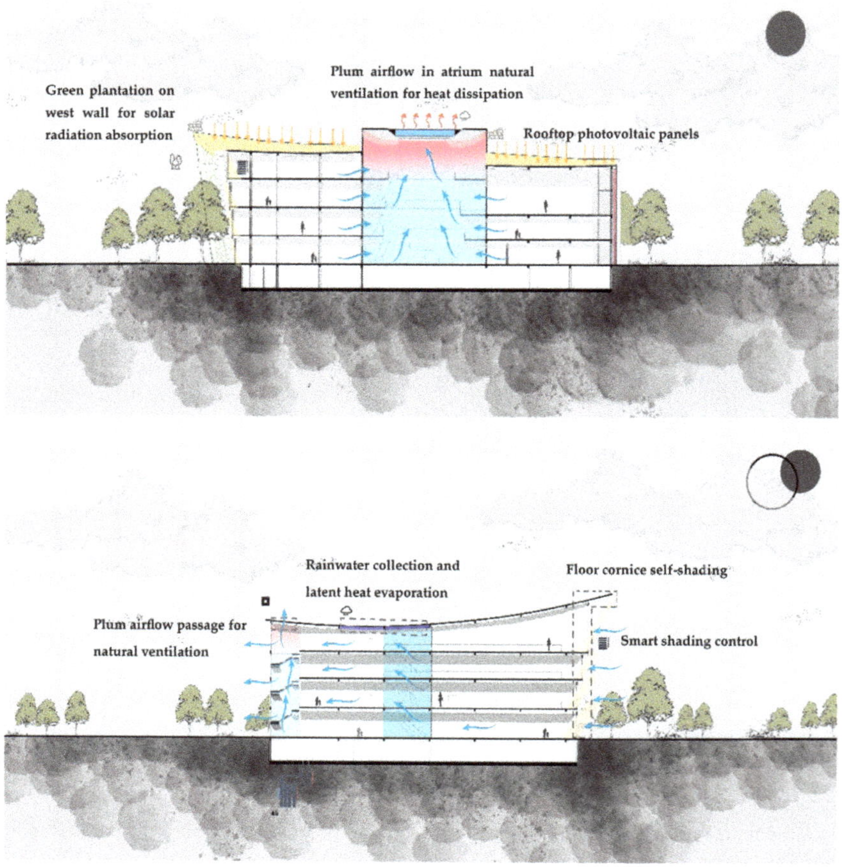

Figure 7. Schematic design mechanisms for passive thermal optimization (drawn by authors) (Color differences represent different bioclimatic technologies).

4.2. Natural Ventilation Simulation

The main focus of this study is investigating the cooling potential of natural ventilation and its energy-saving effects. Given that Guangzhou experiences a hot summer and warm winter climate, with monthly average temperatures consistently exceeding 10 °C, particularly reaching 28 °C in the summer, the proportion of air conditioning cooling consumption in total building energy consumption is significant. Therefore, simulations are set under summer conditions, with the external wind direction at that time being southeast.

Given that the primary focus of this study is on the isolated impact of natural ventilation on cooling and energy conservation, despite the utilization of various bioclimatic techniques in the building, they do not directly affect natural ventilation. Therefore, the energy simulation did not account for the influence of bioclimatic techniques on energy consumption.

Figure 8a represents a simulation of the outdoor wind environment around the building. Outdoor wind speeds are indicated in ascending order from blue to red. As observed in Figure 8a, when the wind is blowing from the southeast, no areas within the human activity region are marked with wind speeds less than 0.2 m/s, thus concluding that there are no windless zones within the building.

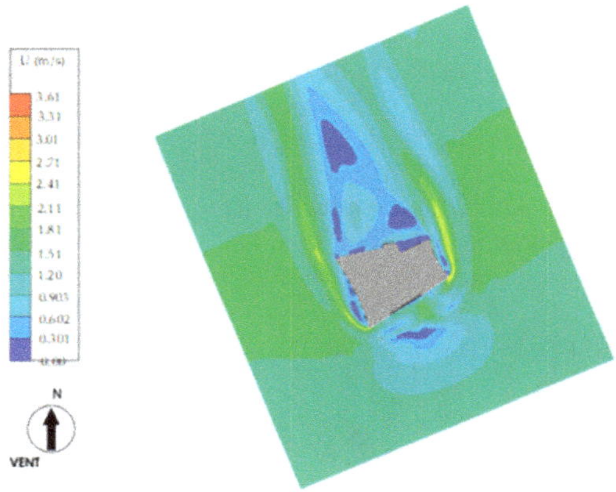

(**a**)

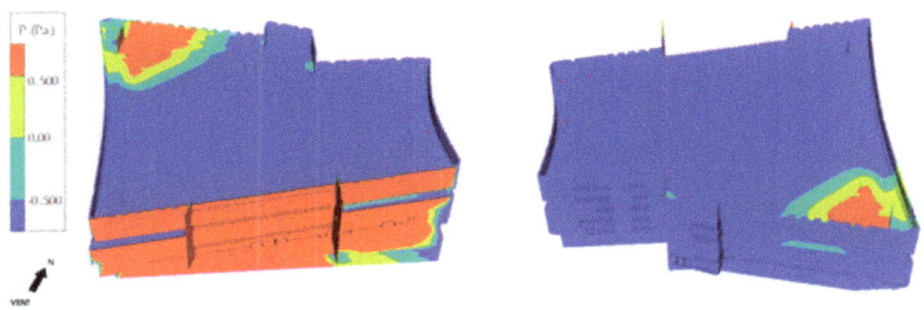

(**b**)

Figure 8. Outdoor wind environment chart. (**a**) outdoor wind speed. (**b**) building external wall surface air pressure.

Only when the absolute wind pressure on the exterior surfaces of the windows is sufficiently high can a good effect of window-based ventilation be ensured, leading to the creation of a favorable indoor airflow environment. Figure 8b displays the wind pressure distribution on the external window surfaces of the windward side of the building in summer conditions. This graph clearly conveys the wind pressure distribution on the windward side of the building. From Figure 8b, combined with the values in the illustration, it can be clearly seen that the wind pressure on the surface of the external window is less than 0.5 Pa in the external window area.

Figure 8b depicts the wind pressure distribution on the exterior window surfaces of the leeward side of the building in summer conditions, clearly presenting the wind pressure distribution pattern on the leeward side of the building. From Figure 8, combined with the values in the illustration, it can be clearly seen that the wind pressure on the surface of the external window is less than 0.5 Pa in the external window area. The building is equipped with a total of 102 operable external windows, with 92 of them being affected by an outdoor wind pressure differential exceeding 0.5 Pa. Consequently, the window-based ventilation in this building exhibits excellent performance, contributing to improved indoor airflow quality. This provides a solid foundation for harnessing the potential of natural cooling through indoor ventilation.

Figure 9 shows a schematic diagram of the indoor wind environment. Figure 9a,b are the indoor wind speed map and indoor wind speed vector map of the building, respectively. Combined with the legend, it can be seen that most of the wind speeds in the ventilation corridor are greater than 0.5 m/s, which proves that the ventilation corridor increases the good effect on the amount of natural ventilation. Figure 9c is the natural wind flow line map of the interior of the building, which clearly reflects the excellent ventilation effect of the ventilation corridor. Figure 9d shows the addition of east–west windows to the original building, which clearly shows that the increase in the indoor east–west wind is larger, affecting the ventilation effect of the ventilation corridor. Figure 9b,c are compared, and combined with the outdoor southeast wind direction, it can be concluded that the practice of catering to the outdoor wind direction of the building's internal ventilation corridor effectively enhances the indoor ventilation volume and controls the direction of the natural wind into the room well.

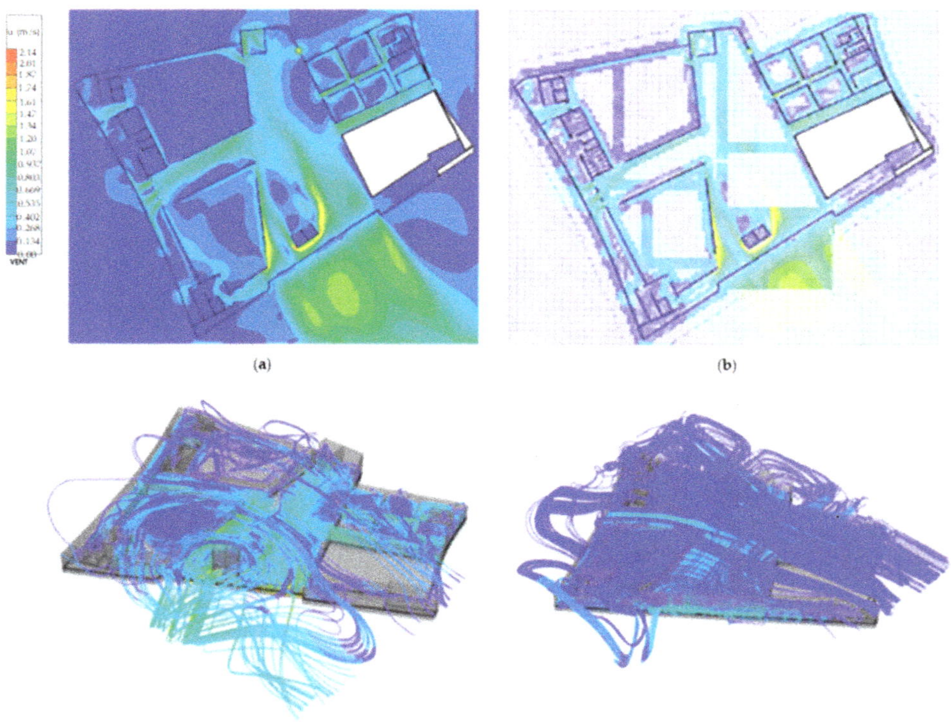

Figure 9. Indoor airflow simulation results under natural ventilation. (**a**) indoor air velocity. (**b**) indoor wind speed vector. (**c**) indoor airflow streamline. (**d**) Indoor airflow streamline (for comparison).

5. Discussion

The outdoor weather conditions of the office building are based on conditions in Guangzhou, China, which is in a hot summer and cold winter zone. The characteristic temperature method (CTM) is utilized for the research. Based on the building energy gene theory [30], the dynamic load and energy consumption of buildings are simulated by CTM, and the relationship between load or energy consumption and various other factors can be revealed. According to CTM, if solar radiation gain is considered, the indoor characteristic temperature can be expressed by

$$T_{\text{in}} = \frac{\sum K_i F_i T_{si} + \sum F_i I\left(\eta_i + \frac{\alpha_i}{\alpha_o}\rho_G\right)C_i \mp Q_{\text{AC}}}{\sum K_i F_i} \tag{13}$$

where F_i is the building envelope area, m^2; T_{si} is the equivalent solar-air temperature, °C; K_i represents heat transfer coefficient, W/(m$^2 \cdot$ K); I represents solar radiation, W/m^2; η_i is the transmittance; ρ_G is the absorption ratio; and C_i is the shading coefficient. Q_{AC} is the cooling or heating capacity provided by the air conditioner to maintain the indoor setting temperature within the thermal comfort zone [10]. On the other hand, if not considering solar radiation, I is equal to zero and T_{si} is approximately the outdoor temperature T_{out}, °C. So, the expression is changed into

$$T_{\text{in}} = \frac{\sum K_i F_i T_{air} \mp Q_{\text{AC}}}{\sum K_i F_i} \tag{14}$$

This CTM approach for indoor temperature prediction and building load simulation has been validated in previous studies and verified by commercial building simulation software including DOE-2 (v 2.2, U.S. Department of Energy), EnergyPlus (v 23.2.0, U.S. Department of Energy) and DeST (v DeST-h, Tsinghua University, China) [11,19,31,32]. For the other working conditions of building simulation, such as indoor air exchange rate (ACH) through a mechanical ventilation system, internal heat gains from indoor lighting, occupancy and devices, etc., key parameters are obtained based on the benchmark values according to the U.S. Department of Energy (DOE) Reference Building and ASHRAE AS (2013) Standard of thermal environmental conditions for human occupancy (building's daily office hours are 9:00–17:00) [33–35]. For the indoor thermal comfort evaluation, the integrated uncomfortable degree I_{year} is used as the index, which is defined as:

$$I_{\text{year}} = \int_{\substack{\text{year} \\ T_L > T_{\text{in}}}} (T_L - T_{\text{in}})d\tau + \int_{\substack{\text{year} \\ T_{\text{in}} > T_H}} (T_{\text{in}} - T_H)d\tau \tag{15}$$

where T_{H} (26 °C) and T_{L} (18 °C) are the upper and lower indoor temperature limits of the thermal comfort zone, respectively. This evaluation index depicts the integrated degree of deviation of the indoor temperature from the thermal comfort zone [10,11].

Based on the aforementioned building thermal performance modeling, the average indoor air temperature variations can be obtained with comparison under natural ventilation optimization (Figure 10). According to the thermal comfort demands, the upper temperature limit is set at 26 °C and the lower temperature limit at 18 °C. When indoor temperatures fall within this range, individuals experience a state of warmth and comfort, without feeling excessively cold or hot. The blue line represents the indoor temperature variations under natural ventilation conditions. Based on annual hour statistics, indoor temperatures remain within the thermal comfort range (19–26 °C) for a total of 4760 h, occurring primarily in the time intervals of 0–3000 h and 7000–8760 h. Conversely, the green line illustrates indoor temperature fluctuations in the absence of natural ventilation. Similarly, based on annual hour statistics, indoor temperatures persist within the thermal comfort zone for a total of 3100 h, mainly falling within the time intervals of 1000–3000 h and 6700–7800 h. By comparing the durations of thermal comfort under these two conditions, it becomes evident that under natural ventilation, indoor temperatures remain within the thermal comfort zone for an additional 1660 h compared to conditions without natural ventilation, indicating a significant improvement in indoor thermal comfort. Furthermore, as observed in Figure 10, under natural ventilation conditions, the indoor minimum temperature is relatively higher compared to conditions without natural ventilation, with a maximum temperature difference of 10 degrees.

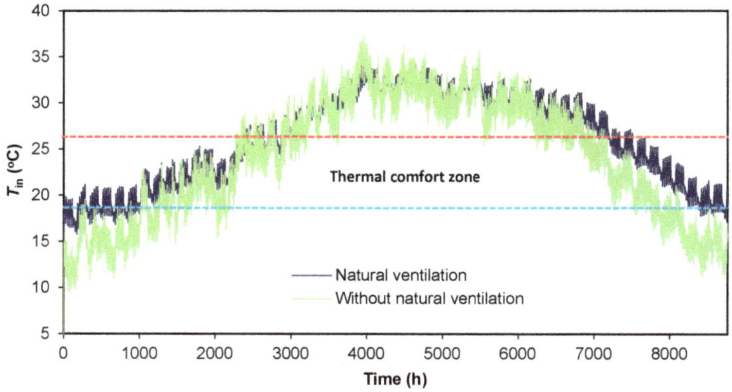

Figure 10. Indoor air temperature variations with thermal comfort comparison (red and blue lines represent the upper and lower temperature values for indoor thermal comfort zone).

Figure 11 compares the energy consumption of the designed building with that of the baseline building. In the simulated science museum building, the energy consumption primarily includes cooling, heating, lighting, and domestic hot water. Among these, lighting energy consumption constitutes 68% of the total building energy consumption, making it the highest among all energy consumption categories. Heating energy consumption has the lowest proportion, at only 2%, a difference attributed to the geographical location of the simulated building. Notably, in the simulated building located in Guangzhou, particular attention should be given to cooling energy consumption, which accounts for 29% of the total energy consumption.

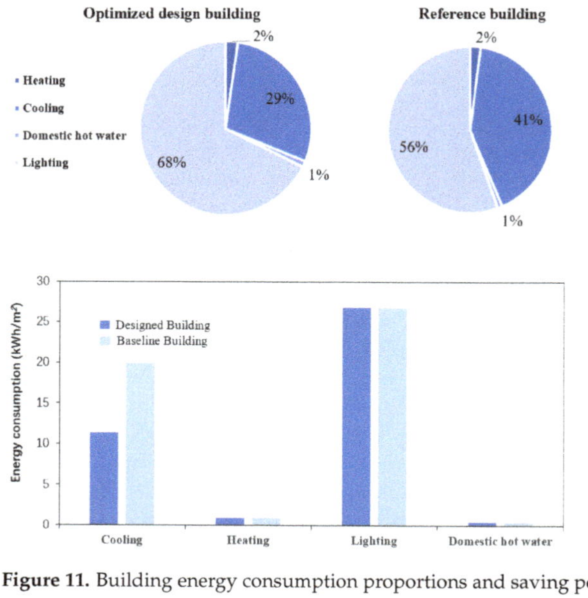

Figure 11. Building energy consumption proportions and saving potential evaluation.

In the energy composition of the baseline building, lighting also has the highest share of energy consumption in the baseline building, accounting for 56% of the total energy consumption. Heating accounts for 2% and domestic hot water is the lowest at 1% as in the simulated building.

Since natural ventilation mainly affects air conditioning energy consumption in terms of energy saving, the design building reduces cooling energy consumption by 8.54 kWh/m^2 and heating energy consumption by 0.1 kWh/m^2 compared to the baseline building. Since the case is located in Guangzhou, China, a hot-summer and warm-winter zone, the increase in heating energy consumption is small, even though the temperature is not in the comfort temperature range in winter, the actual temperature is not much different from the minimum comfort temperature, and the heating demand is not large. The cooling energy consumption and total energy consumption of the design building are 11.37 kWh/m^2 and 39.55 kWh/m^2, respectively, and the cooling energy consumption accounts for 29% of the total energy consumption. Meanwhile, the cooling energy consumption and total energy consumption of the baseline building are 19.91 kWh/m^2 and 48.19 kWh/m^2, respectively, and the cooling energy consumption accounts for 41% of the total energy consumption. It can be seen that the proportion of refrigeration energy to total energy consumption in the design building is 12% lower than that in the baseline building. This disparity indicates that the simulated science museum building, through architectural design measures such as increasing ventilation corridors and enhancing natural ventilation, fully exploits the natural cooling potential, resulting in a significant 42.9% reduction in cooling energy consumption, thus achieving notable energy savings.

In comparison to the baseline building, the designed structure's air conditioning consumption stands at 12.3 kWh/m^2, while the baseline building's consumption reaches 20.94 kWh/m^2. This denotes a 41.2% reduction in air conditioning consumption for the designed structure. Simultaneously, natural ventilation has extended the duration of indoor thermal comfort by 1660 h, significantly enhancing the indoor thermal comfort conditions.

The study investigates the indoor air exchange rates during the transitional seasons due to their optimal natural ventilation potential. Table 2 illustrates the air exchange rates within the building during these transitional seasons under natural ventilation conditions. As depicted in the figures, the area within the building experiencing air exchange rates exceeding 2 h^{-1} amounts to 12,124.48 m^2, constituting 63.77% of the total building area of 19,012.41 m^2. Hence, optimized designs of buildings can achieve air exchange rates surpassing two times per hour under natural ventilation conditions.

Table 2. Table of the number of air changes under typical working conditions in summer.

Area Ratio of Air Changes Greater Than 2 per Hour		
Area with More than 2 Air Changes per Hour (m^2)	**Total Area (m^2)**	**Area Ratio (%)**
12,124.48	19,012.41	63.77

6. Conclusions

Natural ventilation shows high application potential in public buildings because of its highly efficient indoor environment quality enhancement and large building energy-saving expectations. How to determine the land use and space planning with building natural ventilation considerations for energy saving is an important research area from the architectural perspective in the early building design stages. In this paper, taking a practical new construction project of a science museum, located in Guangzhou, China, as a typical illustrative example, the detailed architecture and structure design is conducted and optimized with natural ventilation considerations under local climatic conditions. Based on the modeling and simulation, the preliminary results show that the atrium-centered architectural design is favorable for natural ventilation, in terms of the building main opening having an orientation consistent with local dominant wind directions. Therefore, this typical case provides a design approach for large public buildings in the early design stages to enhance indoor thermal comfort and reduce building energy consumption. Specifically, during the initial design phase of large public buildings, incorporating the design of ventilation corridors and aligning the main openings of the building with the prevailing local wind direction emerges as an effective design strategy. Such optimization design could facilitate an average air exchange rate over 2 h^{-1} driven by city wind pressure. Moreover, through building performance simulation and comparison, the natural ventilation in this case building can contribute to about 41.2% air conditioning energy saving ratio due to the free cooling effect. Simultaneously, natural ventilation significantly improved the indoor thermal comfort.

This study presents a typical case in which, during the initial design phase of public building construction, the cooling effects and energy-saving potential of natural ventilation are considered from an architectural perspective. The case aims to elucidate design methodologies and strategies for incorporating natural ventilation factors in the early stages of public building design. However, in the practical operation of buildings, the effectiveness of natural ventilation for cooling and its energy-saving potential are influenced by the external climatic conditions of the building site. Therefore, applying the specific design details introduced in this paper to cooling effects and energy-saving potential achieved under different climatic conditions may result in variations.

Nevertheless, the planning concepts and design methods proposed in this paper, which involve incorporating ventilation corridors in the initial design of public buildings to enhance the cooling effects of natural ventilation and increase energy-saving potential, are universally applicable. This implies that similar design strategies may achieve comparable results even under diverse climatic conditions. Thus, this study provides a universal design reference and application paradigm for projects considering the cooling effects and energy-saving potential of natural ventilation in the early stages of public building design.

Author Contributions: Conceptualization, M.Z. and Y.Z.; methodology, W.H.; software, Y.H. and J.X.; validation, Y.H., M.Z. and Y.Z.; formal analysis, W.H.; investigation, J.X.; resources, W.H.; data curation, Y.Z.; writing—original draft preparation, M.Z. and W.H.; writing—review and editing, J.X. and Y.Z.; visualization, Y.H.; supervision, Y.Z. All authors have read and agreed to the published version of the manuscript.

Funding: This research was funded by the National Natural Science Foundation of China (no. 52108032) and the National Social Science Foundation of China (no. 23BGL283).

Institutional Review Board Statement: Not applicable.

Informed Consent Statement: Not applicable.

Data Availability Statement: Climatic data are available on the Guangzhou Meteorological Information Community Service Network website (http://www.tqyb.com.cn/sqfw/climate/index.html, accessed on 26 October 2023).

Conflicts of Interest: The authors declare no conflicts of interest.

References

1. Invidiata, A.; Lavagna, M.; Ghisi, E. Selecting Design Strategies Using Multi-Criteria Decision Making to Improve the Sustainability of Buildings. *Build. Environ.* **2018**, *139*, 58–68. [CrossRef]
2. Yang, L.; Yan, H.; Lam, J.C. Thermal Comfort and Building Energy Consumption Implications—A Review. *Appl. Energy* **2014**, *115*, 164–173. [CrossRef]
3. Ma, H.; Du, N.; Yu, S.; Lu, W.; Zhang, Z.; Deng, N.; Li, C. Analysis of Typical Public Building Energy Consumption in Northern China. *Energy Build.* **2017**, *136*, 139–150. [CrossRef]
4. Liu, X.; Liu, X.; Luo, X.; Fu, H.; Wang, M.; Li, L. Impact of Different Policy Instruments on Diffusing Energy Consumption Monitoring Technology in Public Buildings: Evidence from Xi'an, China. *J. Clean. Prod.* **2020**, *251*, 119693. [CrossRef]
5. Letelier, V.; Bustamante, M.; Munoz, P.; Rivas, S.; Ortega, J.M. Evaluation of mortars with combined use of fine recycled aggregates and waste crumb rubber. *J. Build. Eng.* **2021**, *43*, 103226. [CrossRef]
6. Noori, N.; de Jong, M.; Hoppe, T. Towards an integrated framework to measure smart city readiness: The case of Iranian cities. *Smart Cities* **2020**, *276*, 676–704. [CrossRef]
7. Gracias, J.S.; Parnell, G.S.; Specking, E.; Pohl, E.A.; Buchanan, R. Smart cities-A structured literature review. *Smart Cities* **2023**, *265*, 1719–1743. [CrossRef]

8. Solano, J.C.; Caamaño-Martín, E.; Olivieri, L.; Almeida-Galárraga, D. HVAC Systems and Thermal Comfort in Buildings Climate Control: An Experimental Case Study. *Energy Rep.* **2021**, *7*, 269–277. [CrossRef]
9. Duca, G.; Trincone, B.; Bagamanova, M.; Meincke, P.; Russo, R.; Sangermano, V. Passenger dimensions in sustainable multimodal 302 mobility services. *Sustainability* **2022**, *14*, 12254. [CrossRef]
10. Qi, X.Y.; Zhang, Y.; Jin, Z.N. Building energy efficiency for indoor heating temperature set-point: Mechanism and case study of mid-rise apartment. *Buildings* **2023**, *13*, 1189. [CrossRef]
11. Ding, P.; Li, J.; Xiang, M.L.; Cheng, Z.; Long, E.S. Dynamic heat transfer calculation for ground-coupled floor in emergency temporary housing. *Appl. Sci.* **2022**, *12*, 11844. [CrossRef]
12. Indraganti, M. Adaptive Use of Natural Ventilation for Thermal Comfort in Indian Apartments. *Build. Environ.* **2010**, *45*, 1490–1507. [CrossRef]
13. Li, W.; Chen, Q. Design-Based Natural Ventilation Cooling Potential Evaluation for Buildings in China. *J. Build. Eng.* **2021**, *41*, 102345. [CrossRef]
14. Zhang, C.; Pomianowski, M.; Heiselberg, P.K.; Yu, T. A review of integrated radiant heating/cooling with ventilation systems—Thermal comfort and indoor air quality. *Energy Build.* **2020**, *223*, 110094. [CrossRef]
15. González-Cruz, E.M.; Krüger, E.L. Experimental study on a low energy radiant-capacitive heating and cooling system. *Energy Build.* **2022**, *255*, 111674. [CrossRef]
16. Krusaa, M.R.; Hviid, C.A. Combining suspended radiant ceiling with diffuse ventilation-Numerical performance analysis of low-energy office space in a temperate climate. *J. Build. Eng.* **2021**, *38*, 102161. [CrossRef]
17. Ji, W.J.; Zhao, K.J.; Zhao, B. The trend of natural ventilation potential in 74 Chinese cities from 2014 to 2019: Impact of air pollution and climate change. *Build. Environ.* **2022**, *218*, 109146. [CrossRef]
18. Gil-Baez, M.; Barrios-Padura, Á.; Molina-Huelva, M.; Chacartegui, R. Natural Ventilation Systems in 21st-Century for near Zero Energy School Buildings. *Energy* **2017**, *137*, 1186–1200. [CrossRef]
19. Rosato, A.; Ciervo, A.; Ciampi, G.; Scorpio, M.; Guarino, F.; Sibilio, S. Impact of solar field 30 design and back-up technology on dynamic performance of a solar hybrid heating network 31 integrated with a seasonal borehole thermal energy storage serving a small-scale residential 32 district including plug-in electric vehicles. *Renew. Energy* **2020**, *154*, 684–703. [CrossRef]
20. Prieto, C.; Cimmino, M. Thermal interactions in large irregular fields of geothermal 25 boreholes: The method of equivalent boreholes. *J. Build. Perform. Simul.* **2021**, *14*, 446–460. [CrossRef]
21. Guo, S.R.; Yang, H.Y.; Li, Y.R.; Zhang, Y.; Long, E.S. Energy saving effect and mechanism of cooling setting temperature increased by 1 °C for residential buildings in different cities. *Energy Build.* **2019**, *202*, 109335. [CrossRef]
22. Chen, L.; Shi, Q. Experimental study and performance analysis on a closed-cycle rotary dehumidification air conditioning system in deep underground spaces. *Case Stud. Therm. Eng.* **2022**, *37*, 102245. [CrossRef]
23. Zhang, Y.; Zhang, X.; Guo, S.R.; Long, E.S. Thermal plume simulation of VRF air conditioners for cooling system in high-rise buildings: A case study in China. *J. Eng. Res.* **2019**, *7*, 48–61.
24. Javad, K.; Navid, G. Thermal comfort investigation of stratified indoor environment in displacement ventilation: Climate-adaptive building with smart windows. *Sustain. Cities Soc.* **2019**, *46*, 101354. [CrossRef]
25. Afroz, Z.; Shafiullah, G.M.; Urmee, T.; Shoeb, M.A.; Higgins, G. Predictive modelling and optimization of HVAC systems using neural network and particle swarm optimization algorithm. *Build. Environ.* **2022**, *209*, 108681. [CrossRef]
26. Nematchoua, M.K.; Sadeghi, M.; Reiter, S. Strategies and scenarios to reduce energy consumption and CO2 emission in the urban, rural and sustainable neighbourhoods. *Sustain. Cities Soc.* **2021**, *72*, 103053. [CrossRef]
27. Sezen, K.; Gungor, A. Performance analysis of air source heat pump according to outside temperature and relative humidity with mathematical modeling. *Energy Convers. Manag.* **2022**, *263*, 115702. [CrossRef]
28. Yerdesh, Y.; Abdulina, Z.; Aliuly, A.; Belyayev, Y.; Mohanraj, M.; Kaltayev, A. Numerical simulation on solar collector and cascade heat pump combi water heating systems in Kazakhstan climates. *Renew. Energy* **2020**, *145*, 1222–1234. [CrossRef]
29. Jin, Z.; Zheng, Y.; Zhang, Y. A Novel Method for Building Air Conditioning Energy Saving Potential Pre-Estimation Based on Thermodynamic Perfection Index for Space Cooling. *J. Asian Archit. Build.* **2023**, *22*, 2348–2364. [CrossRef]
30. Long, E.S. Research on the influence of air humidity on the annual heating or cooling energy consumption. *Build. Environ.* **2005**, *40*, 571–578.
31. Xu, L.T.; Long, E.S.; Wei, J.C. Study on the limiting height of rooftop solar energy equipment in street canyons under the cityscape constraints. *Sol. Energy* **2020**, *206*, 1–7. [CrossRef]
32. Ding, P.; Li, Y.R.; Long, E.S.; Zhang, Y.; Liu, Q.J. Study on heating capacity and heat loss of capillary radiant floor heating systems. *Appl. Therm. Eng.* **2020**, *165*, 114618. [CrossRef]
33. Mazzeo, D.; Kontoleon, K.J. The role of inclination and orientation of different building roof typologies on indoor and outdoor environment thermal comfort in Italy and Greece. *Sustain. Cities Soc.* **2020**, *60*, 102111. [CrossRef]

34. Zhang, W.; Wu, Y.; Calautit, J.K. A review on occupancy prediction through machine learning for enhancing energy efficiency, air quality and thermal comfort in the built environment. *Renew. Sustain. Energy Rev.* **2022**, *167*, 112704. [CrossRef]
35. *ANSI/ASHRAE Standard 169-2013*; Climatic Data for Building Design Standards. American Society of Heating, Refrigerating, and Air-Conditioning Engineers, Inc.: Atlanta, GA, USA, 2013.

applied sciences

MDPI

Article

Study of the Impact of Indoor Environmental Quality in Romanian Schools through an Extensive Experimental Campaign

Tiberiu Catalina [1,2,*], Andrei Damian [1] and Andreea Vartires [1]

[1] Faculty of Building Services, Technical University of Civil Engineering, 021414 Bucharest, Romania; adamian7@yahoo.com (A.D.); vartires2@gmail.com (A.V.)

[2] National Institute for Research-Development in Construction, Urbanism and Sustainable Territorial Development—INCD URBAN-INCERC, 021652 Bucharest, Romania

* Correspondence: tiberiu.catalina@utcb.ro; Tel.: +40-763-915-461

Abstract: Decentralized ventilation systems in schools are becoming more important due to the focus on indoor air quality and energy economy. The research aims to explore how these technologies affect classroom air quality, thermal comfort, and noise. The study examined four decentralized ventilation systems in a real-world school using field measurements and data analysis. This included measuring the CO_2, temperature, noise, and thermal comfort using the Predicted Mean Vote (PMV) index. All systems greatly improved the air quality, keeping CO_2 levels within suggested limits. They failed to control indoor humidity, often lowering it to below optimal levels. Noise surpassed the 35 dB(A) criteria at maximum operation but was acceptable at lower airflows. Noise and air drafts did not bother residents. The study found that decentralized ventilation systems improve air quality and are easy to adapt to, although they need humidity control and noise management at higher operational levels.

Keywords: indoor air quality; decentralized ventilation; school environment

Citation: Catalina, T.; Damian, A.; Vartires, A. Study of the Impact of Indoor Environmental Quality in Romanian Schools through an Extensive Experimental Campaign. *Appl. Sci.* **2024**, *14*, 234. https://doi.org/10.3390/app14010234

Academic Editors: Daniel Sánchez-García, David Bienvenido Huertas and Dikaia E. Saraga

Received: 18 November 2023
Revised: 22 December 2023
Accepted: 22 December 2023
Published: 27 December 2023

1. Introduction

In the last decades, many studies have been performed to investigate the indoor environmental quality (IEQ) in educational buildings, such as primary and secondary schools or even colleges and universities. All of these studies outlined the importance of a good IEQ, in order to protect pupil's health and academic performance, knowing the fact that children spend 8 h per day or more during workdays in these types of buildings.

The idea of Indoor Environmental Quality (IEQ), as described in the scientific literature for various indoor environments, becomes more significant when considering schools due to evidence showing that children are more susceptible than adults to indoor conditions [1,2]. Some authors linked the IEQ concept to the TAIL concept, meaning the following: T—Thermal comfort, A—Acoustics, I—Indoor air quality, and L—Lighting [3,4], the four main criteria that influence the health and wellbeing of classroom pupils and educational staff. For the design stage of the HVAC systems for all buildings, including the educational ones, the indoor environmental parameters, with respect to the TAIL concept, are defined in the European standard EN 16798-1 [5].

Catalina et al. [6] investigated the IEQ concept in eight Romanian schools, five located in urban regions and three in rural ones, by means of extended surveys among pupils and academic staff. It was shown that the proximity to intense road traffic led to increased indoor air concentrations of carbon dioxide (CO_2) and suspended particles (PMs) in the classrooms from urban areas, compared to the schools located in a rural environment.

2. Literature Review

In relation to the assessment of indoor pollutant levels in schools, there is still significant variability among the available studies. However, Tran et al. [2] recently conducted a comprehensive review to assist future researchers in developing a standardized rating system and measurement protocols for the consistent evaluation of indoor environmental quality (IEQ) in educational facilities.

While the thermal comfort in schools has been investigated frequently by using the well-known PMV-PPD model proposed by Fanger in the early 80s, the indoor air quality in schools concept remains a challenge today, due to different influencing factors, such as the type of indoor pollutants to be investigated, the school location (in a rural or urban environment), the most appropriate ventilation system, or the local climate.

Galicic et al. [7] investigated the indoor air quality factors and natural ventilation in 454 primary Slovenian schools, by means of extended surveys, performed for one winter month (February 2020). The authors outlined the importance of the school micro location, such as the proximity to main roads or other polluting sources (industrial areas), on the IAQ level in classrooms.

Concerning the importance of the natural ventilation, it has been shown that the lack of outdoor filtration and the strong dependence of the ventilation airflow on the meteorological factors (wind speed and direction and outdoor temperatures) make this system unreliable, and a mechanical ventilation system should be mandatory.

Regarding the indoor air pollutants in educational buildings, the majority of scientists assumed that some of them could be considered less harmful than others. While the carbon dioxide (CO_2) is considered a marker of human presence and a maximal indoor average concentration of 1000 ppm could be considered a threshold for "un-ventilated" air [8–10], larger concentrations (around 2000 ppm), frequently seen in poorly ventilated spaces, including classrooms, lead to a decrease in the cognitive and academic performance of pupils, as shown in a recent study [11].

Apart from CO_2, there are many other indoor pollutants in schools that could be harmful at elevated concentrations and with long-term exposure, taking into account that pupils spend around 8 to 12 h in the indoor school environments. It has been shown that TVOC (Total Volatile Organic Compounds), a category of pollutants released indoors by building materials, furnishings, acrylic paints, permanent markers, and some cleaning products [12], have measured concentrations markedly elevated compared to those in the outdoor air and lead to severe adverse health effects among school children [13–16]. A toxic influence on pupils' health is also present as PAHs (Polycyclic Aromatic Hydrocarbons), transported to the indoor school environments via ventilation airflow, when schools are in the proximity of heavy road traffic [17].

Concerning airborne suspended particles, already known in the scientific literature as Particulate Matters (PMs), their harmful effect on children's health could be important, especially for the smaller ones ($PM_{2.5}$ and PM_1, with the down index corresponding to the largest particle diameter of this class, in μm), which could penetrate the inner parts of the pulmonary tissue after inhalation.

Consequently, numerous studies have examined the sources of particulate matter (PM) in schools, whether they originate from indoors or outdoors, and if they are influenced by factors, such as the school location and local climate (e.g., the built environment). These studies have also explored the harmful effects of PM on students' health and proposed ventilation solutions to mitigate the concentration of PM in the classroom air [18–21] A recent systematic literature review on the IAQ related to aerosol dispersion in indoor spaces with a high occupancy degree (schools included) showed the necessity to establish a robust optimization tool to assess the correlation between several influencing factors, such as ventilation, infection risk, people'sbehavior, and design comfort parameters [22]. The authors of this research concluded that such an optimized tool is lacking at present and should be clearly defined based on a more holistic approach. Regarding the influence of the traffic air pollution nearby the schools, Gartland et al. [23] showed the major impact of elevated PMs (notably PM_{10} and $PM_{2.5}$) and nitrogen oxide (NO_x) outdoor concentrations on the corresponding indoor air concentrations, which directly affect the executive function and academic performance of primary-school-aged children.

An important contribution to the ventilation strategies to be adopted for schools to reduce the risk of airborne viral infections was presented by Almaimani et al. [24], which searched an optimal ventilation strategy to minimize the air transmission of biological pathogens in classrooms, with a particular focus of the SARS-CoV-2 virus, responsible for the COVID-19 infection [25]. In their study, the authors demonstrated that a 2 ACH (air changes per hour) ventilation rate, combined with a 2-meter physical distance between two persons in the class should be an optimal strategy to reduce the POI (Probability Of Infection) with SARS-CoV-2, at a decent energy cost. The increase in the ventilation rate above this value was found to be more effective than the increase in the social distance, in terms of a POI reduction, but with higher cooling energy costs.

It appears obvious, as a result of numerous studies investigating the IAQ in schools, that ventilation is the main factor influencing the magnitude of indoor pollutant concentrations. Moreover, mechanical ventilation is much more appropriate to ensure a good IAQ (e.g., concentrations below threshold values given by reference regulations), because it does not depend on the variability of meteorological factors (wind and outdoor temperature) or on the occupants behavior (windows opening periods). Regarding the necessary ventilation airflow for the dilution of indoor pollutants, an interesting comparison between the values recommended by the American norm (ASHRAE standard 62.1 [26]) and European norm (EN standard 16798-1 [5]) was conducted by Kuramochi et al. [27]. The authors showed that, for a classroom with 10–12 years old pupils, the ventilation airflow recommended by the EN 16798-1, equal to 9.8 l/s*person, is larger than the corresponding airflow recommended by ASHRAE 62.1 (e.g., 6.71 l/s*person). Moreover, Kuramochi et al. [27], performed in this study, a meta-analysis judging the effect of ventilation on intellectual productivity in schools, based on a large survey with 3679 participants, by varying ventilation airflows. They demonstrated that an airflow of 10.7 l/s*person would be an ideal choice for an optimum IAQ in classrooms, when searching for high intellectual productivity of pupils. This value is closer to the ventilation airflow recommended by the European norm than to the corresponding value recommended by the American norm.

In the literature, many other studies relying on the IAQ in schools and ventilation strategies for improving IAQ have been performed. Jendrossek et al. [28] analyzed the influence of the ventilation solutions on the airborne risk of infection in schools. Calama-Gonzalez et al. [29] compared different ventilation scenarios (constant ventilation based on reference airflow by person, constant ventilation based on CO_2 concentration limitations, and demand-controlled ventilation), while Chang et al. [30] investigated the air-conditioning operation strategies for thermal comfort and IAQ in Taiwan's elementary schools.

Not all the studies found in the literature lead to the same conclusions regarding the optimum ventilation strategy for a school building, due to different influencing factors, such as the building architecture and the corresponding constraints, the local climate, and the proximity to a polluted environment (road traffic, industry, or other). We can conclude that every case study implying an educational building needs a particular focus and a preliminary experimental campaign for assessing IAQ, which should be mandatory, in order to judge the optimal ventilation strategy, with respect to energy savings for the system that is possible to implement. Moreover, the building owner, in conjunction with the architect,

unicity of the school. The best choice should be, therefore, the decentralized and compact ventilation systems rather than the centralized ones, which could affect the esthetical image of the building. Recent studies on the decentralized ventilation systems for classrooms could be found in the literature, the great majority of them presenting compact systems using air-to-air heat exchangers [31–33].

3. Indoor Air Quality Solutions

A decentralized ventilation system is one where individual ventilation units are placed in the rooms or areas to be ventilated, unlike a traditional ventilation system, which includes a centralized system from which air is distributed through ducts from a single central unit.

Decentralized ventilation systems require regular maintenance to achieve their efficiency and prevent the accumulation of dust and dirt in the ventilation units. The filter must be changed regularly. Additionally, they may not be as effective as centralized systems for ventilating large areas or very tall buildings. This constraint does not apply to our practical case because the classrooms are of a reasonable volume.

Decentralized ventilation systems can be controlled separately, depending on the room temperature, humidity level, or CO_2 concentration of the indoor air. They can be equipped with sensors that automatically adjust the ventilation rate based on these parameters.

Implementing single-flow controlled mechanical ventilation (CMV) enables the ventilation of a space while effectively addressing humidity levels. A single-flow CMV system can result in the wasteful consumption of energy due to the extraction vents creating negative pressure in the ventilated area. This negative pressure causes outside air to be drawn in, often through window grilles located at the top of the frame. The heating must then operate more to compensate for the intake of cold air into the ventilated space due to the infiltration of cold air, particularly in winter or in cold countries. These energy losses can lead to the overconsumption of around 20% of heating.

Decentralized double-flow ventilation keeps the same principle as classic double-flow CMV, except that the heat exchanger and the ventilation are in a single box, which is installed at the level of a facade through the wall or through from a window recess. Just like a classic double-flow CMV, the heat exchanger is equipped with filters to filter pollutants and allergens. It is therefore recommended to change these same filters regularly.

There are two principles of ventilation operation inside the decentralized ventilation box:

- The single fan alternates the direction of rotation in cycles of 50 to 120 s to reverse the flow, for extraction, then for blowing;
- The box is equipped with two fans and two air networks to ensure continuous ventilation.

Finally, a humidity-controlled CMV is a controlled mechanical ventilation solution that adjusts the air flow according to the relative humidity of the ambient air. Humidity-controlled ventilation systems are not suitable for use in schools for the main reason that the occupancy of the premises requires a constant minimum air flow to guarantee a suitable CO_2 level. This parameter is not linked to the relative humidity level in the air.

In Romania, there is NP 010–2022 (regulation on the design, construction, and operation of buildings for schools and high schools) [34] with a binding character, recently updated, which provides clarifications regarding the ventilation of spaces in educational buildings and the quality of indoor air in classrooms.

All occupied spaces in schools must be mechanically ventilated, locally or centrally, with ventilation systems equipped with heat recovery units that carry out the heat exchange between the exhausted air and the incoming air. In the classrooms, a flow of fresh air will be ensured according to the requirements of the technical Romanian norm I5 [35] to comply with the air quality category IDA1 (indoor air).

The incoming air is filtered with ePM efficiency filters in correlation with the ODA outdoor air quality class (minimum F7 to F9 is recommended).

It is recommended to use installations with variable air flow, which operate in a controlled manner according to the difference in the CO_2 concentration between indoor and outdoor air. Thus, a maximum CO_2 concentration difference of 400 ppm is allowed in classrooms.

Building materials and furniture items that do not contain or emit formaldehyde or other volatile organic compounds will be used. Radon from building materials and soil must not exceed the concentration of 200 Bq/m^3 on average per year.

If the ventilation units are located directly in the classroom, then the project will be accompanied by an acoustic impact study. The air ducts used in common spaces are made of non-combustible materials.

For air distribution inside the rooms occupied by students, the ventilation system is used by mixing or by displacement, with air vents specific to each chosen ventilation system. Textile air ducts can also be used. The air vents are made so that the air speed in the occupied area does not exceed the limits indicated for the average air movement speeds in the rooms in the occupied area, correlated with the IDA 1 environment category.

It is recommended to respect the values of the air changes per hour depending on the destination of the room, for example for 6–8 h^{-1}, laboratories 8–10 h^{-1}, sport halls 2–3 h^{-1}, canteens, university restaurants 8–12 h^{-1}, kitchens 5–8 h^{-1} (the air change rate is defined as the ratio between the total air flow rate introduced into the room and the air volume of the room).

In laboratories or other spaces where local exhaust devices are provided, the general ventilation system includes special measures to organize the introduction of compensation air.

In the case of centralized ventilation installations, it is recommended that the mechanical ventilation installation be made so that it can also be used to exhaust smoke and hot gases in the case of fire.

Inside the occupied spaces, the principles of user comfort are respected according to EN 16798-1/2019 [5], technical regulations I5 [36] and I 13 [37]. In the classrooms, at least the criteria of the ambience category II (IEQ2), in terms of thermal and acoustic comfort, and the criteria of the air quality category IDA1 are respected. Ambient category I (IEQ I) and air quality category IDA1 are recommended for these.

The indoor air temperature is set according to the purpose of the rooms, in the warm season (classrooms, laboratories 23–25 °C, libraries 24–27 °C, dining areas 23–27 °C, gyms 20–26 °C) and in the cold season (classrooms, laboratories 18 °C, libraries 20 °C, dining areas 18 °C, gyms 18 °C). For cooling, the indoor air temperature is chosen according to the specified values, provided that the difference between the outdoor and indoor temperature does not exceed 10 °C.

Regardless of the heating source, the maximum allowable temperature of the inlet pipe will be 70 °C. In situations where this requirement cannot be ensured, the heaters will be provided with protective grills to avoid accidents.

From the point of view of energy performance, whether they are newly built or rehabilitated buildings, these buildings must have an energy consumption almost equal to zero, by creating a suitable envelope, by providing high-performance technical systems, and by covering the energy requirement with energy from renewable sources in a proportion of at least 30%.

In Table 1, a short comparison between these systems is presented.

Table 1. Comparison of different ventilation solutions.

Ventilation System	Single-Flow CMV	Decentralized Double-Flow CMV	Classic Double-Flow Ventilation
Advantages	Very affordable price, between 400 and 800 Euros. Simple to install and maintain, can be carried out by an individual. Very affordable price, between 400 and 800 Euros.	Appropriation of the machine locally by class users, mainly teachers. Allows equipment to be managed separately, class by class, according to the specific needs of each. This modulation allows for significant energy savings. Allows for investment room by room. Renovation projects or projects with small budgets will therefore be able to benefit more easily.	These are technologies that are already a few years old. Maintenance work is centralized in a single location, at the heart of the installation. Purifies the air.

Table 1. *Cont.*

Ventilation System	Single-Flow CMV	Decentralized Double-Flow CMV	Classic Double-Flow Ventilation
Disadvantages	The system creates a depression in the room, causing cold air to enter and excess heating consumption. Cold drafts in rooms.	Maintenance must be carried out on each piece of equipment. It is still a very recent technology, making the question of repairability difficult to assess. Requires multiple openings in the facade to supply the groups.	These are complex installations that require careful management. Schools often call on outside companies for maintenance because they do not want to take on this responsibility, which represents significant costs. The air duct system requires a lot of space. Groups and ducts that are located outdoors quickly perish due to bad weather. Higher consumption than simple CMV due to the motors. Possibility of noise in the ducts.

The study team proceeded on a comprehensive investigation to investigate several forms of decentralized ventilation systems, acknowledging the current lack of sufficient studies in this area. The unique aspect of our research paper is its examination of different decentralized ventilation models through a comparative analysis. This particular aspect has not been widely investigated in previous scholarly works, thus making a valuable and original contribution to the area.

4. Methodology

To provide a comprehensive comparison of ventilation systems according to the aforementioned quality standards, it was necessary to devise a standardized process that ensures uniformity across all systems. This protocol was implemented to obtain the most precise and reliable data for analysis. The ventilation systems, specifically System#1, System#2, System#3, and System#4, underwent testing in a primry school located in Bucharest. The installation of all equipment took place on the ground floor of the building, namely in classrooms that have comparable capacities. A multitude of sensors was deployed within these rooms. In order to conduct data analysis, our attention will be directed towards the sensors located in the central area of each classroom, specifically in close proximity to the video projector. Other CO_2 sensors were placed on the walls and in the back of the room at normal height (approx. 1.5 m). For the measurements, the ARANET CO_2 monitors have been used as an advanced and accurate device specifically engineered for the purpose of detecting carbon dioxide concentrations in diverse settings. The device has a large measuring range ranging from 0 to 9999 ppm, rendering it appropriate for a diverse range of applications, encompassing educational settings, as well as industrial situations. The monitor's elevated resolution of 1 part per million (ppm) enables the precise and comprehensive monitoring of carbon dioxide (CO_2) levels. The device exhibits a level of precision with regards to accuracy, adhering to a standard of ±30 ppm or ±3% of the measured value, thereby guaranteeing the dependability of the collected data. The time constant τ, which represents 63% of the total response time, is set at 100 s. This number represents the sensor's capacity to rapidly detect and communicate changes in CO_2 levels, hence guaranteeing timely and precise information about air quality. Furthermore, the ARANET CO_2 monitor is equipped with the ability to measure temperature, including a range from 0 to 50 °C and boasting a precise resolution of 0.1 °C. The temperature readings exhibit a high level of precision, with a margin of error of ±0.3 °C. The device is capable of quantifying relative humidity within a specified range of 0–85%. It achieves this with a resolution of 1% and an accuracy of ±3%. To evaluate the efficacy of the equipment, an additional sensor was set up within a classroom without ventilation, alongside an external sensor to quantify the ambient air quality surrounding the building. The present study will examine the variations in humidity, temperature, and CO_2 levels over a duration of one month. This period coincides with the presence of children in the classroom for five consecutive days each week, while they remain at rest during the weekends. We shall do a comparative analysis of the data both internally and in relation to the external sensor.

To assess the level of comfort in a room and consequently the level of contentment among occupants, a subsequent set of measurements was carried out on 15 February 2023 in the morning. This was performed in four ventilated rooms utilizing a Testo440 sensor along with its corresponding probes. Additionally, this sensor was capable of measuring several environmental parameters, such as temperature, air velocity, relative humidity, and concentrations of carbon dioxide (CO_2). One advantage of this approach is the availability of data regarding the number of students in each room and their respective activities. This information enables us to make more accurate estimations of the carbon dioxide (CO_2) levels in the air, considering the influence of human occupancy.

5. Case Study: Cuza School Pilot School

The school building is in Bucharestand has a height regime of Basement + Ground floor + 2 Floors. From the information received from the manager of the building, it appears that it was built in 1967, the project being one that was often replicated during that communist period. The building has a rectangular shape with sides of 18.80 m and 45.50 m, respectively, and a footprint of approximately 765 m^2. The level heights are as follows: 1.80 m for the technical basement (partially) and 3.40 m for the ground floor, the 1st floor, and the 2nd floor, while the constructed area is 2435 m^2.

From a functional point of view, the building includes 17 classrooms, laboratories for chemistry, biology and informatics, chancellery, library, bathrooms, administrative spaces (director's office), a medical office, technical spaces, annexes, and storages, and it is considered as a typical school for Romania.

The structure of the outer envelope is made of solid brick walls (fired clay elements) with a thickness of 37.5 cm. Inside the building are structural brick walls 25 cm thick. The exterior PVC joinery and double glazing is partially degraded. The building has a gable roof with a wooden frame. The construction was designed according to the seismic design norm P13 from 1963.

The building is heated by steel radiators, a panel type, mounted on the walls with the help of support brackets and protected with masks to prevent injury to children by touching hot surfaces. The radiators are connected to the cogeneration system of the city of Bucharest, operating with a thermal agent with parameters of 80/60 °C, in quantitative regulation according to the outside temperature. The distribution of the thermal agent for heating and hot water for consumption is performed through the technical basement of the building and through columns to the floors.

The lighting of the classrooms is performed with lighting fixtures equipped with compact fluorescent lamps, depending on the purpose of the rooms.

The building is not equipped with mechanical ventilation and air cooling systems; thus, we expected high indoor pollution.

In the period 2010–2011, thermal rehabilitation works took place with the aim of increasing the energy performance of the educational unit, respectively reducing the energy consumption for space heating, under the conditions of ensuring and maintaining indoor thermal comfort. Figure 1 illustrates the exterior façade and a typical classroom from this school.

The thermal rehabilitation works consisted of the following:

- Thermal insulation of the external walls with fireproof expanded polystyrene of a 10 cm thickness.
- Replacement of the pipes for the distribution of the thermal heating agent with polypropylene pipes, in the areas where failures have occurred.
- Restoring the thermal insulation of the distribution pipes in the basement or replacing them entirely.

- All the sanitary groups were equipped with exhaust fans, and to avoid overpressure, transfer grids are installed in the doors; the exhaust air flow from each compartment of the sanitary group is 100–120 m^3/h.

The first ventilation system was installed in Room (1) and it is heat recovery ventilation equipment that is usually mounted in a false ceiling or in a space adjacent to the classroom, thus allowing the floor to remain free; the module is connected to the piping necessary for the supply of fresh air and the exhaust of stale air to the outside, but also to the network of ventilation ducts for the supply/exhaust of air from the classroom.

(a) (b)

Figure 1. (**a**) External photo of the pilot school. (**b**) Typical classroom.

It is also used in the free-cooling version when the outside temperature is lower than the inside temperature (for example, during the night). It can be used as an individual ventilation unit or can be integrated into air conditioning systems (air-refrigerant type). There is a varied range of units, for an air flow between 150 and 2000 m^3/h, but for this specific application, the model with 800 m^3/h was selected.

Medium and fine dust filters (M6, F7, F8) are optional for this system, to meet customer requirements or applicable legislation. The installation time was short due to the easy adjustment of the nominal air flow—in our case a 1 day installation; therefore, fewer air dampers are required than in the case of traditional installations. It can work in conditions of overpressure and depression.

The system is installed with two filters, G3 + F7, with automatic switching of heat-recovery units to a free cooling mode in the event of high heat, and has a cost of 3400 €/unit (without VAT) + €147 for filters and has operating temperatures from −15 to 50 °C. The introduction of fresh air was performed using textile piping (see Figure 2), while the extraction was performed using steel classic ventilation ducts.

The second system (see Figure 3) is also a decentralized ventilation unit with a high heat-recovery efficiency (up to 93% according to technical specifications), low noise levels, low installed power consumption, and minimal installation requirements. The equipment is equipped with an aMotion control module to operate all the necessary functions and consists of flexibly mounted inlet and suction fans, a countercurrent heat-recovery exchanger, a sliding supply air filter, a bypass flap for the heat exchanger, autonomously operated closing flaps, and a control box. The condensation tray (no drain) is heated using an electric cell with an automatic switching function. The upper section is equipped with noise attenuators with separation, grilles for the exhaust and intake of air, a filter for the extracted air, and an external CO_2 concentration sensor. The lower part of the unit has a spacer frame made of anti-vibration rubber. This system was also installed in one day and was connected to Wifi for control. Compared to the previous solution, it has the disadvantage that it is occupying around 1 m^2 in the back of the room, but on the other hand, it is well fixed to the wall and in case of earthquakes (Bucharest has a high risk), the system is secured with no danger for the children.

(**a**) (**b**)

Figure 2. (**a**) Heat-recovery unit placed in the school hallway. (**b**) Textile introduction of fresh and extraction with regular HVAC duct.

Figure 3. Photo with the second heat recovery ventilation unit placed in the school.

The third system is a decentralized mechanical ventilation system consisting of two HRV ventilation units and one ERV ventilation unit installed in a classroom number (3)—see Figure 4. The systems are smaller and are decentralized ventilation systems, including a heat-recovery system from the exhausted air, thus reducing energy consumption for heating or cooling the outside air. Integrated CO_2 sensors enable the continuous monitoring of carbon dioxide concentrations in the classroom. When the CO_2 concentration exceeds a preset level, the ventilation system automatically activates, ensuring a constant supply of fresh air. The outside air introduced is preheated by means of an electric resistance and reheated with the help of a battery fed with hot thermal water (supplied by the heating network of the city of Bucharest). The systems not only allow for the introduction of fresh filtered air but also act as radiators for the heating. Clearly, its main advantages is the fact that it does not reduce the area and is also fixed to the wall, making it safe.

While two HRV units were placed left and right of the room, in the middle, the ERV unit was installed and equipped with an advanced energy-recovery system, known as enthalpy recovery, which allows for the transfer of not only heat, but also moisture between the exhaust air and the air introduced into the classroom. Through this process, the ERV Decentralized Ventilation System ensures more the efficient regulation of humidity inside the classroom. Thus, the state of thermal comfort is improved, not only by maintaining a constant temperature, but also by controlling the humidity, which has a positive impact on the performance of the occupants. Also, proper humidity management can prevent indoor air quality problems, such as the growth of mold and other microorganisms that prefer moist environments.

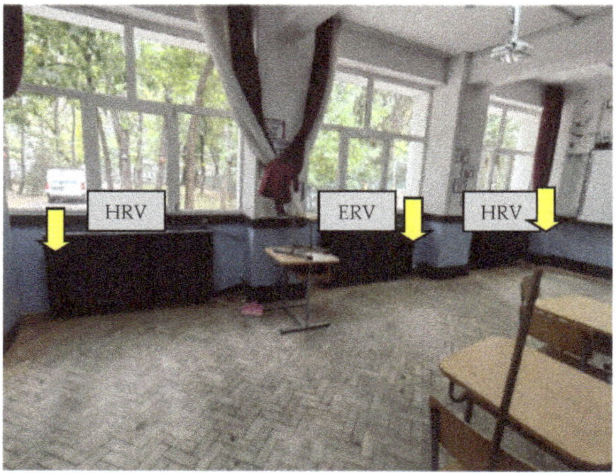

Figure 4. Photo with the third heat recovery ventilation unit placed in the school.

The installed ventilation equipment has a nominal air flow of 250 m^3/h/unit, so 750 m^3/h total, a declared noise level 32.6 dB, an electric battery for frost protection with a power of 0.54 kW, a reheating battery that works with hot water, a heat recovery in countercurrent type HRV or type ERV, depending on the model, and two centrifugal fans.

Moving to the last study system, called System #4, it is a decentralized ventilation system with heat recovery 700 m^3/h and is specially designed for classrooms (see Figure 5).

Figure 5. Photo with the second heat-recovery ventilation unit placed in the school.

The plastic heat exchanger, with countercurrent hexagonal plates, guarantees, according to the manufacturer, a thermal efficiency of up to 90%. The fans are self-powered with EC motors that save electricity and operate silently. The built-in carbon dioxide (CO_2) sensor can automatically control the unit; when a set CO_2 level is reached in a classroom, the device shuts down. The metal casing contains double walls and insulating material in the space of 25 mm between the panels, which ensures both a high thermal performance and low noise levels. There are also special silencers on both the intake and supply ports to ensure a quiet environment in classrooms.

The ventilation unit uses the ModBus protocol to connect to and communicate with the building management system and to report any malfunctions or periodic maintenance. The ventilation system can be monitored and controlled using a computer or a central system. The unit uses three filters, and the air is filtered both at the inlet and outlet (M5 filter for dust and other pollutants, G4 and F7 filters for particles up to 1 micron and allergens, spores, and pollen). In addition, when the filters are loaded with a certain number of particles, the system issues a warning.

6. Results and Analysis

The graph below shows the evolution of CO_2 levels and temperature in the classroom studied ⑤ without a ventilation system. There are wide variations in CO_2 levels over the course of the day, generally fluctuating between 750 and 1500 ppm, with a peak between 7 and 10 a.m., at above 1500 ppm. As a reminder, the ideal CO_2 level in a room is recommended to be less than 1000 ppm. The levels on this day are not satisfactory and show the need for a ventilation system to evacuate CO_2 air to the outside.

The level of CO_2 in the air can also be used to assess classroom activity. Indeed, we know that CO_2 is released by the people in the room as they breathe. Therefore, the more people there are in a room and the greater their physical activity, the faster the CO_2 levels in the air will rise, saturating the air with CO_2.

As can be seen from Figure 6, the CO_2 levels reach alarming values of more than 3000 ppm, while the air temperature, due to high internal gains, goes up to even 26 °C. We remind the reader that the measurements took place in the middle of March 2023. For the entire set of measurements from 11 March 2023 to 11 April 2023, the average temperature was 23 °C, with a maximum of 26 °C and a minimum of 13.3 °C (during the nighttime on the weekend). For the CO_2, the data showed maximum values of even 7036 ppm (20 March 2023 at 9:23 a.m.) while the average (including nighttime) was 845 ppm.

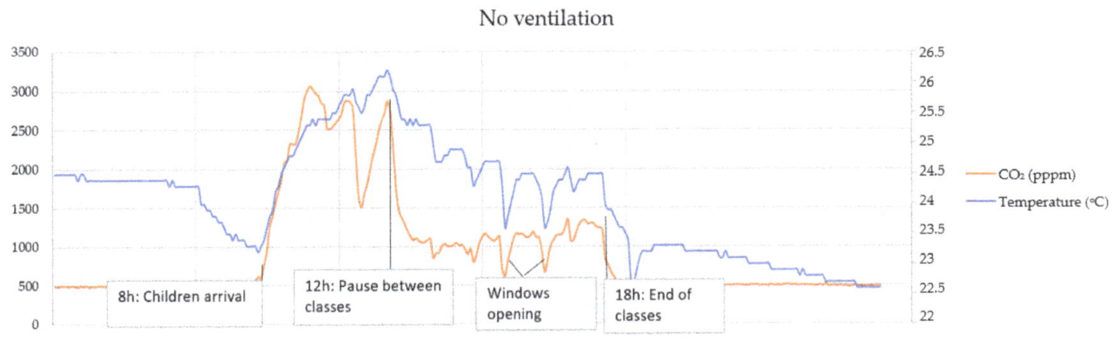

Figure 6. Typical day in the non-ventilated classroom.

The data shown in the graph (see Figure 6), which illustrate the change in carbon dioxide (CO_2) levels and temperature within a classroom lacking a ventilation system, provide compelling evidence to support the necessity of conducting thorough investigations regarding the implementation of ventilation in educational settings. The variations in observed carbon dioxide (CO_2) levels, namely the peaks that reach, in most of the cases, 3000 parts per million (ppm), greatly exceed the recommended maximum threshold of 1000 ppm. This divergence not only signifies inadequate air quality but also emphasizes the potential health and cognitive consequences for individuals in educational settings. The presence of values over 3000 ppm, occasionally peaking at 7036 ppm, suggests an environment that has the potential to significantly impede cognitive functions and impair learning efficiency. This is particularly worrisome considering that these measures were obtained in a conventional educational environment during standard school hours.

Elevated temperatures, particularly when combined with elevated amounts of carbon dioxide, have the potential to induce discomfort, impair concentration, and ultimately diminish academic performance. The presence of these conditions in March, a month often associated with lower temperatures, suggests that the severity of the situation may be significantly amplified during periods of higher temperatures.

This highlights the pressing necessity for the implementation of efficient ventilation systems in educational environments, with the aim of guaranteeing the comfort, health, and cognitive well-being of all present. The research presented in this study makes a substantial contribution to the ongoing academic discussion surrounding indoor environmental quality in educational settings, as in this paper, we are focusing on possible solutions.

The study of four distinct ventilation systems inside a classroom environment has yielded noteworthy findings about their efficacy in preserving indoor air quality and controlling the temperature (see Figure 7).

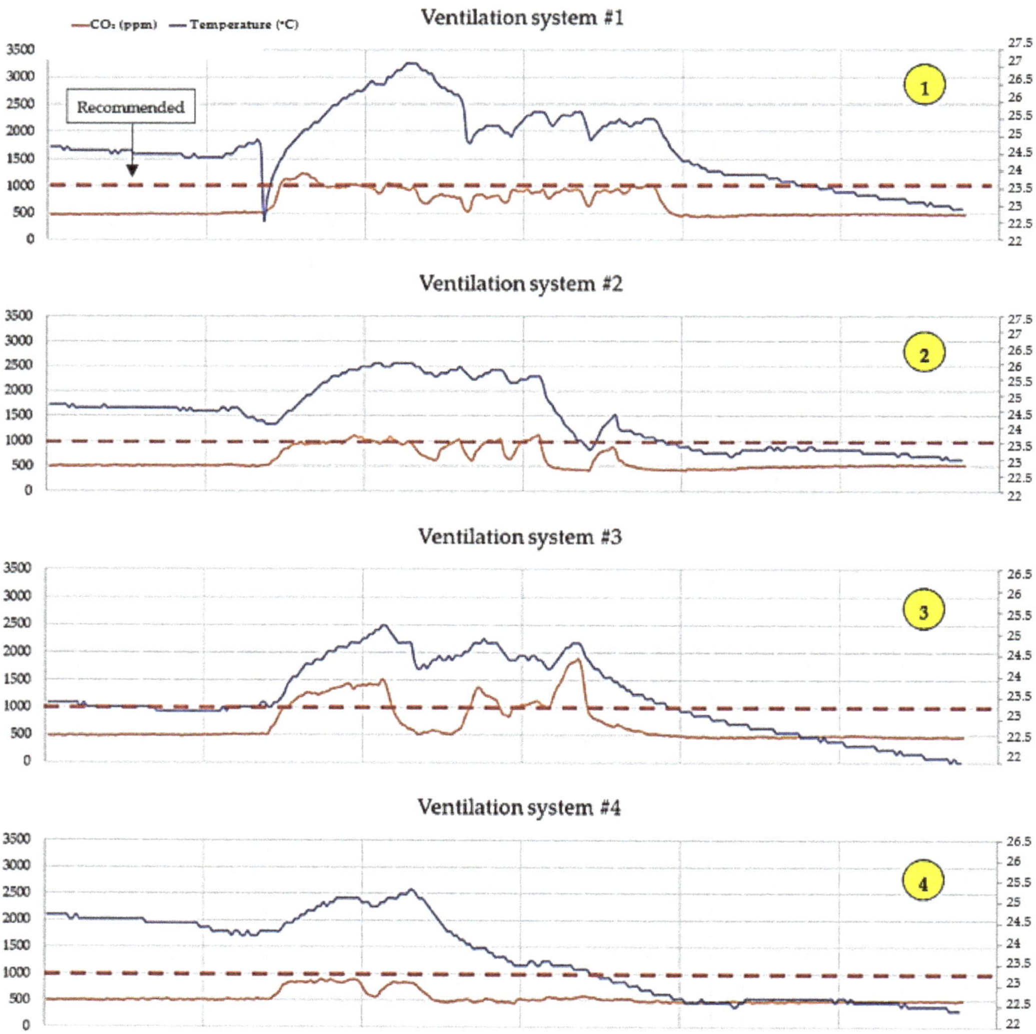

Figure 7. Typical day in the ventilated classrooms.

System 1 has shown remarkable efficacy by consistently upholding CO_2 concentrations below 1000 ppm for 97% of the duration, thereby serving as a robust metric for assessing favorable indoor air quality. Nevertheless, it is important to highlight that the temperature within the room increased to 27 °C, which is marginally higher compared to that with the other systems. Nevertheless, the system's capacity to reliably control the concentration of CO_2 highlights its efficacy.

The findings of System 2 demonstrated even more noteworthy outcomes, as it successfully maintained carbon dioxide (CO_2) levels below 1000 parts per million (ppm) for 99% of the duration of occupancy. This outcome serves as a strong indicator of the system's exceptional control of air quality. The temperature within the enclosed space reached a maximum of 26 °C, a value that falls within the allowed range for creating appropriate educational settings.

The third system, despite maintaining air quality levels below 1000 ppm for 65% of the time, did not employ its boost mode. The utilization of this function during periods of rest has the potential to significantly improve its effectiveness of removing pollutants. Significantly, this classroom, characterized by the presence of two external walls, showed a lower maximum temperature of 25.2 °C in contrast to the preceding two systems.

System 4 consistently demonstrated exceptional air quality, as seen by its ability to consistently control carbon dioxide (CO_2) levels below 1000 parts per million (ppm) during its operation. The classroom, experiencing reduced solar heat gain, had a temperature of 25.3 °C, which is beneficial for creating a comfortable learning environment.

In general, it can be observed that all four systems demonstrated satisfactory performance by successfully delivering fresh air and ensuring the maintenance of suitable indoor conditions. The notable disparity in air quality and temperature regulation between the ventilated and non-ventilated classroom serves to emphasize the effectiveness of these systems and emphasizes the pressing necessity for their implementation in educational settings. The present investigation not only showcases the efficacy of these systems in upholding ideal learning settings but also provides significant data to the domain of indoor environmental quality in educational structures.

The analysis of carbon dioxide (CO_2) concentrations in different ventilation systems, in comparison to a situation without any ventilation, provides valuable findings regarding their efficacy in controlling indoor air quality (see Figure 8). Without adequate ventilation, carbon dioxide (CO_2) levels experience substantial variations, ranging from a minimum of 440 parts per million (ppm) during nighttime to a maximum of 7036 ppm. Consequently, the average CO_2 concentration calculated only for the occupation period (8:00–18:00) in this situation amounts to 1296 ppm.

System 1 demonstrates a notable enhancement with a more favorable concentration of 697 ppm (46% reduction compared to with no ventilation). System 2 demonstrates superior performance compared to the other systems, as it effectively achieves a harmonious equilibrium between the lowest maximum concentration of 1495 ppm and an appropriate average concentration of 636 ppm (50.9% reduction compared to with no ventilation). Moreover, System 2 constantly maintains outstanding air quality throughout its operation. System 3, however, is less efficient than System 2 and demonstrates a notable reduction in the maximum concentration of CO_2 to 2072 ppm and maintains an average level of 753 ppm (41.8% reduction compared to with no ventilation). Finally, System 4 demonstrates strong performance by maintaining a maximum carbon dioxide (CO_2) level of 3234 parts per million (ppm) (the system shut down at that moment) and an average level of 673 ppm, thereby preserving favorable air quality for a significant duration.

In general, it can be observed that each ventilation system exhibits a notable improvement in air quality when compared to that with the absence of ventilation.

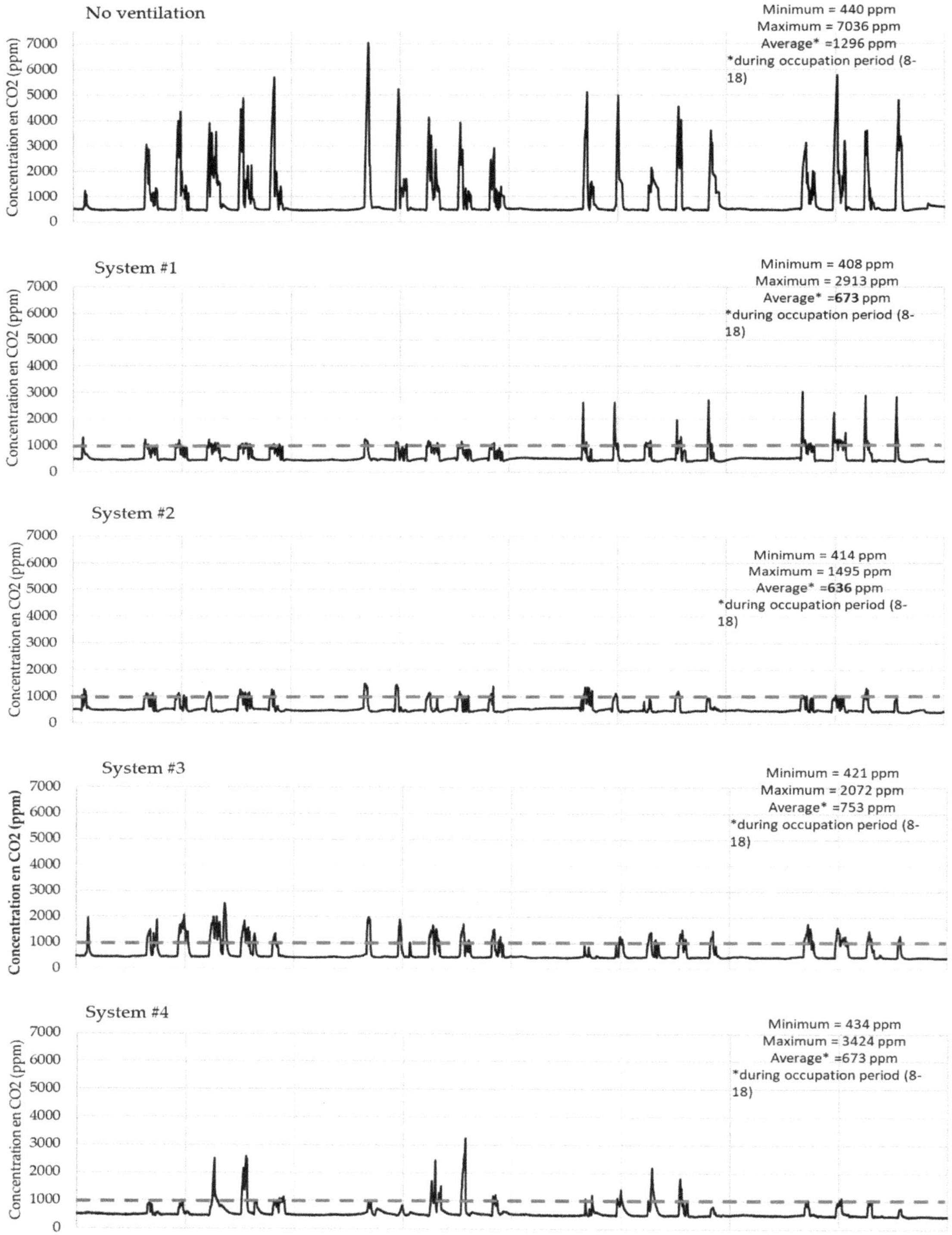

Figure 8. Long-term analysis of CO_2 levels for the heat recovery ventilation units (11 March 2023 to 11 April 2023).

Similar results were obtained in other studies that analyzed decentralized ventilation (e.g., like SYSTEM 2) and we can state that the experimental data are validated with this cross-reference [33] on CO_2 and temperature values. In [6] it was also found that CO_2 levels increase gradually but can be kept to values of around 1200 ppm with decentralized ventilation systems.

The investigation of air temperature in several ventilation systems, including a scenario without ventilation, demonstrates a significant degree of uniformity in maintaining comparable temperature levels among educational spaces. The mean ambient temperature within the enclosed classroom, lacking proper ventilation, was measured as 23.5 °C. When comparing the data, it can be observed that Systems 3 and 4 exhibited a similar average temperature of 23.1 °C, which nearly resembled the temperature of the non-ventilated environment. System 2 exhibited a marginally lower mean temperature of 23.4 °C, but System 1, which was installed in a room featuring two external walls, registered a little higher average temperature of 24.1 °C.

The noteworthy aspect lies in the negligible temperature difference observed across the various scenarios, regardless of the presence or absence of ventilation systems. The findings indicate that the ventilation systems are successfully sustaining a consistent thermal environment, a critical factor for ensuring comfort and cognitive performance in educational environments. The fact that Systems 1 and 3, located in rooms with two external walls, showed no statistically significant temperature fluctuations compared to the other systems is particularly noteworthy. This observation suggests that the insulation and architectural features of the rooms are successfully mitigating the risk of thermal energy transfer across the external walls.

Based on the consistent room volumes and the approximate ratio of inhabitants (about 23 kids and 1 professor), it is apparent that the ventilation systems are not inducing substantial variations in the thermal conditions within the classrooms. The consistency of temperature, irrespective of the type of ventilation system utilized, is advantageous in terms of education, as it guarantees a uniform and pleasant learning atmosphere for both students and educators.

The control of humidity within educational settings is a crucial element in establishing an optimal learning environment, and ventilation systems are vital in facilitating this endeavor. Maintaining humidity levels within the optimal range of 30% to 50% is crucial for ensuring comfort, promoting good health, and supporting cognitive function. Elevated levels of humidity can contribute to the growth and spread of mold and mildew, posing risks to both the physical stability of the building and the well-being of individuals inhabiting it. Such conditions have the potential to induce breathing problems and induce allergic responses. On the contrary, decreased levels of humidity can lead to many forms of discomfort, including the manifestation of dryness in the skin and eyes. Moreover, it can also elevate the probability of contracting respiratory diseases and facilitate the transmission of airborne viruses. Ventilation systems play a crucial role in maintaining equilibrium in interior humidity levels. By facilitating the circulation of new air and eliminating stagnant, damp air, these systems have the capacity to inhibit the accumulation of excessive moisture, thereby reducing the likelihood of mold proliferation and upholding an indoor environment conducive for human health. Moreover, the influence of humidity on thermal comfort is substantial. Despite the implementation of precise temperature regulation, unsuitable levels of humidity might result in individuals perceiving the environment as excessively warm or chilly.

The investigation of the researched ventilation systems, which lack integrated humidity management, uncovers notable difficulty in upholding ideal indoor humidity levels, particularly in winter when the exterior air is cold and dry. The introduction of outside air into classrooms through these systems may result in a notable decrease in interior humidity levels, which could potentially give rise to perceptions of discomfort and health-related concerns. The phenomenon is particularly accentuated when airflows are increased, as ventilation systems have the potential to severely dry the indoor air.

The data obtained from the systems highlight this particular risk (see Figure 9). System 1 shows an average humidity level of 30.87%, which is closely trailed by System 3 at 30.76% and System 2 at 30.43%. The observed results exhibit a proximity to the lower threshold of permissible indoor humidity, suggesting a modest yet constant decrease in moisture levels. System 4 exhibits a little improvement, as it consistently sustains an average humidity level of 33.45%. Although this value is closer to the target range, it still resides towards the lower end. On the other hand, the room without ventilation has a comparatively elevated average humidity level of 35.04%. This can be attributed to the absence of external air circulation and the inherent moisture produced within the space.

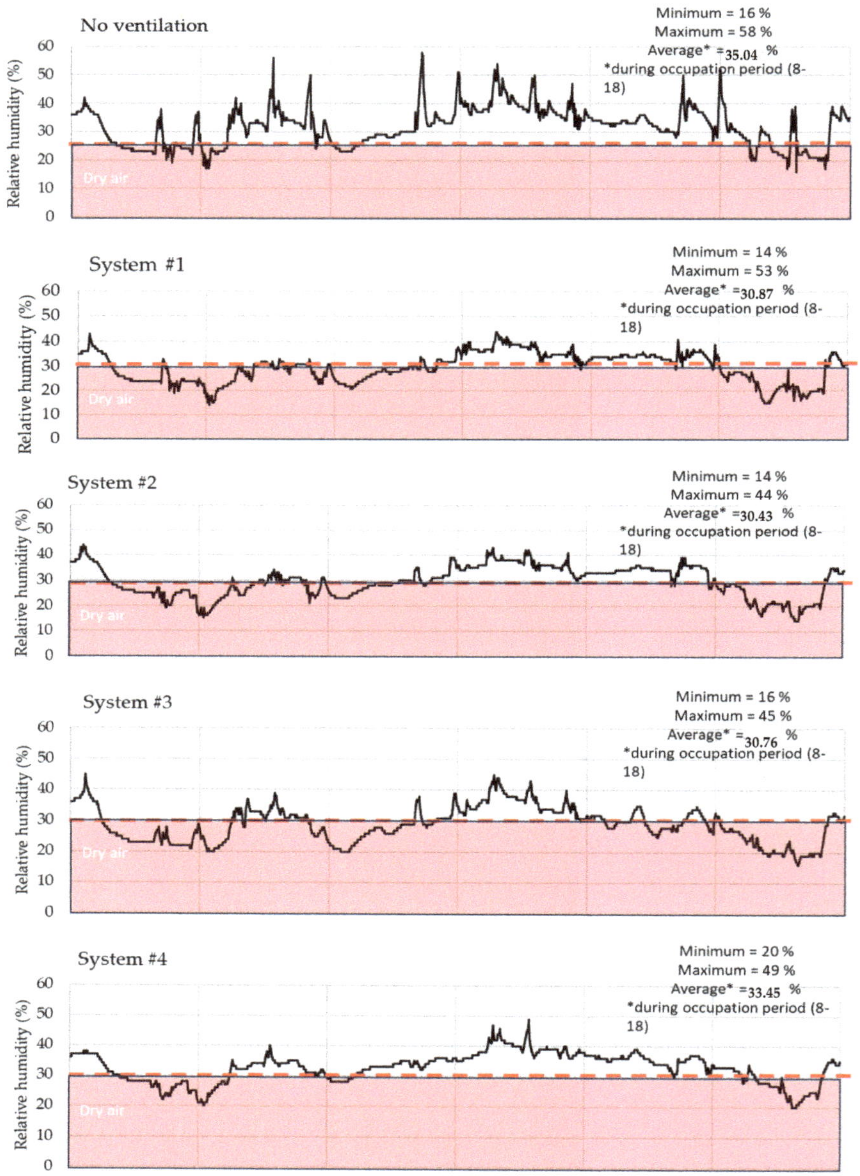

Figure 9. Long-term analysis of humidity levels for the heat recovery ventilation units (11 March 2023 to 11 April 2023).

The analysis of the ventilated and non-ventilated scenarios indicates that ventilation plays a vital role in maintaining the air quality. However, it is important to note that the implementation of ventilation systems can mistakenly result in decreased humidity levels. The concern is notably apparent in systems lacking the means for controlling humidity levels. To address this issue, the implementation of humidifiers in conjunction with ventilation systems could be a feasible approach, aiding in the maintenance of a balanced and pleasant interior environment. The data unequivocally demonstrate that although ventilation systems are efficient for controlling air quality, they do require supplementary measures, such as humidification, to guarantee comprehensive indoor environmental quality.

The Predicted Mean Vote (PMV) index is a widely recognized method for calculating thermal comfort in indoor environments. Developed by P.O. Fanger in the 1970s, the PMV model estimates the average thermal sensation of a large group of people. It is based on the heat balance of the human body and incorporates six key factors: air temperature, mean radiant temperature, air velocity, humidity, clothing insulation, and metabolic rate. The PMV index serves to forecast the mean rating of a large number of people on a seven-point scale measuring the thermal feeling, which spans from −3 (indicating chilly) to +3 (indicating hot), while a rating of 0 signifies a state of neutral thermal comfort. The inclusion of this scale in the PMV model is crucial as it allows for the measurement of the subjective aspect of thermal comfort, recognizing that different individuals may have varying responses to identical environmental conditions. In our study, assessing the thermal comfort in classrooms equipped with different ventilation systems, we employed the Testo 400 IAQ and Comfort Kit to measure the Predicted Mean Vote (PMV) index. The evaluation of thermal comfort using the Predicted Mean Vote (PMV) index for the four different ventilation systems provides valuable insights regarding their efficacy in sustaining a pleasant indoor environment. System 1 showed a PMV satisfaction percentage of 79.6%, whereas the PMV values ranged from −0.5 to +0.5. This finding suggests that a significant proportion of individuals reported a thermal sensation that was close to neutral, a state that is generally regarded as comfortable in the context of indoor environments (see Figure 10).

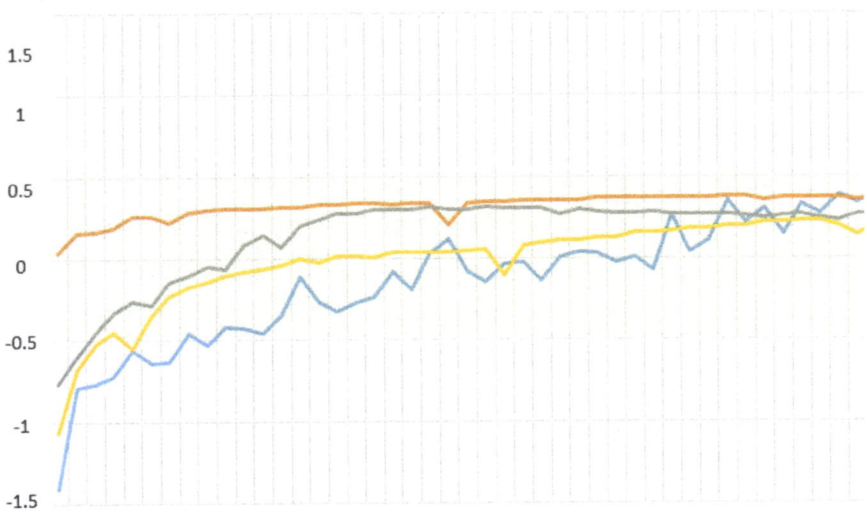

Figure 10. PMV index measurements for the analyzed systems.

System 2 provides a remarkable satisfaction rating of 100%, indicating that all individuals occupying the space had thermal conditions that fell within the ideal Predicted Mean Vote (PMV) range. The observed level of satisfaction is significant and suggests that System 2 is highly proficient in maintaining an optimal thermal environment.

System 3 had a commendable performance, with a satisfaction percentage of 94.1%. The substantial proportion mentioned indicates that a large majority of individuals were situated inside the desirable range of the Predicted Mean Vote (PMV), hence suggesting the efficient management of temperature conditions. In a similar vein, it is worth noting that System 4 demonstrated a satisfaction rate of 92.2%, thereby reinforcing the effectiveness of these ventilation systems in delivering thermal comfort.

In comparison, it can be observed that Systems 3 and 4 exhibit marginally lower satisfaction rates in contrast to System 2, yet they consistently uphold elevated levels of thermal comfort. The observation that the satisfaction rates of all systems, except for System 1, surpassed 90% suggests a notable level of success across the board.

In conclusion, each of the four systems exhibits a notable capacity to effectively maintain optimal thermal comfort, since there have been no documented occurrences of air drafts. The impact of air drafts on perceived thermal comfort is a critical element to consider. The observed high levels of satisfaction across all systems serve to underscore their effectiveness in establishing and sustaining a pleasant and stable indoor thermal environment, a crucial factor in promoting the well-being and productivity of occupants.

The control of noise levels is a crucial feature of the design and operation of decentralized ventilation systems, especially in environments, such as schools or offices. HVAC systems, by their nature, can generate noise, which if not adequately controlled, can become a source of distraction or discomfort. If not effectively managed, this noise can potentially lead to distractions or discomfort. In such circumstances, the prescribed standards for permissible noise levels generally advocate a maximum threshold of 35 dB(A). The purpose of establishing this threshold is to guarantee that the background noise generated by the ventilation system does not disrupt the main functions of the area, such as educational activities in classrooms or concentrated work in offices.

For decentralized systems, their inherent advantages in terms of flexibility and energy economy necessitate meticulous design considerations to mitigate noise. This includes the careful selection of equipment that fulfills low-noise standards, the strategic placement of this equipment to minimize sound transmission, and the potential integration of sound-dampening materials or designs. The challenge is to balance the acoustic comfort with the system's effectiveness in air circulation and quality.

Surpassing the established standard of 35 dB(A) can result in heightened levels of tension, diminished ability to concentrate, and a general decline in pleasure with the interior environment. Hence, while considering the design and execution of decentralized HVAC systems, it is imperative to recognize that acoustic performance holds equal significance to factors, such as thermal comfort and air quality. It is imperative to maintain the prescribed noise levels to establish an indoor atmosphere that is conducive to comfort, well-being, and productivity. Measurements of acoustic levels were measured for all four systems at different scenarios (e.g., maximum air flow or different air flow stages) using a class 1—sound meter SVANTEK 970. The measurements were realized during the night to minimize the background noise. For all measurements, the background noise was recorded and subtracted for the results. The measurement of data was performed based on the frequency (31.5 Hz to 8000 Hz).

Figure 11 shows that System 1 and 3 are by far the noisiest among all four. For System 1, this is justified as it does not have a sound attenuator. At maximum operation, it emits an average noise level of 64.15 dB(A), which is far too high for a classroom. As a reminder, a background noise level of 65 dB corresponds to the noise heard in a moving car. Even when operating at very low speeds (2/6 for System 3), the system fails to meet the standard, with a noise level of 39.07 dB(A).

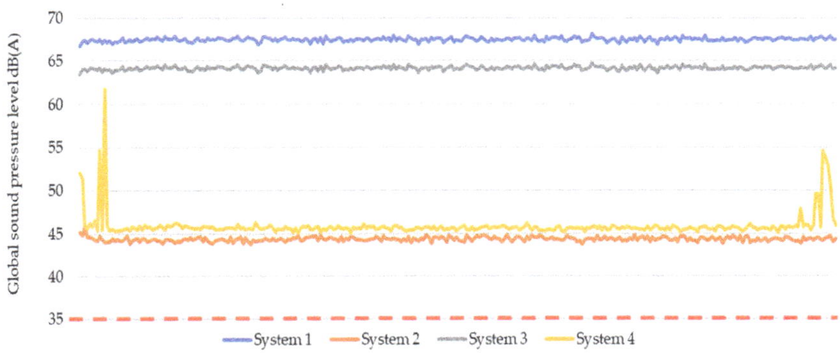

Figure 11. Sound pressure level in dB(A) at maximum air flow.

System #2 and System #4 achieved similar results, with maximum noise levels of around 45 dB(A) when delivering maximum air flow of 700 m^3/h. Although this value is higher than the standard, no complaints were issued during the entire period, thus confirming that even 45 dB(A) is acceptable for the occupants. On the other hand, we note that System #4 operated at 55%, or 385 m^3/h, and System #2 at only 40%, or 320 m^3/h, with maximum values of 35–36 dB(A), thus respecting the norms.

7. Discussion

In the absence of proper ventilation, the concentration of carbon dioxide (CO_2) surpassed the established threshold of 1500 parts per million (ppm) for a significant duration of 86.5 h. The observed length is notably elevated, suggesting a substantial likelihood of compromised air quality, which has the potential to seriously impact the cognitive abilities and well-being of both children and educators. System 1 exhibits a notable enhancement, as seen by the CO_2 levels exceeding 1500 ppm after a mere duration of 7.95 h. The observed decrease in air quality, when compared to that with the absence of ventilation, serves as compelling evidence for the efficacy of the system in managing air quality. Among the many systems considered, System 2 demonstrates superior efficacy by consistently maintaining carbon dioxide (CO_2) levels below the threshold of 1500 parts per million (ppm) during the entire duration of the month. This finding suggests that the system achieves a high level of effectiveness in maintaining stable indoor air quality, therefore serving as a commendable benchmark for ventilation practices in educational environments. System 3, albeit an improvement compared to the absence of ventilation, resulted in carbon dioxide (CO_2) concentrations surpassing 1500 parts per million (ppm) for a duration of 23.5 h. While this improvement is noteworthy compared to the absence of ventilation, it implies that there is potential for further optimization in this system. System 4 exhibits a moderate level of efficacy in preserving air quality, but is less efficient than Systems 1 and 2, as evidenced by its sustained exposure to carbon dioxide (CO_2) concentrations above 1500 parts per million (ppm) for a duration of 10.66 h.

The analysis of indoor humidity levels in classrooms equipped with different ventilation systems yields a significant observation: although these systems effectively regulate CO_2 levels, they tend to decrease humidity, frequently reaching the minimum permitted threshold. Systems 1–3 have a mean humidity level in the proximity of 30%; however, System 4 has a little superior performance with an average humidity level of 33.45%. Nevertheless, as compared to the average of 35.04% in the non-ventilated space, all other measurements exhibit inferior results. This implies that although ventilation plays a critical role in maintaining air quality, it can also result in decreased humidity levels, which may have implications for both comfort and health. Hence, the incorporation of humidification solutions into these systems may be imperative to maintain an ideal indoor environment.

The space occupied by the system might limit classroom functionality, and unacceptable noise levels could be disruptive, affecting the learning environment and overall student and staff comfort. Addressing these challenges is crucial for evaluating the feasibility and overall effectiveness of decentralized ventilation systems in educational settings.

8. Conclusions and Future Work

The comparative analysis of carbon dioxide (CO_2) levels in classrooms equipped with various ventilation systems, as opposed to a scenario without any ventilation, over a period of one month, offers a distinct signal regarding the efficacy of these systems in upholding satisfactory indoor air quality throughout school hours.

Our study led to a few significant results regarding the assessment of decentralized ventilation systems in educational environments. First and foremost, these systems exhibit a high level of efficacy in improving air quality, successfully controlling CO_2 levels to ensure that they remain below acceptable thresholds. This is of crucial significance for the physical and cognitive wellbeing of both students and staff members. One significant constraint that has been identified is the absence of adequate humidity regulation. In specific instances, this phenomenon has led to a decrease in indoor humidity levels that fall below the optimal range, which may have potential implications for both comfort and health.

In relation to noise levels, it can be observed that the systems do not consistently maintain compliance to the 35 dB(A) noise threshold when operating at their maximum air supply capacity. However, these systems are able to sustain satisfactory acoustic conditions when run within a moderate range of levels, specifically at 55% or between 320 to 380 m^3/h. The absence of any complaints from occupants regarding noise or air drafts is a notable characteristic, suggesting a general satisfaction with the ambient conditions facilitated by these technologies.

The systems' ease of installation and non-invasive characteristics makes them a highly favorable choice when retrofitting existing school structures. The incorporation of this characteristic, alongside with its cost-efficiency, taking into account the significant advantages for the well-being of children and decreased energy usage, puts up a compelling rationale for its implementation. The expenses associated with implementing these systems in classrooms, varying from 4000 to 6500 Euros, can be considered very affordable when considering the positive impact on health and the environment. This is particularly significant as these systems effectively eliminate the necessity for window ventilation.

An effective ventilation strategy that has emerged from our study is the use of boost mode during break times or periods of high background noise. This methodology aims to enhance air quality while minimizing any possible adverse effects on classroom activity. Nevertheless, it is apparent that humidifiers are essential throughout the winter season in order to uphold indoor humidity levels at above 40%, thereby effectively resolving the problem of low humidity induced by these systems.

Similar measurement campaigns in schools could be found in the literature, concerning the evolution of indoor CO_2 concentrations, temperature, and relative humidity. For instance, Lazovic et al. [38] performed an extensive experimental study in four schools from Serbia (three located in urban areas and one in a rural area), with the results showing a good correlation in terms of CO_2 concentration profiles with the results outlined in this paper (Figure 6). The authors showed a strong correlation between the indoor temperature, humidity, and CO_2 concentration for naturally ventilated classrooms, while for the mechanical ventilation case, all these parameters are under control, and this correlation is much weaker. Gladyszewska-Fiedoruk [39] arrived at a similar conclusion in a study of IAQ and thermal comfort performed for a kindergarten case. Similar results were found by Sanchez-Fernandez et al. [8] in a study of the evaluation of different natural ventilation strategies using CO_2 sensors to monitor IAQ in classrooms and by Almaimani et al. [24] when assessing several optimal ventilation strategies for a school in Saudi Arabia, in order to minimize the risk of the transmission of airborne viral infections indoors.

In summary, our study provides significant contributions in terms of understanding the application and constraints of decentralized ventilation systems in the retrofitting process of educational facilities. Although there are certain challenges associated with humidity control and noise management at higher operational levels, the overall advantages of these systems in terms of air quality, installation convenience, occupant satisfaction, and energy efficiency establish them as a feasible and efficient solution for enhancing indoor environmental quality in educational settings.

Limitations of the current study include its focus on a limited number of schools, potentially limiting the generalizability of the findings. Also, the study primarily concentrates on physical measurements of IEQ parameters without delving deeply into subjective perceptions or health outcomes of students and teachers, which could be addressed in future research.

Future studies should consider longitudinal approaches, observing the long-term impacts of decentralized ventilation systems over various seasons and years. This extended timeframe would enable a deeper understanding of the consistency and sustainability of these systems in enhancing IEQ. Expanding the scope to encompass a broader range of schools across different geographical and socio-economic spectrums within Romania is recommended. Such diversity in the study sample would shed light on the influence of external environmental factors and socio-economic conditions on IEQ. Investigating the direct relationship between enhanced IEQ and educational outcomes, such as student academic performance, attendance, and health, is crucial. This would establish a more concrete link between environmental quality within schools and tangible educational benefits. Future research should adopt a more holistic approach to IEQ, examining the combined effects of various IEQ components (air quality, thermal comfort, lighting, and acoustics) and their collective impact on occupants' health and the learning environment.

Author Contributions: Conceptualization, T.C. and A.D.; methodology, T.C.; validation, T.C., A.D. and A.V.; formal analysis, T.C.; investigation, T.C., A.D. and A.V.; resources, T.C.; data curation, T.C.; writing—original draft preparation, T.C., A.D. and A.V.; writing—review and editing, T.C.; visualization, T.C. and A.V.; supervision, T.C.; project administration, T.C.; funding acquisition, T.C. All authors have read and agreed to the published version of the manuscript.

Funding: This work was supported by National Research Grants of the UTCB (ARUT 2023), project number UTCB-31 and Cnfis-FDI-2023-F-0655.

Institutional Review Board Statement: Not applicable.

Informed Consent Statement: Not applicable.

Data Availability Statement: The data presented in this study are available in the article.

Acknowledgments: We acknowledge the help all the partners and collaborators from this unique project in Romania—Healthy School Project https://scolisanatoase.ro/ accessed on 1 October 2023.

Conflicts of Interest: The authors declare no conflict of interest.

References

1. Catalina, T.; Iordache, V. IEQ assessment on schools in the design stage. *Build. Environ.* **2012**, *49*, 129–140. [CrossRef]
2. Tran, M.T.; Wei, W.; Dassonville, C.; Martinsons, C.; Ducruet, P.; Mandin, C.; Héquet, V.; Wargocki, P. Review of Parameters Measured to Characterize Classrooms' Indoor Environmental Quality. *Buildings* **2023**, *13*, 433. [CrossRef]
3. Wargocki, P.; Wei, W.; Bendžalová, J.; Espigares-Correa, C.; Gerard, C.; Greslou, O.; Rivallain, M.; Sesana, M.M.; Olesen, B.W.; Zirngibl, J.; et al. TAIL, a New Scheme for Rating Indoor Environmental Quality in Offices and Hotels Undergoing Deep Energy Renovation (EU ALDREN Project). *Energy Build.* **2021**, *244*, 111029. [CrossRef]
4. Wei, W.; Wargocki, P.; Ke, Y.; Bailhache, S.; Diallo, T.; Carré, S.; Ducruet, P.; Maria Sesana, M.; Salvalai, G.; Espigares-Correa, C.; et al. PredicTAIL, a Prediction Method for Indoor Environmental Quality in Buildings Undergoing Deep Energy Renovation Based on the TAIL Rating Scheme. *Energy Build.* **2022**, *258*, 111839. [CrossRef]
5. *EN 16798-1*; Energy Performance of Buildings. Ventilation for Buildings. Indoor Environmental Input Parameters for Design and Assessment of Energy Performance of Buildings Addressing Indoor Air Quality, Thermal Environment, Lighting, and Acoustics. British Standards Institution: London, UK, 2019.

6. Catalina, T.; Ghita, S.A.; Popescu, L.L.; Popescu, R. Survey and Measurements of Indoor Environmental Quality in Urban/Rural Schools Located in Romania. *Int. J. Environ. Res. Public Health* **2022**, *19*, 10219. [CrossRef]
7. Galicic, A.; Rožanec, J.; Kukec, A.; Carli, T.; Medved, S.; Eržen, I. Identification of Indoor Air Quality Factors in Slovenian Schools: National Cross-Sectional Study. *Processes* **2023**, *11*, 841. [CrossRef]
8. Sánchez-Fernández, A.; Coll-Aliaga, E.; Lerma-Arce, V.; Lorenzo-Sáez, E. Evaluation of Different Natural Ventilation Strategies by Monitoring the Indoor Air Quality Using CO_2 Sensors. *Int. J. Environ. Res. Public Health* **2023**, *20*, 6757. [CrossRef]
9. Schibuola, L.; Tambani, C. Indoor Environmental Quality Classification of School Environments by Monitoring PM and CO_2 Concentration Levels. *Atmos. Pollut. Res.* **2020**, *11*, 332–342. [CrossRef]
10. Persily, A.K. *Evaluating Building IAQ and Ventilation with Indoor Carbon Dioxide*; American Society of Heating, Refrigerating and Air-Conditioning Engineers, Inc.: Atlanta, GA, USA, 1997.
11. Wargocki, P.; Porras-Salazar, J.A.; Contreras-Espinoza, S.; Bahnfleth, W. The Relationships between Classroom Air Quality and Children's Performance in School. *Build. Environ.* **2020**, *173*, 106749. [CrossRef]
12. Mishra, N.; Bartsch, J.; Ayoko, G.A.; Salthammer, T.; Morawska, L. Volatile Organic Compounds: Characteristics, Distribution and Sources in Urban Schools. *Atmos. Environ.* **2015**, *106*, 485–491. [CrossRef]
13. Baloch, R.M.; Maesano, C.N.; Christoffersen, J.; Banerjee, S.; Gabriel, M.; Csobod, É.; de Oliveira, F.E.; Annesi-Maesano, I. Indoor air pollution, physical and comfort parameters related to schoolchildren's health: Data from the European SINPHONIE study. *Sci. Total Environ.* **2020**, *739*, 139870. [CrossRef]
14. Huang, L.; Wei, Y.; Zhang, L.; Ma, Z.; Zhao, M. Estimates of emission strengths of 43 VOCs in wintertime residential indoor environments, Beijing. *Sci. Total Environ.* **2021**, *793*, 148623. [CrossRef]
15. Becker, P.N.; Calesso, T.E.; Agudelo-Castaneda, D.M.; da Silva, C.M.; Oliveira Silva, L.F.; Vigo, A.; Kumar, P. Indoor-outdoor relationships of airborne nanoparticles, BC and VOCs at rural and urban preschools. *Environ. Pollut.* **2021**, *268*, 115751.
16. De Lima, B.D.; Kautzmann, R.M.; da Silveira, F.R.; da Silva, C.M.; de Vargas, F.C.; Taffarel, S.R. Quantitative evaluation of total volatile organic compounds in urban and rural schools of southern Brazil. *Environ. Monit. Assess.* **2020**, *192*, 634. [CrossRef]
17. Oliveira, M.; Slezakova, K.; Madureira, J.; de Oliveira Fernandes, E.; Delerue-Matos, C.; Morais, S.; do Carmo Pereira, M. Polycyclic Aromatic Hydrocarbons in Primary School Environments: Levels and Potential Risks. *Sci. Total Environ.* **2017**, *575*, 1156–1167. [CrossRef]
18. Fernandes, A.; Ubalde-López, M.; Yang, T.C.; McEachan, R.R.C.; Rashid, R.; Maitre, L.; Nieuwenhuijsen, M.J.; Vrijheid, M. School-Based Interventions to Support Healthy Indoor and Outdoor Environments for Children: A Systematic Review. *Int. J. Environ. Res. Public Health* **2023**, *20*, 1746. [CrossRef]
19. Tran, T.D.; Nguyen, P.M.; Nghiem, D.T.; Le, T.H.; Tu, M.B.; Alleman, L.Y.; Nguyen, V.M.; Pham, D.T.; Ha, N.M.; Dang, M.N.; et al. Assessment of Air Quality in School Environments in Hanoi, Vietnam: A Focus on Mass-Size Distribution and Elemental Composition of Indoor-Outdoor Ultrafine/Fine/Coarse Particles. *Atmosphere* **2020**, *11*, 519. [CrossRef]
20. Viana, M.; Rivas, I.; Querol, X.; Alastuey, A.; Sunyer, J.; Alvarez-Pedrerol, M.; Bouso, L.; Sioutas, C. Indoor/outdoor relationships of quasi-ultrafine, accumulation and coarse mode particles in school environments in Barcelona: Chemical composition and Sources. *Atmos. Chem. Phys.* **2013**, *13*, 32849–32883.
21. Hussein, T.; Alameer, A.; Jaghbeir, O.; Albeitshaweesh, K.; Malkawi, M.; Boor, B.E.; Koivisto, A.J.; Löndahl, J.; Alrifai, O.; Al-Hunaiti, A. Indoor Particle Concentrations, Size Distributions, and Exposures in Middle Eastern Microenvironments. *Atmosphere* **2020**, *11*, 41. [CrossRef]
22. Hobeika, N.; García-Sánchez, C.; Bluyssen, P.M. Assessing Indoor Air Quality and Ventilation to Limit Aerosol Dispersion—Literature Review. *Buildings* **2023**, *13*, 742. [CrossRef]
23. Gartland, N.; Aljofi, H.E.; Dienes, K.; Munford, L.A.; Theakston, A.L.; van Tongeren, M. The Effects of Traffic Air Pollution in and around Schools on Executive Function and Academic Performance in Children: A Rapid Review. *Int. J. Environ. Res. Public Health* **2022**, *19*, 749. [CrossRef]
24. Almaimani, A.; Alaidroos, A.; Krarti, M.; Qurnfulah, E.; Tiwari, A. Evaluation of Optimal Mechanical Ventilation Strategies for Schools for Reducing Risks of Airborne Viral Infection. *Buildings* **2023**, *13*, 871. [CrossRef]
25. World Health Organization. *Infection Prevention and Control during Health Care When COVID-19 Is Suspected*; World Health Organization: Geneva, Switzerland, 2020.
26. *ASHRAE Standard 62.1 2019*; Ventilation for Acceptable Indoor Air Quality. ASHRAE: Peachtree Corners, GA, USA, 2019.
27. Kuramochi, H.; Tsurumi, R.; Ishibashi, Y. Meta-Analysis of the Effect of Ventilation on Intellectual Productivity. *Int. J. Environ. Res. Public Health.* **2023**, *20*, 5576. [CrossRef]
28. Jendrossek, S.N.; Jurk, L.A.; Remmers, K.; Cetin, Y.E.; Sunder, W.; Kriegel, M.; Gastmeier, P. The Influence of Ventilation Measures on the Airborne Risk of Infection in Schools: A Scoping Review. *Int. J. Environ. Res. Public Health* **2023**, *20*, 3746. [CrossRef]
29. Calama-González, C.M.; León-Rodríguez, A.L.; Suárez, R. Indoor Air Quality Assessment: Comparison of Ventilation Scenarios for Retrofitting Classrooms in a Hot Climate. *Energies* **2019**, *12*, 4607. [CrossRef]
30. Chang, L.-Y.; Chang, T.-B. Air Conditioning Operation Strategies for Comfort and Indoor Air Quality in Taiwan's Elementary Schools. *Energies* **2023**, *16*, 2493. [CrossRef]
31. Pfluger, R.; Längle, K.; Rojas, G. Minimal Invasive Ventilation Systems with Heat Recovery for School Buildings. In Proceedings of the 36th AIVC Conference "Effective Ventilation in High Performance Buildings", Madrid, Spain, 23–24 September 2015.

32. De Schepper, P. Case study: Comparison between a central and a decentral ventilation unit in a school building from the 80's. In Proceedings of the 39th AIVC Conference "Smart Ventilation for Buildings, Antibes Juan-les-Pins, France, 18–19 September 2018.
33. Catalina, T.; Istrate, M.A.; Damian, A.; Vartires, A.; Dicu, T.; Cucoş, A. Indoor Air Quality Assessment in a Classroom Using a Heat Recovery Ventilation Unit. *Rom. J. Phys.* **2019**, *9–10*, 14.
34. NP 010-2022—Romanian Regulation on the Design, Construction, and Operation of Buildings for Schools and High Schools. Available online: https://www.mdlpa.ro/pages/reglementare24 (accessed on 8 September 2023).
35. *I5-2022;* Romanian Norm for the Design, Execution and Operation of Ventilation and Air-Conditioning Systems Air Conditioning. Available online: https://drimand.ro/download/43%20NORMATIV%20I%205%20-%202022.pdf (accessed on 8 September 2023).
36. *I5-2022;* Standard for the Design, Execution and Operation of Ventilation and Air-Conditioning Installations. ASRO: Bucharest, Romania, 2022.
37. *I13-2022;* Standard for the Design, Execution and Operation of Central Heating Installations. ASRO: Bucharest, Romania, 2022.
38. Lazovic, I.; Stevanovic, Z.M.; Jovasevic-Stojanovic, M.V.; Zivkovic, M.M.; Banjac, M.J. Impact of CO2 concentration on indoor air quality and correlation with relative humidity and indoor air temperature in school buildings in Serbia. *Therm. Sci.* **2016**, *20*, 297–307. [CrossRef]
39. Gladyszewska-Fiedoruk, K. Correlations of air humidity and carbon dioxide concentration in the kindergarten. *Energy Build.* **2013**, *43*, 45–50. [CrossRef]

applied sciences

MDPI

Article

Extending the IFC-Based bim2sim Framework to Improve the Accessibility of Thermal Comfort Analysis Considering Future Climate Scenarios

Veronika Elisabeth Richter *, Marc Syndicus, Jérôme Frisch and Christoph van Treeck

Institute of Energy Efficiency and Sustainable Building (E3D), RWTH Aachen University, Mathieustr. 30, 52074 Aachen, Germany; syndicus@e3d.rwth-aachen.de (M.S.)
* Correspondence: richter@e3d.rwth-aachen.de

Abstract: Future weather scenarios significantly affect indoor thermal comfort, influencing people's well-being and productivity at work. Thus, future weather scenarios should be considered in the design phase to improve a building's climate change resilience for new constructions as well as renovations in building stock. As thermal comfort is highly influenced by internal and external thermal loads resulting from weather conditions and building usage, only a dynamic building performance simulation (BPS) can predict the boundary conditions for a thermal comfort analysis during the design stage. As the model setup for a BPS requires detailed information about building geometry, materials, and usage, recent research activities have tried to derive the required simulation models from the open BIM (Building Information Modeling) Standard IFC (Industry Foundation Classes). However, even if IFC data are available, they are often faulty or incomplete. We propose a template-based enrichment of the BPS models that assists with imputing missing data based on archetypal usage of thermal zones. These templates are available for standardized enrichment of BPS models but do not include the required parameters for thermal comfort analysis. This study presents an approach for IFC-based thermal comfort analysis and a set of zone-usage-based templates to enrich thermal comfort input parameters.

Keywords: Building Information Modeling; IFC to simulation; thermal comfort; building performance simulation; climate change

Citation: Richter, V.E.; Syndicus, M.; Frisch, J.; van Treeck, C. Extending the IFC-Based bim2sim Framework to Improve the Accessibility of Thermal Comfort Analysis Considering Future Climate Scenarios. *Appl. Sci.* **2023**, *13*, 12478. https://doi.org/10.3390/app132212478

Academic Editors: David Bienvenido Huertas and Daniel Sánchez-García

Received: 17 October 2023
Revised: 13 November 2023
Accepted: 16 November 2023
Published: 18 November 2023

1. Introduction

Climate change necessitates future weather scenarios that will significantly affect humans' everyday lives and pose challenges to housing design and construction. These challenges emphasize the need for measures such as shading, temperature peak reduction through thermal mass (if available), or night cooling ventilation [1]. Bell et al. [2] estimate an increase in cooling demand for buildings at up to 35% for the year 2050 and stress that BPS must be improved and advanced in order to serve as a reliable planning and prediction tool, for example, by properly incorporating not only warmer average temperatures due to climate change but also short yet extreme weather events. People in Western societies spend 90% of their time indoors throughout the day [3,4], and thermal comfort influences well-being and productivity at work, in schools, and at home [5–7]. Therefore, future annual weather conditions should be considered in a building's design phase to improve the building's climate change resilience for new constructions and renovations in existing buildings.

Since thermal comfort is strongly influenced by internal and external loads resulting from weather conditions and building usage, only a dynamic building performance simulation (BPS) can predict the boundary conditions for a thermal comfort analysis during the design stage. As the model setup for a BPS requires detailed information on building geometry, materials, and usage, recent research activities have attempted to derive the required simulation models from the open BIM (Building Information Modeling) Standard IFC (Industry Foundation Classes) [8]. However, even when the IFC data are available, it is often erroneous or incomplete. A template-based enrichment of the BPS models helps fill in missing data based on archetypal usage of thermal zones (e.g., for defining internal loads and schedules based on the usage of thermal zones). These templates are available for standardized enrichment of BPS models, but do not include the required parameters for thermal comfort analysis.

We propose a basic setup to integrate automated thermal comfort analysis into an IFC-based BPS setup. This setup enables the user to perform an IFC-based thermal comfort analysis in the design process with minimal additional effort. It assists in rapidly evaluating building designs and even small changes without major remodeling efforts. The impact of climate change on building design can be easily considered when comparing the impact of design decisions. Furthermore, this setup cannot only be used to minimize building operational costs but also consider, e.g., embodied carbon for different design strategies when optimizing the building's design for maximum thermal comfort design. For a quick template-based model enrichment, we propose zone-usage-based thermal comfort parameters. These parameters are rough estimates but enable users to consider usage-specific setups and can be further specified for individual project needs.

After an introduction to the related research on thermal comfort and climate change, IFC-based BPS, and future weather scenarios, we present our approach for IFC-based thermal comfort simulation. This approach extends the existing IFC-based BPS approach *bim2sim* [9,10]. A set of thermal comfort parameters for template-based BPS model enrichment supports the presented bim2sim extension. The proposed methods and templates are evaluated on a case study IFC building applying a TMYx (2007–2021) and SSP5-8.5 (2050, 2080) weather scenarios for Cologne, Germany.

1.1. Related Research

The related research for our study is grouped into the three areas of (1) thermal comfort with respect to climate change, (2) general IFC-based methods for BPS, and (3) related research on future weather scenarios for the use in BPS.

1.1.1. Thermal Comfort and Climate Change

Thermal comfort in general and the Fanger [11] model in particular have been used in different simulation scenarios, but oftentimes for a specialized purpose or situation, such as thermal comfort in trains [12], the prediction of thermal comfort in indoor swimming pools [13], the heat exposure in a kitchen environment [14], or the investigation of rural heating systems' efficacy [15]. In early design stages, suitable models for occupant behavior allow the designers to evaluate design decisions with respect to their combined influence on comfort and the building's energy consumption [16]. Gritzki and Rösler [17] describe a simulation approach where BPS and HVAC system simulations were coupled with Computational Fluid Dynamics (CFD) in order to investigate the feasibility of Net-Zero Energy concepts in office buildings. Although their approach is quite extensive, including thermal and draught simulation as well as prediction of CO_2 distribution, it might prove beneficial to interface this approach with IFC to facilitate integration into the BIM process.

Barbosa et al. [18] describe an approach to investigate the vulnerability of Portuguese residential buildings, focusing on occupancy and insulation. Their multi-step procedure involves geometry and constructive data acquisition, acquisition of monitoring data, modeling in EnergyPlus as well as calibration against weather data and the final comfort assessment. The results emphasized the beneficial role of external insulation in increasing the adaptive capability of buildings. As stated by the authors, future replications should also include air-tightness of the building or additional shading devices in the simulation.

Applying the Simulation-based Large-scale uncertainty/sensitivity Analysis of Building Energy performance (SLABE) methodology, Escandón et al. [19] projected the impact of climate change on a building category with a case study object located in Seville, Spain.

Generalization took place by Monte Carlo resampling of the EnergyPlus input parameters. Among the nine geometrical parameters, floor height, form ratio, floor area, and orientation had considerable influence. Among the 17 envelope parameters, the absorptance of the wall's external layer showed the highest standard rank regression coefficient in that category. User behavior was identified as the major influence within the three analyzed operation parameters, but also among all categories, and its relevance was estimated to even increase with the progress of climate change. In line with previous studies, the authors project an overall worsening of the indoor thermal conditions towards the year 2050, resulting in elevated cooling demands (up to 250 % increase) and an increase in discomfort hours of about 36 %. Escandón and colleagues suggest and emphasize the usefulness of incorporating the approach as a plugin in BIM tools.

Aiming to improve multi-type building performance prediction as well as optimization, Yan et al. [20] propose a machine learning (ML)-based procedure, which was applied to dwellings by the Singaporean Housing Development Board. Transfer learning techniques, as well as multi-objective genetic optimization, were utilized to derive optimal performance and design parameters for daylight performance, energy efficiency, and thermal comfort from short to long-term future climate conditions.

Various approaches to evaluate the thermal resilience of buildings were reviewed by Siu et al. [21]. According to the authors, one of the main issues to be resolved is a lack of standardized procedures for the prediction and evaluation of a building's thermal resilience with simulations. This holds true not only for the BPS part but also for the methods applied to create extreme weather data. Many of the studies projecting thermal comfort changes caused by climate change oftentimes focus on regions where heat is already present [22–24]. As no region will remain untouched by climate change or its consequences, it might prove beneficial to include a thermal comfort simulation and prediction component in BIM models.

The heat-balance guided thermal comfort model by P. O. Fanger [11] is implemented in current comfort-related standards such as the ASHRAE 55 [25] or DIN EN ISO 7730 [26]. Although the model has been criticized because of some shortcomings compared to adaptive approaches—such as its application focus on mechanically conditioned buildings—it is still one of the most common models applied to derive predictions for thermal comfort [27]. Besides the theoretically sound foundation on human heat exchange with the environment, the model is fairly easy to apply based on four physical (air temperature, radiant temperature, air velocity, and relative humidity) as well as two person-related (physical activity, clothing) variables. Both of the model's metrics, the predicted mean vote (PMV) and the predicted percentage dissatisfied (PPD), are easy to interpret and present themselves as suitable variables for simulation purposes.

1.1.2. IFC-Based BPS

The use of Building Information Modeling (BIM) as a collaborative approach for building design is widely used in the European Union, but the implementation is heterogeneous [28]. IFC [8] is used as an open data exchange format for BIM data. The IFC-based setup of BPS models has been widely discussed in related research, but the IFC-based process still leads to errors (e.g., syntax, geometry, semantics, consistency) [29–31]. So far, the processes are not sufficiently supported and integrated by available software tools, preventing them from being more widely adopted in industry.

The basic concept of translating IFC-based geometry into BPS models utilizes the second level space boundaries (IFC4: *IfcRelSpaceBoundary2ndLevel*), initially introduced by Bazjanac [32]. These space boundaries describe the virtual heat-transmitting surfaces in-between thermal zones and the building's surroundings, considering adjacent thermal zones and building constructions. While some studies discuss methods for generating second-level space boundaries for building simulation [33–39] or directly integrate space boundary generation methods in their IFC-based BPS approaches [39–41], other studies rely on the IFC4 schema-conform definition of space boundaries for the (semi-)automatic generation of BPS models [9,42–45]. To be able to rely on the quality of provided space boundaries, a validation and/or a correction process may need to be applied, as proposed by Ying and Lee [29], Richter et al. [30]. At least for minor corrections (e.g., surface normal orientation), the correction process, in most cases, has less computational costs than generating new space boundaries and should be preferred over generating a full set of new space boundaries.

With its modular and open-source approach for IFC-based BPS simulations, the bim2sim tool [9,10] provides a solid base for adapting the EnergyPlus-based BPS methods toward thermal comfort analysis. The bim2sim tool aims to support engineers with a geometric and semantic simulation model setup using IFC and template-based data. This drastically reduces the manual model setup time, while only minor manual model corrections are required after applying the bim2sim tool [10].

The structure of this open-source framework is displayed in Figure 1, showing the central data processing and enrichment steps for the two BPS plugins PluginEnergyPlus and PluginTEASER, using EnergyPlus and Modelica as simulation backends, respectively. A similar approach is available for HVAC simulations with the PluginAixLib, also using Modelica as a simulation backend. The modular approach of the bim2sim tool and the integrated HVAC simulation support the future implementation of a dynamic co-simulation setup of BPS and HVAC, which is not part of the proposed methods but could also be extended for the application of thermal comfort. The initial implementation of the bim2sim framework also includes plugins for exporting CFD boundary conditions for use in the commercial CFD-Software ANSYS FLUENT (Version 2022 R2) and an initial implementation supporting Life Cycle Assessment (LCA) [10].

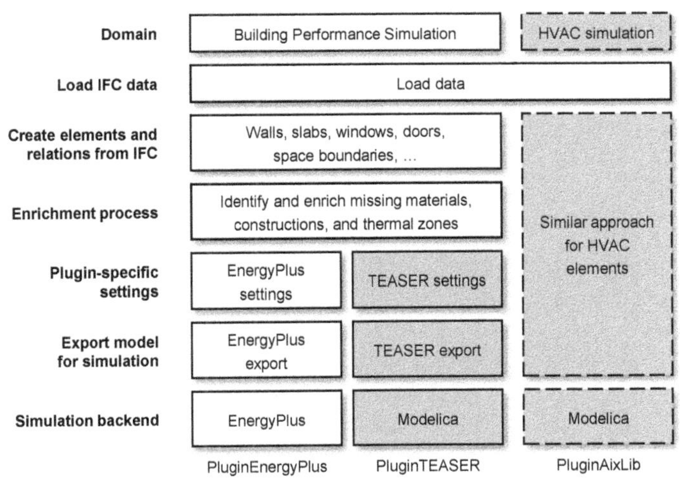

Figure 1. Simplified representation of the main bim2sim workflow; for Building Performance Simulation including the PluginEnergyPlus (white) and the PluginTEASER (gray, solid line), and for HVAC simulations the PluginAixlib (gray, dashed line).

All plugins start with the shared process of loading the IFC data. In the next step, the required elements for the BPS domain are created for the respective process (e.g., walls, slabs, windows, doors, and space boundaries). The available element properties within the IFC are identified, evaluated, and enriched (user input- or template-based) if properties are missing. This process is followed by plugin-specific settings, the export for the simulation backend, and the simulation itself.

The template-based enrichment in bim2sim builds on archetypal templates (e.g., for zone usage), which are derived within TEASER [46]. These use conditions represent the boundary conditions for zone types defined in DIN V 18599-10 [47] and VDI 2078 [48] and are further enriched by internal load profiles defined in SIA 2024 [49] (TEASER Documentation: https://rwth-ebc.github.io/TEASER//master/docs/code/teaser.logic.buildingobjects.html#teaser.logic.buildingobjects.useconditions.UseConditions, accessed on 12 November 2023).

Mahecha Zambrano et al. [16] state that the representation of occupant behavior in BPS, i.e., occupancy and occupancy-related schedules (internal loads due to lighting and equipment), is oversimplified, and that the development of a generalized model for occupant behavior is impossible due to the occupants' diversity. For the simulation of thermal comfort, the hierarchy of occupant behavior needs to be evaluated, e.g., how the occupants decide on changing room set point temperatures and adjusting their clothing when feeling thermally uncomfortable [16]. The choice of the modeling approach for occupant behavior in related research is based on the type of behavior, the building design stage, or the spatial scale of the simulation. Depending on the scale of the simulation, stochastic occupancy models can be used to derive suitable occupancy profiles for the buildings. To minimize the (manual and computational) effort in considering occupant behavior, Mahecha Zambrano et al. [16] suggest using a pre-processed set of schedules for occupant behavior with an estimation of the probability of these scenarios. This schedule set could support the designer in the decision-making process when considering extreme events. However, as the available occupant models lack validation, standardized validation approaches are required to be able to compare the results of different studies [16].

The bim2sim tool acts as basis for further development of IFC-based thermal comfort methods within the present study. However, thermal comfort parameters are yet not sufficiently represented within the usage-based TEASER templates.

1.1.3. Future Weather Scenarios

For the generation of future weather scenarios, General Circulation Models (GCMs) are downscaled from their worldwide scale of typically 1–5° latitude and longitude [50] (i.e., about 111–555 km) to regional scale high-resolution Regional Climate Models (RCMs) with a resolution of 4 km or less [51]. Wilby and Wigley [52] introduced four statistical downscaling approaches: regression, weather pattern approaches, stochastic weather generators, and limited-area climate models. Belcher et al. [53] further refined these approaches to dynamical downscaling, stochastic weather generation, interpolation, and introduced morphing. While dynamical downscaling is computationally expensive, stochastic weather generation requires large input data sets, and interpolation may lead to biased resulting data; the proposed morphing technique has low computational cost and builds upon real climate data [53]. However, even though this technology produces consistent future weather data based on future climate predictions, the resulting characteristics are still mainly influenced by the input weather data and, thus, do not reflect changed characteristics and variability (e.g., heat waves) in future climate [53].

Zeng et al. [54] present a recent critical review on these generation approaches of future weather data for building performance simulation, giving advice for the choice of future weather files according to the application (i.e., energy analysis, thermal resilience, HVAC design, utility analysis). For the analysis of thermal resilience, they recommend the use of future extreme weather data instead of typical year weather data. However, morphed weather data (unable to reflect actual future weather variability) was still used in related research [54].

Nielsen and Kolarik [55] presented a review on existing climate research. They discovered that more than half of their 47 analyzed studies (2015 and newer) used the outdated weather data of CMIP3 (Phase 3 of the Coupled Model Intercomparison Project (CMIP), supporting the fourth assessment report of the Intergovernmental Panel on Climate Change (IPCC) based on weather data mostly generated in 2005 and 2006: https://pcmdi.github.io/mips/cmip3/, accessed on 12 November 2023), partially due to the availability through the CCWorldWeatherGen Tool that easily generates EnergyPlus Weather files. Only five out of the analyzed studies underlined why they chose the selected climate model even though the resulting weather files show high variance and may lead to an erroneous interpretation of the simulation results. The ways to deal with solar radiation data for BPS are manifold in related research, as global horizontal irradiance from GCM and RCM have to be converted to direct normal irradiance and diffuse horizontal irradiance for the use in BPS [55]. They published the results of their study on a continuously updated webpage (FutureWeatherBPS: www.futureweatherbps.com, accessed on 12 November 2023), currently (November 2023) including data from 82 studies and a total of 210 locations.

Rodrigues et al. [56] propose an open-source morphing tool for future weather data, as the existing tools (CCWorldWeatherGen [50], Weather Morph [57], WeatherShift (https://weathershift.com/, accessed on 17 November 2023)) have limited accessibility (i.e., not open-source), rely on outdated data models, or are not free to use and, thus, may not be accessible for researchers with limited funding.

Hong et al. [58] discuss ten questions on building's and occupants adaptation to climate changes. Considering changes in the outdoor environment and their impact on the buildings, they list general climate trends and local weather conditions, urban microclimate and heat island effects and hazards as influential on thermal resilience of buildings.

1.2. Aim of Study

The current study integrates thermal comfort analysis into an existing IFC-based BPS approach. In this way, we address the need for an automated thermal comfort model setup to avoid remodeling when considering design alternatives [21]. Zone usage-based thermal comfort parameters were derived for template-based enrichment of thermal zones, aiming to extend the existing templates. The proposed methods were evaluated on a case study building represented in IFC4. From the BPS-based thermal comfort analysis results, thermal zones with critical thermal comfort situations were identified. The case study assessed the applicability of state-of-the-art comfort metrics and the resilience of the building in future weather scenarios.

2. Materials and Methods

This section starts with an introduction to the thermal comfort approaches used in our study for thermal comfort analysis. In the following Section 2.2, we propose the implementation of our bim2sim *PluginComfort*, which extends the existing bim2sim EnergyPlus methods by an additional thermal comfort analysis. Section 2.3 addresses the extension of the TEASER templates by activity (Section 2.3.1) and clothing parameters (Section 2.3.2) used for thermal comfort analysis for each archetypal thermal zone. Finally, in Section 2.4, we describe the weather data used in our study.

2.1. Evaluate Thermal Comfort Models

EnergyPlus provides the implementation of the adaptive comfort analysis according to DIN EN 15251 [59], which has been withdrawn and replaced by DIN EN 16798-1 [60], which is not yet implemented in EnergyPlus. With the standard's transition, the methods for thermal comfort still rely on ISO 7730 [26]. While methods to calculate the running mean outdoor temperature have not changed, the standards for acceptable temperatures have changed (cf. Table 1), i.e., the lower boundary for each comfort category has been reduced by one degree Celsius, and thus, slightly cooler temperatures are defined to be acceptable. However, since the calculation of temperatures itself did not change with the transition to DIN EN 16798-1, the available EnergyPlus implementation of the DIN EN 15251 can still be used with minor modifications in the evaluation of results.

The template values we derived for the clothing parameters are static throughout the whole year as a simplification for template-based model enrichment. In further research, we could use ASHRAE Standard 55 [25] dynamic clothing model in comparison to our static template-based clothing parameter set. ASHRAE 55 enables dynamic clothing based on outdoor air temperatures. This may not be applicable to office and other formal situations (e.g., in formal office or meeting situations), so zone-usage-based clothing may be more applicable in this case.

Table 1. Differences in running mean outdoor temperatures as set points for thermal comfort categories in DIN EN 15251 [59], DIN EN 16798-1 [60].

	DIN EN 15251	DIN EN 16798-1
Category I, upper threshold		24.1...30.7 °C
Category I, lower threshold	21.75...26.7 °C	19.1...25.7 °C
Category II, upper threshold		25.1...31.7 °C
Category II, lower threshold	20.75...25.7 °C	18.1...24.7 °C
Category III, upper threshold		26.1...32.7 °C
Category III, lower threshold	18.1...24.7 °C	17.1...23.7 °C
Applicable for running mean temperature:		
Upper threshold		10...30 °C
Lower threshold	15...30 °C	10...30 °C

2.2. *Implementation of PluginComfort for bim2sim*

The proposed *PluginComfort* is an extension of the bim2sim *PluginEnergyPlus* (cf. Figure 1). This new plugin requires comfort parameters for each thermal zone, introduced as simplified usage-based parameters in Section 2.3.

Figure 2 visualizes the structure of the proposed thermal comfort plugin. This plugin builds upon the existing implementation of the PluginEnergyPlus within the bim2sim framework, as described in Figure 1. While the general tasks like loading IFC data and general element setup from IFC are the same as for all other plugins within the bim2sim environment, additional template-based data are loaded within the enrichment process of the thermal zones, adding personal comfort parameters for activity and clothing. If this data are available from the given IFC data, these personal parameters could also be loaded directly from IFC. However, since IFC does not yet provide standardized property sets for these personal parameters, the current implementation only considers a template-based enrichment for personal parameters.

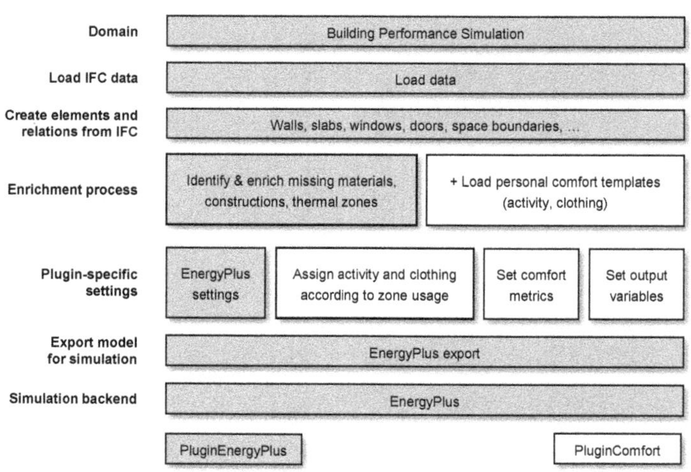

Figure 2. Simplified representation of the workflow of the bim2sim-based thermal comfort plugin PluginComfort; the new plugin builds upon the existing parts of the PluginEnergyPlus (gray) by loading additional data in the enrichment process and comfort-related settings in the plugin-specific settings (white).

Within the plugin-specific settings, the implementation of the PluginEnergyPlus is extended by adding the personal comfort parameters for activity and clothing to the individual thermal zones according to their zone usage. Furthermore, the applicable comfort metrics are selected, and the related output variables are set. The export of the EnergyPlus simulation model is the same as for the PluginEnergyPlus. The simulation results of the PluginComfort can be evaluated with suitable visualizations to support design decisions within the design stages of the building.

With this IFC-based thermal comfort plugin, BPS models for design variants can quickly be created, minimizing the manual effort to remodel the geometry. Applying future weather data within the IFC-based BPS assists in evaluating the impact of climate change and the thermal resilience of the design variants. For this evaluation, the heating and cooling loads are related to the resulting thermal comfort of the building. Considering the modular structure of the bim2sim framework, an additional analysis of the embodied carbon of the design variants could easily be added to further support the design decisions by extending the existing implementation of the LCA Plugin *PluginLCA* [10].

2.3. Enrichment of Thermal Comfort Parameters for Archetypal Zone Usage

Richter et al. [61] outline the input requirements for an IFC-based thermal comfort analysis. Although many parameters for thermal comfort analysis, as specified in ISO 7730, can be derived at runtime from a BPS simulation (e.g., air/radiant/floor temperature, air velocity, humidity), parameters for clothing insulation and metabolic rates must be defined prior to simulation. This parameter definition is specific for the conditions of the simulated environment. They also suggest that these parameters can be included in the underlying IFC or be defined in room-specific templates, similar to the approach used in TEASER for archetypal simulations.

In our study, we expand on and refine the existing archetypal TEASER templates (cf. Section 1.1.2) for these activity degree and clothing parameters. These parameters are derived from established standards and are mapped to TEASER's archetypal zones. The following sections provide a detailed description of this extending and mapping process.

2.3.1. Enrichment of Activity Degree Parameters for Archetypal Zone Usage

The existing TEASER templates already include archetypal data for different usage of thermal zones, including occupant activity degree data and occupant heat flow in W/person. These parameters are derived from values provided by the Swiss standard SIA 2024 [49], which provides data to calculate the energy demand of buildings in early building design stages. However, these available activity degrees are standardized (mostly either 1.2 met or 2.0 met) and do not reflect the expected variations in activity due to individual zone usage. Therefore, these parameters are unsuitable for evaluating zone-usage-based thermal comfort, as shown by a first comparison of the existing TEASER activity data with standards for thermal comfort analysis [26,62].

There are several national and international standards for defining the activity degree of occupants. DIN EN ISO 8996 [63] provides data to accurately determine the metabolic rate in working environments. This standard also provides calculation methods to determine the metabolic rates for various activities, taking into account human conditions such as gender, age, and body weight. However, since we extend a template-based approach that provides approximate data to enrich archetypal zone setups, we require more generalized metabolic data at this stage. For this purpose, DIN EN ISO 8996 [63] categorizes metabolic rates in its Annex A, ranging from (0) Resting (100–125 W/person) to (4) Very high metabolic rate (>465 W/person). As these categories are broad, each covering a range of metabolic rates of about 100 W/person, we rely on other international standards and guidelines (i.e., ISO 7730 [26], ASHRAE Fundamentals [62]) for more specific metabolic data related to particular activities.

ISO 7730 [26] provides metabolic rates and heat flows for different activities in Table A.5 and Table B.1. The ASHRAE Fundamentals [62] give a detailed introduction to thermal comfort analysis in Chapter 9, providing typical metabolic rates for an average adult person (person's body surface area $A = 1.8\,\mathrm{m}^2$) in Table 4.

Table 2 displays an excerpt of the activity values for the archetypal room types in the existing TEASER template (cf. Table A1 for full set of parameters for all zone types), along with the corresponding derived activity values obtained from ISO 7730 and ASHRAE Fundamentals. To make the data derivation process transparent, we list the activity types mapped to each room type from both sources (columns (4) and (5)) and their related individual metabolic rates. The metabolic rates from ISO 7730 and ASHRAE Fundamentals show only minor deviations (mostly $0 - 0.2\,\mathrm{met}$), although the matched activities vary due to differences in specifications. The newly derived (and combined) activity degree is calculated as the mean of columns (6) and (7), rounded up to the nearest decimal place. This value is rounded up because a higher activity degree results in higher cooling loads and an increased risk of occupant overheating during summer. On the basis of this combined activity degree, the heat flow (W/person) is displayed in column (10), calculated with $1\,\mathrm{met} = 58.1\,\mathrm{m}^2$ and the average adult surface area of $A = 1.8\,\mathrm{m}^2$ [62].

Table 2. Deriving activity parameters for archetypal enrichment of thermal zones based on ASHRAE Fundamentals and ISO 7730, including a comparison to existing TEASER template values. Columns: (1) Room type according to TEASER templates, (2) Activity degree from TEASER templates (met), (3) Heat flow from TEASER templates (W/person), (4) Chosen activity type according to ASHRAE Fundamentals, Chapter 9 Table 4, (5) Chosen activity type according to ISO 7730, Table A.5 and B.1, (6) ASHRAE activity degree (met), (7) ISO 7730 activity degree (met), (8) Resulting combined activity degree (met), (9) Absolute deviation from TEASER activity degree (met), (10) Resulting heat flow (W/person), (11) Absolute deviation from TEASER heat flow (W/person).

(1) Room Type	(2)	(3)	(4) ASHRAE Fundamentals	(5) ISO 7730	(6)	(7)	(8) Activity Degree (met)	(9)	(10) Heat Flow (W/Person)	(11)
Single office	1.2	70	Office, Typing	A.5, Single office	1.1	1.2	1.2	0.0	125	55
Bed room	1.2	70	Resting, Sleeping	B.1, Reclining	0.7	0.8	0.8	0.4	84	14
Kitchen in nonresidential buildings	1.2	70	Cooking	B.1, Standing, medium activity	1.8	2	1.9	0.7	199	129
WC and sanitary rooms in non-residential buildings	1.2	70	Resting, Seated, quiet	B.1, Seated, relaxed	1	1	1	0.2	105	35
Traffic area	1.2	70	Office Walking	B.1, Walking, 2 km/h	1.7	1.9	1.8	0.6	188	118
Living	1.2	70	Resting, Seated, quiet	B.1, Sedentary activity	1	1.2	1.1	0.1	115	45

Columns (9) and (11) show the absolute difference to the original activity parameters from the TEASER Templates given in columns (2) and (3), respectively. While some of the metabolic rates correspond to the previous TEASER values, the new values for the heat flow per person greatly exceed the previous values (mostly between 79 and 169 %, referring to the full parameter set in Table A1).

Updating the activity data in the TEASER Templates is necessary for thermal comfort analysis, as they show high deviations from the metabolic rates reported in existing thermal comfort standards. Since the activity values provided in ASHRAE Fundamentals and ISO 7730 show only small deviations, the derived combined values give a reasonable estimate for these corrected parameter values.

2.3.2. Enrichment of Clothing Insulation Parameters for Archetypal Zone Usage

Contrary to the derivation of activity parameters, the TEASER templates do not provide pre-existing values for clothing insulation parameters. Thus, these values have to be derived from scratch. These values could be derived from the detailed clothing combinations in the international standard DIN EN ISO 9920 [64], which focuses on detailed descriptions of clothing settings. However, similar to the DIN EN ISO 8996 [63] standard for activity degrees, the DIN EN ISO 9920 [64] describes the clothing insulation with such high detail that it is not easily applicable for deriving standardized clothing insulation settings for extending the pre-existing templates. We only expect these more detailed clothing parameters to increase the accuracy of the simulation results if we perform a massive study on average clothing in buildings.

Similar to the previous section, ISO 7730 [26], Table C.1 and the ASHRAE Fundamentals [62], Chapter 9, Table 7 provide data for predefined clothing sets. Opposite to activity degrees, clothing parameters may vary with the seasons. ISO 7730 states operative temperatures for summer and winter, referring to a general clothing value of 0.5 clo (summer) and 1.0 clo (winter). However, as for the individual archetypal room settings (e.g., offices), specific clothing standards may apply that do not vary with the seasons. The first version of the data set only provides a single clothing parameter since the set point cooling and heating temperatures also do not vary with the seasons in the TEASER templates.

The calculation of the complete clothing insulation I_T consists of multiple parts, e.g., base insulation (clothing insulation I_{cl}), air insulation I_a, and clothing area factor f_{cl} [64]. The position of the human body (e.g., seated, standing) and the surroundings (e.g., chair if seated, bed if sleeping) also affect the person's insulation. ASHRAE Fundamentals [62] state that a factor of up to 0.15 clo should be added to the clo value caused by clothing when a person is seated on a chair. Nevertheless, clothing insulation and air insulation cannot simply be added, as the clothing affects the air layers [64]. The surrounding insulation significantly affects the effective insulation of a person in bedding systems. However, as our study proposes a new set of generalized templates for archetypal zone usage that is used for simple estimation of thermal comfort in the design phase, we decided to add clothing insulation values as additional insulation factors, such as chairs (when seated) or beds (when sleeping). The reduced accuracy of the clothing insulation is covered by the general assumption of an estimation of an average/standard setup (of person, clothing, and activity degrees) per archetypal zone usage.

In the proposed template (see Table 3 for an excerpt and cf. Table A2 for the full parameter set), we split the insulation parameter into two parts: clothing insulation and surrounding insulation. By splitting these parameters, we ensure transparency of our assumptions. We consider the sum of these two parameters for our further thermal comfort analysis.

Table 3. Deriving clothing parameters for archetypal enrichment of thermal zones based on ASHRAE Fundamentals and ISO 7730. Columns: (1) Room type according to TEASER templates, (2) Chosen clothing type according to ASHRAE Fundamentals, Chapter 9 Table 7, (3) Chosen clothing type according to ISO 7730, (4) ASHRAE clothing (clo), (5) ISO 7730 clothing (clo), (6) Resulting combined clothing parameter (clo), (7) Chosen surrounding insulation type, (8) Surrounding insulation (clo).

(1) Room Type	(2) ASHRAE Fundamentals	(3) ISO 7730	(4)	(5)	(6) Clothing Insulation (clo)	(7) Surrounding Insulation Description	(8) Surrounding Insulation (clo)
Single office	Trousers, long-sleeved shirt	Underwear, shirt, trousers, socks, shoes	0.61	0.7	0.66	ISO 7730, C.3 Executive chair	0.15
Bed room	Walking shorts, shortsleeved shirt	Panties, T-shirt, shorts, light socks, sandals	0.36	0.3	0.33	Average based on Zhang et al. [65]	2
Kitchen in nonresidential buildings	Long-sleeved overalls, T-Shirt	Underpants, shirt, trousers, smock, socks, shoes	0.72	0.9	0.81	None	
WC and sanitary rooms in non-residential buildings	Trousers, long-sleeved shirt	Underwear, shirt, trousers, socks, shoes	0.61	0.7	0.66	None	
Traffic area	Trousers, long-sleeved shirt	Underwear, shirt, trousers, socks, shoes	0.61	0.7	0.66	None	
Living	Trousers, long-sleeved shirt	Underwear, shirt, trousers, socks, shoes	0.61	0.7	0.66	ISO 7730, C.3 Executive chair	0.15

Depending on climate zones as well as culture, bedding systems vary with regard to insulation, materials, and general configuration. However, as Zhang et al. [65] describe, the filling of the bedding materials has a minor effect on the resulting insulation of the bedding system. Still, it highly depends on the weight per unit area of the bedding system. Thus, even a study on Chinese bedding systems can be considered for Western Europe if the weight of the bedding system is chosen appropriately. Zhang et al. [65] measured bedding system insulation between 1.53 and 4.89 clo depending on the percentage of body coverage and system. As we aim for a surrounding insulation factor that we can add to the clothing insulation, we estimate an additional surrounding insulation factor of 2.0 clo. This surrounding insulation factor for bedding already incorporates a reduction of the combined clothing and surrounding insulation value (clothing and surrounding insulation factor follow the principles of superposition, such that for the general template value, we decided to keep the clothing value fixed, while we reduce the surrounding insulation value). We keep in mind that this surrounding insulation factor highly depends on the bedding system weight (that may be changed due to weather conditions), body coverage, and even sleeping posture. A more detailed statistical study on average sleeping parameters should be considered in further research. For PMV calculations in ISO 7730, the maximum clo value is limited to 2.0 clo, which the proposed combinations of clothing and additional insulation may exceed. Thus, the applicability of the ISO 7730 needs to be further tested.

As the effect of surrounding insulation is rather small (e.g., for wooden stools), these parameters can be considered optional and used for the fine-tuning of models (e.g., in classroom or meeting room settings, where all people are expected to be seated on a chair). We expect the additional insulation of the bedding systems to be crucial for thermal comfort, while office chairs are negligible. However, the impact of clothing and surrounding insulation factors on the general thermal comfort within the room should be elaborated in more detailed case studies.

2.4. Definition of Weather Data

For evaluating the impact of climate change by using our proposed IFC-based thermal comfort analysis, we use TMYx weather (2007–2021) for Cologne Bonn, as this matches with the German construction setup from TEASER templates. On this weather data, we applied the open-source morphing tool Future Weather Generator [56] to generate future weather predictions for 2050 (i.e., median months from 2036 to 2065) and 2080 (i.e., median months from 2066 to 2095) [56] based on the up-to-date CMIP6 Global Climate Model EC-Earth3 [66]. For simplicity, the years of the future weather scenarios are denoted in the following as 2050 and 2080. For further evaluation of the proposed methods, we choose the IPCC Scenario SSP5-8.5, following the recommendation of Zeng et al. [54] to use extreme weather data for analysis of thermal resilience. We choose the worst-case scenario to underline the potential of thermal comfort analysis during design stage, keeping in mind that all weather data only represent predictions of future scenarios. By using the morphing tool by Rodrigues et al. [56], we address the criticism that much research still relies on the outdated climate model CMIP3 [55]. However, for a detailed analysis of climate change's impact on thermal comfort, CMIP6-weather data should be used with specific predictions that match the actual design purpose (e.g., extreme weather summer periods for determination of overheating risks), which has not been taken into account within our proposed demonstration case.

Figure 3 displays the outdoor temperature for the selected weather scenarios per month. The monthly median temperatures in the SSP5-8.5 scenarios are for 2050 up to 4.37 °C higher than the TMYx baseline and 6.4 °C higher for the corresponding 2080 weather data. Since the weather data was transformed to the future weather scenario using morphing, the general characteristics (e.g., amount of outliers per month) are approximately similar to the baseline weather. As already criticized by Zeng et al. [54], the variability of current weather data is preserved through the morphing methodology, such that changes in the variability (or even developed phenomena like heat islands) are not represented in the morphed data.

Our proposed study focuses on climate trends morphed for local weather conditions. The analysis of urban microclimate, heat islands, and hazards is part of further research and can easily be built upon our proposed IFC-based thermal comfort framework.

To evaluate the use of the IFC-based thermal comfort, we use the SSP5-8.5 weather data for 2080 to further analyze the results. As we do not have specific hot summer design weather data available for the latest IPCC scenario, the 2080 data are the most extreme data that are available for our current comfort analysis.

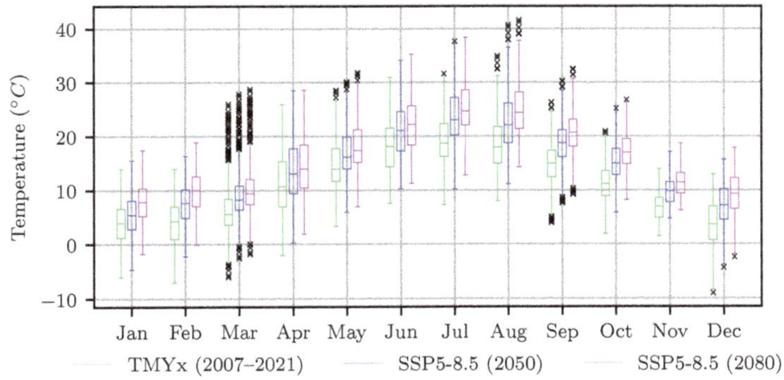

Figure 3. Monthly outdoor temperature for TMYx (2007–2021) and SSP5-8.5 for 2050 and 2080 weather data (Boxplots defined by median and Interquartile Range (IQR) from 25th to 75th percentile and whiskers limited by ±1.5 IQR).

3. Results

The simulation models are generated using the proposed template-based thermal comfort data and the IFC-based bim2sim workflow with the proposed *PluginComfort* for BPS-based thermal comfort analysis. The evaluation in Sections 3.2 and 3.3 analyzes the impact of the future weather scenarios on thermal comfort with and without cooling, compared to the current weather. As the present study addresses IFC-based evaluations of climate change impact on buildings using a case study in Germany, the results focus on the evaluation of summer comfort, since the climate change increases the observed temperatures throughout the year (cf. Figure 3).

3.1. Use Case: FZK-Haus

The use case building for demonstrating the proposed IFC-based thermal comfort setup is the KIT FZK-Haus (FZK-Haus (IFC4), provided by Institute for Automation and Applied Informatics (IAI) / Karlsruhe Institute of Technology (KIT): https://www.ifcwiki.org/index.php?title=KIT_IFC_Examples, accessed on 12 November 2023) (total floor area: 208.55 m^2) displayed in Figure 4. The simulation setup is based on the structure of the archetypal templates used for TEASER [46], which are also used in bim2sim. For the choice of materials and constructions, the construction type is selected as heavy, and the materials and constructions are set for a building construction year between 1995 and 2015. The thermal zone usage templates are modified to match the use case of a residential house and further extended by the derived templates for activity and clothing (cf. Tables 2 and 3). The set points for heating and cooling (cf. Table 4) are derived from SIA 2024 [49] and do not include a night setback. Spaces that do not have cooling requirements according to SIA 2024 [49] are defined to have a maximum indoor temperature of 32 °C. The simulation setup uses ideal loads to meet the desired set points for heating and cooling in Section 3.2, and heating without cooling in Section 3.3.

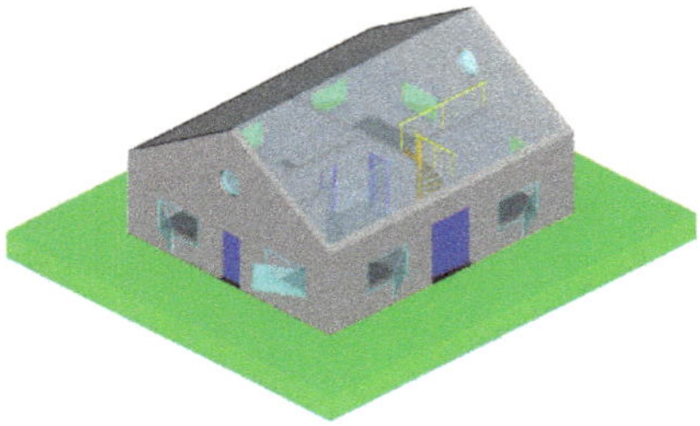

Figure 4. Use Case: FZK Haus.

Table 4. Heating and cooling set points.

	Heating [°C]	Cooling [°C]
Living	21	28
WC Residential	24	32
Kitchen Residential	20	26
Single office	21	26
Traffic area	18	32
Bedroom	21	28

The occupancy schedules are derived and adapted from Mitra et al. [67] and displayed in Figure A1. Schedules and loads for lighting and technical equipment are adapted from Remmen et al. [46] and SIA 2024 [49]. Natural ventilation set points are adapted from Remmen et al. [46] and DIN V 18599-10 [47].

3.2. Template-Based Thermal Comfort with Cooling

To get insights into the impact of future weather scenarios for heating and cooling loads, the use case is first simulated with enabled heating and cooling set points. Figure 5 displays heating and cooling loads for the weather scenarios, showing a major decrease of the heating loads (−36.04 % for 2050 and −51.96 % for 2080). The observed cooling loads drastically increase by 174.60 % for 2050 and 256.15 % for 2080. These calculated energy loads for heating and cooling align with the observations from related research [19,68] for future climate scenarios for the years 2050–2100.

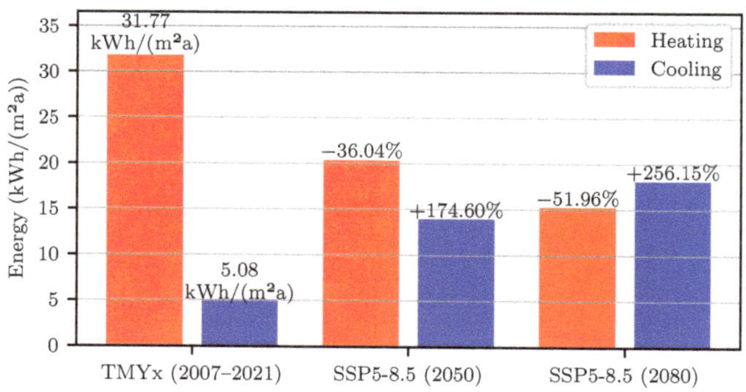

Figure 5. Annual energy demand for heating and cooling per building floor area.

The analysis of the heating and cooling loads shows that in future climate scenarios, significantly higher energy demands occur to meet the cooling set points and likewise the thermal comfort requirements. For an overview on the impact of climate change on thermal comfort under the previously defined conditions, Figures 6 and 7 display the daily mean PMV values per thermal zone (solid lines) and the daily mean outdoor temperature (dashed line) for the summer months between April and October.

For TMYx (2007–2021), the daily mean PMV ranges from −1.21 (Living_2) in December to 1.73 (Bedroom) in July. For SSP5-8.5 (2080), the daily mean PMV ranges from −1.04 (Living_2) in January to 2.06 (Traffic area) in August. The characteristics of the PMV reflect the set points and weather data that have been used for simulation. The cooling set points are reflected in the PMV data, where the PMV reaches a "plateau" in summer (e.g., for the Single office) since the choice of cooling set points indirectly limits the level of PMV. The weather data are reflected in the PMV, where the outdoor temperature has a qualitatively high similarity with the PMV, which is true for the PMV in Figure 6 for the TMYx weather data, but only in April, May, and October for the SSP-5-8.5 weather data in Figure 7. We can observe a stronger correlation between outdoor temperature and PMV for the TMYx data here since the cooling set points are only reached for a smaller period in summer than for the SSP5-8.5 weather scenario.

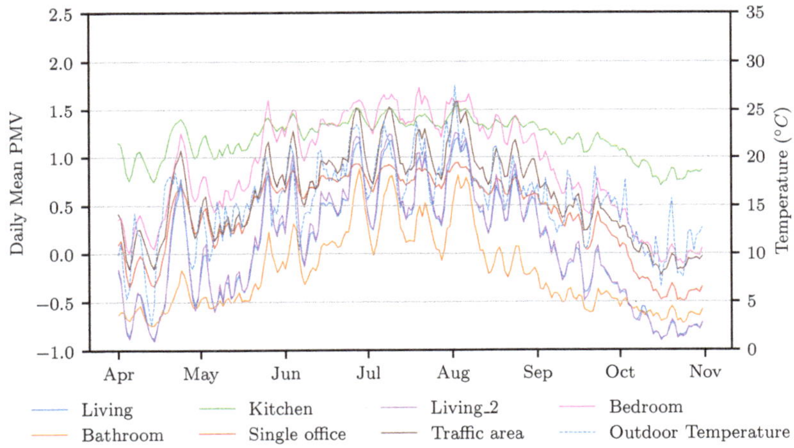

Figure 6. TMYx (2007–2021) mean daily PMV with cooling.

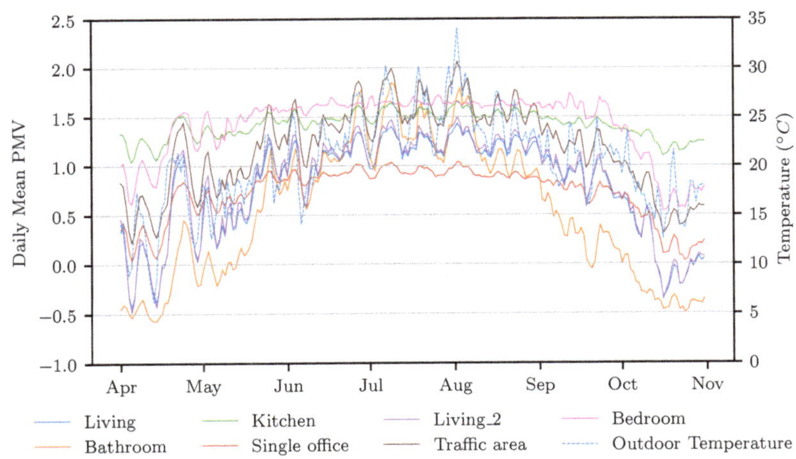

Figure 7. SSP5-8.5 (2080) mean daily PMV with cooling.

For TMYx (2007–2021) data, the low PMV during winter is caused by larger differences between zone mean air temperature (used for set point control) and the actual operative air temperature that has been used to calculate the PMV (i.e., even with a set point of 21 °C, the actual operative temperature was below 20 °C, causing occupants to feel cold). For the SSP5-8.5 (2080) scenario, the least comfortable status can be seen in the thermal zones without cooling (e.g., traffic area). For the analysis of the proposed clothing and activity parameters, a larger range of set points and building types needs to be tested, ideally supported by measured data and validation.

In Germany, most of the building stock (residential but also a large amount of public office buildings and schools) rely on natural ventilation and do not have cooling devices. Thus, as for fully conditioned buildings, thermal comfort can be accomplished by spending more energy (or by choosing a resilient building design during the design stage, which also could be supported by our proposed methods); we perform a more detailed analysis of our approach without cooling.

3.3. *Template-Based Thermal Comfort Analysis without Cooling*

For the use case without cooling, we added natural ventilation, which generally increases the heating load of the building through more added outdoor air. The heating loads for this use case are displayed in Figure 8. Here, since with increasing temperature natural ventilation increases the heat loss through the windows, the decrease of heating demand is lower compared to the previous use case with cooling, reducing the heating energy by 27.51 % until 2050 and by 36.65 % until 2080.

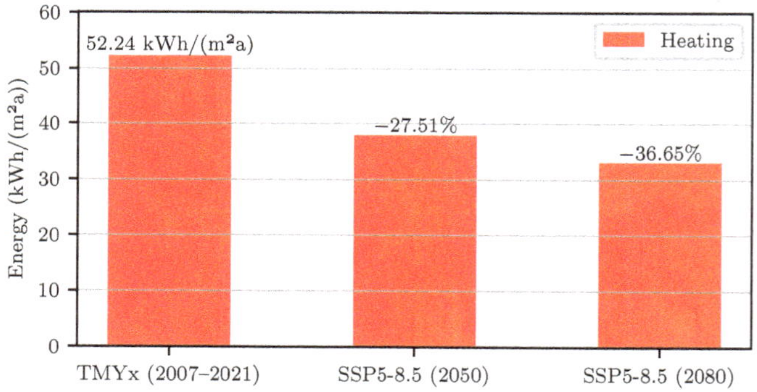

Figure 8. Annual energy demand for heating per building floor area.

As this use case without mechanical cooling allows a more in-depth analysis of the impact of climate change on thermal comfort, further results are visualized for different levels of granularity. Starting in Section 3.3.1 with an annual comparison of the mean PMV values and hours within individual PMV ranges, calendar plots highlight the changes in the daily mean PMV values, while in Section 3.3.2 further diagrams give insights into the adaptive thermal comfort per room.

3.3.1. Analysis of Fanger's PMV

Figure 9 displays an overview of the annual PMVs in the individual thermal zones. This mean annual PMV value does not take into account seasonal differences in the PMV but still outlines that for the given weather data of TMYx and SSP5-8.5 (2080), the annual PMV increases combined over all thermal zones about 0.44 (mean PMV over all thermal zones).

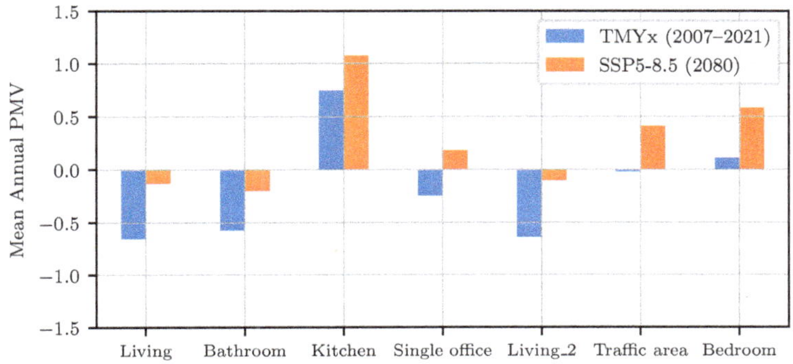

Figure 9. Mean annual PMV per thermal zone comparing the scenarios for TMYx (2007–2021) and SSP5-8.5 (2080) weather data.

Figure 10 displays the changes between the two climate scenarios per thermal zone. Here, the number of hours per thermal zone within a specific PMV range is displayed for TMYx (2007–2021) in Figure 10a, for SSP5-8.5 (2080) in Figure 10c, and the changes between these scenarios in Figure 10b. The results in Figure 10b highlight that the number of hours that are rather cold (PMV smaller 0) drastically shifts to warmer predicted temperature perceptions, resulting in more than 700 hours with a PMV larger than 2 in the kitchen. In the SSP5-8.5 scenario for 2080, all zones show PMVs larger than 2, while none of the thermal zones have shown so much discomfort for the baseline scenario.

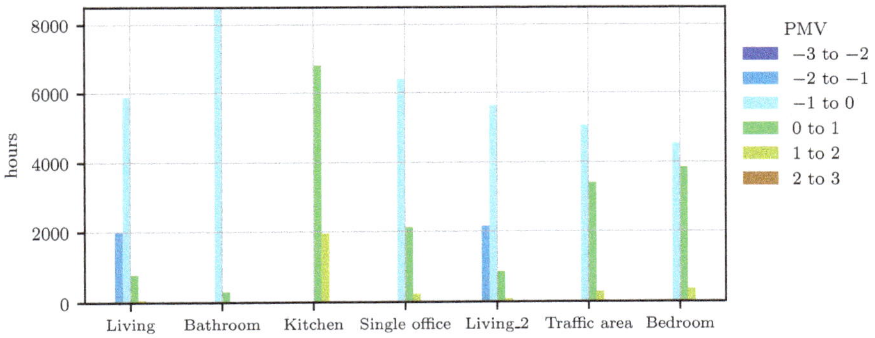

(**a**)

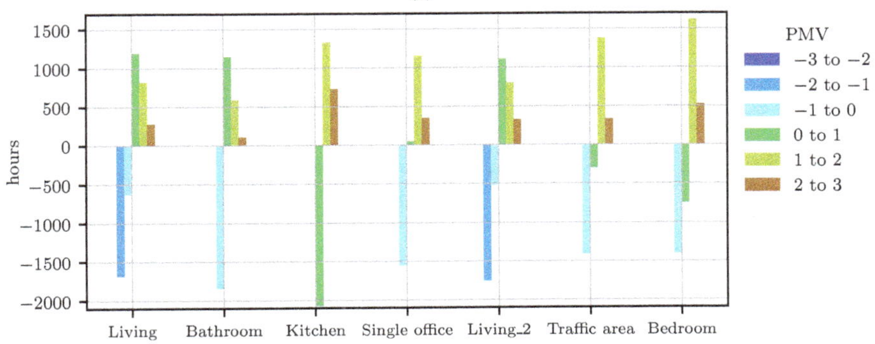

(**b**)

Figure 10. *Cont.*

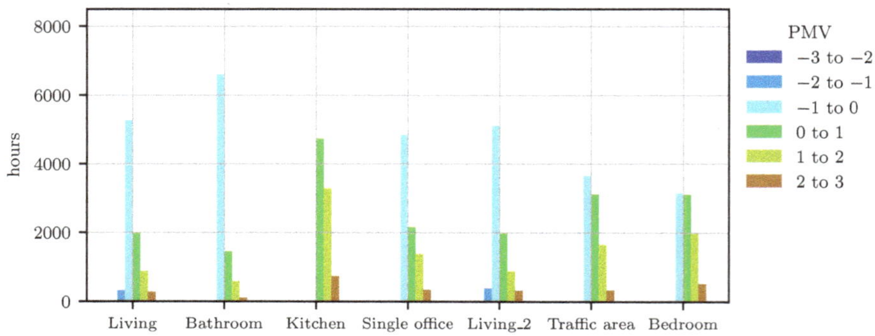

(**c**)

Figure 10. Number of hours of the individual thermal zones within PMV ranges for TMYx (2007–2021) (**a**), SSP5-8.5 (2080) (**c**) and the difference between TMYx (2007–2021) and SSP5-8.5 (2080) (**b**).

To gain more insights on the actual changes of the daily PMV variations through the year, the changes in the PMV data are visualized in Figures 11 and 12 for the exemplary thermal zones *Living* and *Kitchen* using calendar plots highlighting the 24 h mean PMV values per thermal zone. These calendar plots give a quick impression on the seasonal changes in the thermal comfort, highlighting seasonal shifts between the years TMYx (2007–2021) and SSP5-8.5 (2080).

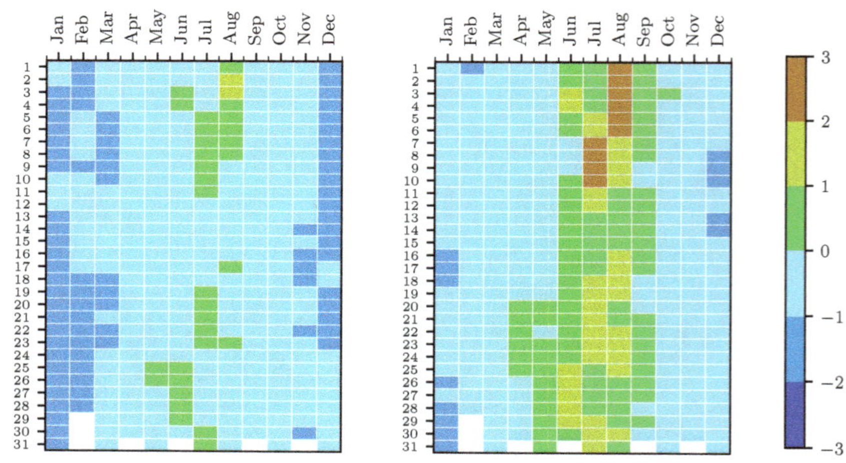

(**a**) TMYx (2007–2021): Living (**b**) SSP5-8.5 (2080): Living

Figure 11. 24 h mean PMV values for thermal zone "Living" for TMYx (2007–2021) and SSP5-8.5 (2080).

For the thermal zone *Living* in Figure 11 with TMYx (2007–2021) weather data, the PMV ranges between −2 and +2 for the whole year, showing several longer periods of PMV below −1 within the months of November through March, mixed with periods of PMV between −1 and 0. This highlights that the heating set points for this thermal zone may not be sufficiently set to meet the comfort requirements or that the proposed clothing or activity values are too low for the winter months. Comparing the daily mean PMV values for the years of TMYx (2007–2021) and SSP5-8.5 (2080), the number of PMV below −1 drastically reduced in the winter months, while the summer months (especially July and August) for SSP5-8.5 (2080) show a longer period of days with a PMV greater 1, and 10 days with a PMV between 2 and 3 and, thus, high discomfort for the occupants.

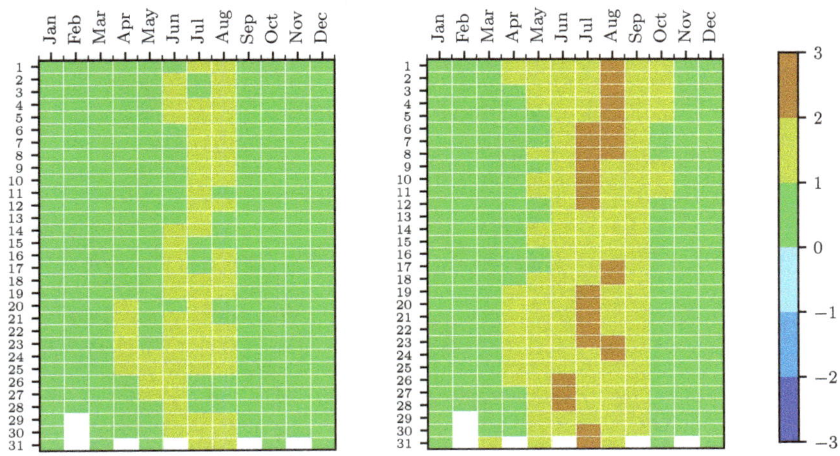

(a) TMYx (2007–2021): Kitchen (b) SSP5-8.5 (2080): Kitchen

Figure 12. 24 h mean PMV values for thermal zone "Kitchen" for TMYx (2007–2021) and SSP5-8.5 (2080).

For the thermal zone *Kitchen* in Figure 12, the PMV index in scenario TMYx (2007–2021) ranges between 0 and 1 (September through March) and a mix between 0 and 2 (April through August). With the 2080 weather data, the PMV between 1 and 2 is no longer mixed with PMV below 1 from May through September, but from June to August, 29 days with a higher PMV than 2 occur, resulting in longer periods of high discomfort.

The heating loads in the kitchen are generally higher than in the other rooms. Additional mechanical ventilation may help reduce the heating loads during kitchen use to prevent overheating. However, due to the predicted climatic changes, a residential kitchen with natural ventilation may just not be able any longer to reach a thermally comfortable status throughout the year.

These calendar plots reflect the daily mean PMV index, including the observed cool-down during the night. The actual maximum heat loads can be higher than the presented values of the calendar plots. Calendar plots for the other thermal zones are placed in Appendix A.4 in Figures A2–A6.

3.3.2. Analysis of Adaptive Comfort

With increasing indoor temperatures due to the lack of cooling, adaptive comfort measures may better represent the actual thermal sensation of occupants. For the evaluation of adaptive thermal comfort we apply the adaptive comfort metrics according to DIN EN 15251 [59] (withdrawn) and DIN EN 16798-1 [60]. As mentioned in Section 2.1, DIN EN 15251 is implemented and available in EnergyPlus, but DIN EN 16798-1 is not. Thus, we combine their results to highlight the differences between these two standards. For the analysis of the acceptable operative temperature, the main difference for DIN EN 16798-1 is that the acceptable lower bound of the operative temperature per category is decreased by 1 °C compared to DIN EN 15251. The upper boundaries are not changed. Both standards include three categories regarding the level of expectation towards thermal comfort within the room, ranging from Category I (high level of expectation) to Category III (low level of expectation).

In Figure 13, the operative indoor temperature (vertical axis) is plotted over the running mean outdoor temperature (horizontal axis) within the outdoor temperature range of 10 °C to 30 °C for thermal zone "Living". Each scattered point in the diagram is colored according to its category in DIN EN 15251. The dashed lines visualize the upper and lower thresholds of the DIN EN 16798-1 categories. For the baseline weather, the operative temperatures are in the lower categories of DIN EN 15251, while with the reduced thresholds in DIN EN 16798-1, the operative temperatures mostly vary within the ranges of Category I and II. For this baseline weather scenario, only a very limited amount of time crosses the upper band of Category I. The range of the mean outdoor temperature is mostly below 24 °C. For The SSP5-8.5 (2080) scenario in the right plot, operative indoor temperatures are widely spread over the mean outdoor temperature range, and the upper thresholds of the Categories are often crossed; some operative temperatures are even beyond the upper threshold of Category III.

Figure 14 displays the adaptive comfort according to DIN EN 15251 and DIN EN 16798-1 for the kitchen zone. Compared to Figure 13, the indoor operative temperatures are higher in general, which is due to the high internal loads of kitchen equipment. For the weather data of TMYx (2007–2021), the operative temperatures are balanced between the thresholds of the categories of DIN EN 16798-1, while for the SSP5-8.5 scenario, again, even the upper band of Category III is exceeded multiple times. As described for Figure 12, natural cooling is not sufficient in this scenario to even keep a low standard of thermal comfort.

For adaptive comfort diagrams for "Living_2" (Figure A7), "Bedroom" (Figure A8), and "Single office" (Figure A9) see the appendix. The thermal zones of the bathroom and traffic area do not show enough occupancy according to the standards to be applicable to the adaptive comfort analysis.

We additionally evaluate the percentage of time within the individual thermal comfort categories for each thermal zone (Figure 15). All percentages of the TMYx (2007–2021) data refer on the total number of 4728 hours within the applicable range of the DIN EN 16798-1 (running outdoor temperature between 10 °C and 30 °C) and 6744 hours for the SSP5-8.5 (2080) scenario, neglecting the actual occupancy time for this evaluation.

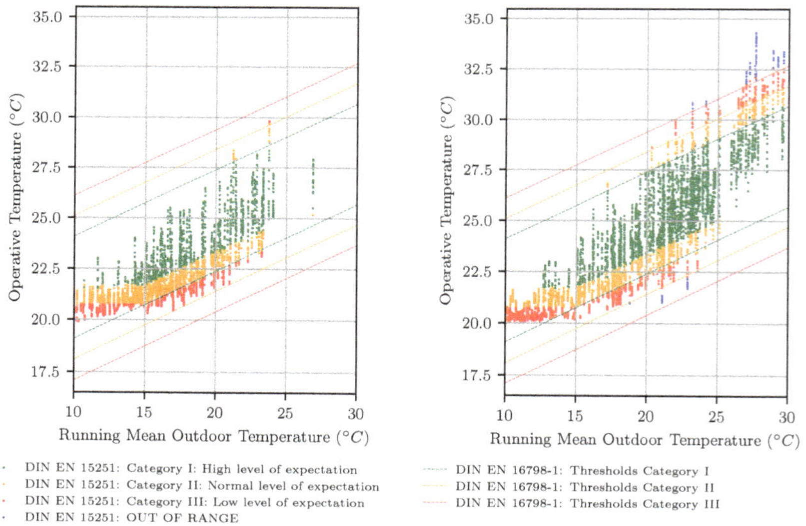

(**a**) TMYx (2007–2021): Living (**b**) SSP5-8.5 (2080): Living

Figure 13. Adaptive comfort according to DIN EN 16798-1/DIN EN 15251 for zone "Living" for TMYx (2007–2021) and SSP5-8.5 (2080).

The evaluation shows that for the base scenario, we have most of the time inside Categories I and II (in Category III: only 1 % for the kitchen and 2 % for traffic area), while we have in the SSP5-8.5 scenario between 0–10 % in Category III and even between 0–4 % worse than Category III. While the conditions of the TMYx (2007–2021) scenario need a further evaluation of the actual occupancy hours to determine if they are still within the acceptable thermal comfort range (max. 1 % of the occupancy time worse than Category II according to DIN EN 16798-1 [60]), the SSP5-8.5 scenario is far out of range.

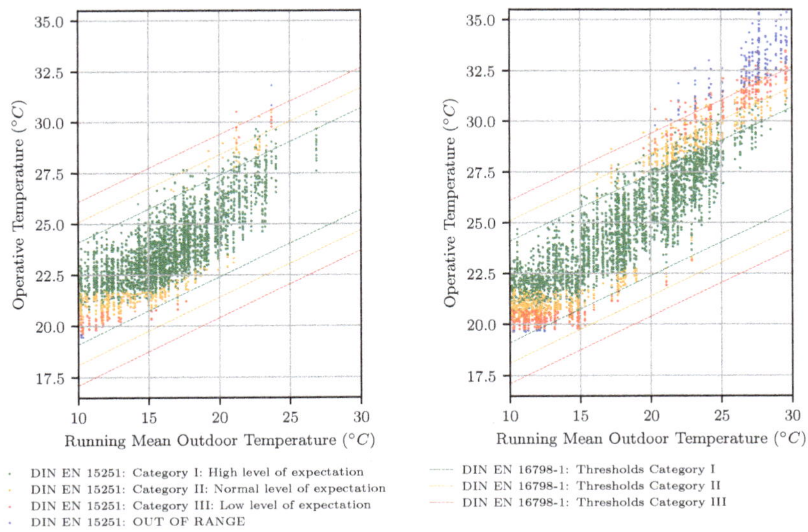

(**a**) TMYx (2007–2021): Kitchen (**b**) SSP5-8.5 (2080): Kitchen

Figure 14. Adaptive comfort according to DIN EN 16798-1/DIN EN 15251 for zone "Kitchen" for TMYx (2007–2021) and SSP5-8.5 (2080).

	Living	Bathroom	Kitchen	Single office	Living_2	Traffic area	Bedroom
CAT I	86%	100%	94%	96%	89%	72%	95%
CAT II	14%	0%	5%	4%	11%	26%	5%
CAT III	0%	0%	1%	0%	0%	2%	0%
> CAT III	0%	0%	0%	0%	0%	0%	0%

CAT I CAT II CAT III > CAT III

(**a**) TMYx (2007–2021)

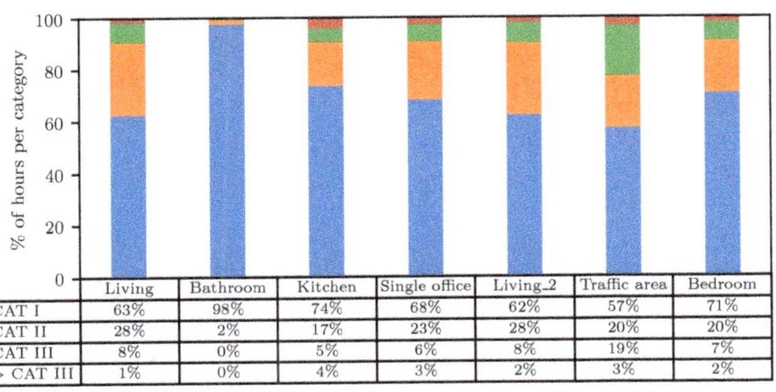

	Living	Bathroom	Kitchen	Single office	Living_2	Traffic area	Bedroom
CAT I	63%	98%	74%	68%	62%	57%	71%
CAT II	28%	2%	17%	23%	28%	20%	20%
CAT III	8%	0%	5%	6%	8%	19%	7%
> CAT III	1%	0%	4%	3%	2%	3%	2%

CAT I CAT II CAT III > CAT III

(**b**) SSP5-8.5 (2080)

Figure 15. Adaptive comfort according to DIN EN 16798-1, percentage of hours per category.

4. Discussion and Limitations

The present study proposes an extension to the IFC-based bim2sim framework to facilitate thermal comfort analysis. The study also introduces an extension to a simplified data set (i.e., TEASER [46]) used to provide the thermal comfort parameters, which are unavailable in the IFC standard. This extended data set is proposed in a manner similar to the original template and builds upon the existing dataset used in bim2sim. This simplified set of thermal comfort parameters has certain limitations that users should consider.

First, the template values provide only an estimate of the number of occupants in each thermal zone. We derived the clothing and activity parameters on the basis of assumed zone usage and with the support of existing standards. To better understand the parameters, we listed the decisions and fundamentals of each individual parameter. This listing enables further users of the templates to determine if the individual parameters are applicable to their case or if they require modification.

The proposed parameters do not account for seasonal variations (or even current outdoor weather conditions also depending on day-/nighttime) regarding clothing, as this reflection would unnecessarily complicate the parameter set. For certain types of zones, such as offices and meeting rooms, seasonal clothing changes may not even be applicable. Further research should test either various sets of clothing parameters, including summer and winter attire, or a range of clothing parameters for each zone on the basis of outdoor temperature. The latter would adapt the dynamic clothing described in ASHRAE 55, but also consider minimum and maximum clothing requirements for specific zone types.

We disregarded regional and cultural differences in clothing in the proposed templates. Since, even within the same culture, clothing standards in offices vary across businesses, the template values need to be adjusted to fit individual circumstances. Furthermore, the template considers only standardized body shapes and activity levels, represented by a body surface area of $1.8\,\text{m}^2$. Gender specifications are not included, nor are individuals who are shorter, taller, underweight, or overweight. To ensure the thermal comfort for underrepresented groups, noise should be introduced to clothing and activity parameters, allowing for the computation of thermal comfort to be repeated and evaluated for a wider range of occupant parameters. Since only standardized clothing has so far been evaluated, no skirts or dresses are included in the proposed clothing template.

We presented a simplified set of parameters for the surrounding insulation. For the sake of simplicity, we suggest adding this value to the occupant's clothing parameter, even though they may not physically add up to the total clo value. Further evaluation is necessary. Additionally, the ISO 7730 PMV only accounts for clothing up to 2 clo, but our mixed furniture approach may result in higher clothing values. To remain within the range of ISO 7730 applicability, we have to limit the PMV to 2 clo, select an alternative comfort index, or expand and validate the ISO 7730 PMV approach for clo values exceeding 2. Similarly to the classic Fanger model, which omits occupant adaptive measures including clothing changes, we also disregard such variability. As predicting individual variations in clothing can be difficult, we suggest adding a clo value increment during colder seasons to account for additional clothing layers worn, and a decrement during warmer seasons to account for reduced clothing. While this approach is crude, it can limit clothing adaptability and assist in incorporating variability driven by human behavior into the simulation.

As stated by Mahecha Zambrano et al. [16], occupants decide to adjust clothing or change the room's set point temperature. Furthermore, they state that developing generalized models for occupant behavior is not possible due to the diversity of occupants. However, the use of standardized occupancy schedules is crucial for computationally efficient decision-making in the early design stages. The approach to generate a set of schedules along with a probability of the occurrence of these scenarios [16] could assist to further improve the accuracy of our proposed IFC-based thermal comfort analysis.

For the evaluation of the results, we modified, as described, more parameters than just the proposed clothing and activity parameters compared to the original TEASER templates [46] to meet the requirements of our residential building. Some parts of the

TEASER templates originate from the DIN V 18599-10 [47], which has been designed for energetic validation and rating of buildings and, thus, may not be fully applicable for engineering tasks within the design stage of buildings. For a further use of our proposed thermal comfort parameters as addition to the TEASER template set, the full set of templates should also be tested for consistency. As an example: in the TEASER templates, the bedroom has full occupancy 24/7 while a hotel room is only occupied during the night.

Taking a closer look at the thermal comfort analysis itself, it has to be noted that BPS using EnergyPlus only calculates a single-node-per-zone PMV measure, neglecting any kind of furniture or people position within the room. The BPS engineer should be aware of the BPS limitations and also of the inaccuracies due to the choice of parameters. For a detailed thermal comfort analysis, the user must apply a much more detailed CFD analysis, which could be focused on building parts that show a high dissatisfaction with thermal comfort discovered in the BPS analysis.

When we take a closer look at the impact of the weather data, the presented results and examples only show the potential of our IFC-based approach. This study does not evaluate or compare actual design alternatives but focuses on proposing the IFC-based thermal comfort approach itself. Other weather files representing other climate change scenarios will lead to different results. However, our approach can be easily repeated using different weather files representing a variety of climate change scenarios. Design alternatives can be evaluated using our IFC-based approach without manual remodeling. However, our simulation results (heating and cooling loads) match the observations from related research [19,68], even though the results can only be qualitatively compared as the building geometry, construction, location, and weather data vary from the cases from related literature. The development of benchmark cases could help to evaluate the accuracy of our proposed methods.

A more general limitation of our proposed methods is that the selected timeframe for weather prediction should be chosen according to the predicted use of the analyzed buildings. The timespan of climate prediction from our use case is only used to highlight the potential of our proposed methods. However, until we reach the SSP5-8.5 (2080) scenario, the building can be refurbished multiple times (HVAC, insulation), or even demolished, or the usage has changed (used as an office of a smaller company vs. used as a holiday apartment), such that the comfort predictions and energy savings do not apply anymore. The construction weight and the resulting resilience of the building due to thermal mass in comparison to the embodied carbon can still be evaluated with our methods even for such large timespans of sixty years, but the occupancy schedules and internal loads can hardly be predicted.

However, as long as the user is aware of the input data limitations when interpreting the thermal comfort results, the reduced model setup time due to enriched IFC data is still highly beneficial to the decision-making within the design process.

5. Conclusions

The IFC-based BPS approach has the capability to reduce the effort for setting up models of design variants for energy and thermal comfort analysis of buildings. The template-based approach assists in rapidly filling in missing thermal comfort parameters. However, the application engineers should be aware of the simplifications that come with the template-based enrichment and should validate the model and its results for plausibility. With the use of our proposed IFC-based thermal comfort analysis plugin for the bim2sim tool, buildings providing IFC data in the design phase can quickly be evaluated for their robustness towards climate change. In this way, design variants can be tested for their resilience regarding different climate change scenarios.

Author Contributions: Conceptualization, V.E.R. and J.F.; methodology, V.E.R. and M.S.; software, V.E.R.; validation, V.E.R.; formal analysis, V.E.R.; investigation, V.E.R.; resources, V.E.R.; data curation, V.E.R.; writing—original draft preparation, V.E.R. and M.S.; writing—review and editing, V.E.R., M.S. and J.F.; visualization, V.E.R.; supervision, J.F. and C.v.T.; project administration, V.E.R.; funding acquisition, V.E.R., J.F. and C.v.T. All authors have read and agreed to the published version of the manuscript.

Funding: This research was funded by the Federal Ministry for Economic Affairs and Climate Action within the project "BIM2Praxis" (grant number 3EN1050A).

Institutional Review Board Statement: Not applicable.

Informed Consent Statement: Not applicable.

Data Availability Statement: The source code of the *bim2sim* tool is available on github: https://github.com/BIM2SIM/bim2sim, accessed on 12 November 2023.

Conflicts of Interest: The authors declare no conflict of interest. The funders had no role in the design of the study; in the collection, analyses, or interpretation of data; in the writing of the manuscript; or in the decision to publish the results.

Abbreviations

The following abbreviations are used in this manuscript:

BPS	Building Performance Simulation
BIM	Building Information Modeling
CFD	Computational Fluid Dynamics
CMIP	Coupled Model Intercomparison Project
IFC	Industry Foundation Classes
IPCC	Intergovernmental Panel on Climate Change
IQR	Interquartile Range
LCA	Life Cycle Assessment
ML	Machine Learning
PMV	Predicted Mean Vote
PPD	Predicted Percentage Dissatisfied

Appendix A. Thermal Comfort Templates

Appendix A.1. Activity Parameters

Table A1. Deriving activity parameters for archetypal enrichment of thermal zones based on ASHRAE Fundamentals and ISO 7730, including a comparison to existing TEASER template values. Columns: (1) Room type according to TEASER templates, (2) Activity degree from TEASER templates (met), (3) Heat flow from TEASER templates (W/person), (4) Chosen activity type according to ASHRAE Fundamentals, Chapter 9 Table 4, (5) Chosen activity type according to ISO 7730, Tables A.5 and B.1, (6) ASHRAE activity degree (met), (7) ISO 7730 activity degree (met), (8) Resulting combined activity degree (met), (9) Absolute deviation from TEASER activity degree (met), (10) Resulting heat flow (W/person), (11) Absolute deviation from TEASER heat flow (W/person).

(1) Room Type	(2)	(3)	(4) ASHRAE Fundamentals	(5) ISO 7730	(6)	(7)	(8) Activity Degree (met)	(9)	(10) Heat flow (W/person)	(11)
Single office	1.2	70	Office, Typing	A.5, Single office	1.1	1.2	1.2	0.0	125	55
Group Office (between 2 and 6 employees)	1.2	70	Office, Typing	A.5, Landscape office	1.1	1.2	1.2	0.0	125	55
Open-plan Office (7 or more employees)	1.2	70	Office, Typing	A.5, Landscape office	1.1	1.2	1.2	0.0	125	55
Meeting, Conference, seminar	1.2	70	Office, Typing	A.5, Conference Room	1.1	1.2	1.2	0.0	125	55
Main Hall, Reception	1.2	70	Office Walking	B.1, Standing, light activity	1.7	1.6	1.7	0.5	178	108
Retail, department store	1.2	70	Office, Filing, Standing	A.5, Department store	1.4	1.6	1.5	0.3	157	87
Retail with cooling	1.2	70	Office, Filing, Standing	A.5, Department store	1.4	1.6	1.5	0.3	157	87
Class room (school), group room (kindergarden)	1.2	70	Office, Writing	A.5, Kindergarten	1	1.4	1.2	0.0	125	55
Lecture hall, auditorium	1.2	70	Office, Typing	A.5, Auditorium	1.1	1.2	1.2	0.0	125	55
Bed room	1.2	70	Resting, Sleeping	B.1, Reclining	0.7	0.8	0.8	0.4	84	14
Hotel room	1.2	70	Resting, Sleeping	B.1, Reclining	0.7	0.8	0.8	0.4	84	14
Canteen	1.2	70	Office, Filing, Seated	A.5, Cafeteria/restaurant	1.2	1.2	1.2	0.0	125	55
Restaurant	1.2	70	Office, Filing, Seated	A.5, Cafeteria/restaurant	1.2	1.2	1.2	0.0	125	55
Kitchen in non-residential buildings	1.2	70	Cooking	B.1, Standing, medium activity	1.8	2	1.9	0.7	199	129
Kitchen—preparations, storage	2	90	Cooking	B.1, Standing, medium activity	1.8	2	1.9	0.1	199	109
WC and sanitary rooms in non-residential buildings	1.2	70	Resting, Seated, quiet	B.1, Seated, relaxed	1	1	1	0.2	105	35
Further common rooms	1.2	70	Office, Typing	B.1, Sedentary activity	1.1	1.2	1.2	0.0	125	55
Auxiliary areas (without common rooms)	1.2	70	Office Walking	A.5, Department store	1.7	1.6	1.7	0.5	178	108
Traffic area	1.2	70	Office Walking	B.1, Walking, 2 km/h	1.7	1.9	1.8	0.6	188	118
Stock, technical equipment, archives	2	90	Office, Filing, Standing	B.1, Standing, light activity	1.4	1.6	1.5	0.5	157	67
Data center	1.2	70	Office, Filing, Standing	B.1, Standing, light activity	1.4	1.6	1.5	0.3	157	87
Commercial and industrial Halls—heavy work, standing activity	2	90	Machine work, heavy	B.1, Walking, 5 km/h	4	3.4	3.7	1.7	387	297

Table A1. *Cont.*

(1) Room Type	(2)	(3)	(4) ASHRAE Fundamentals	(5) ISO 7730	(6)	(7)	(8) Activity Degree (met)	(9)	(10) Heat flow (W/person)	(11)
Commercial and industrial Halls—medium work, standing activity	1.6	80	Machine work, light (electrical industry)	B.1, Standing, medium activity	2.2	2	2.1	0.5	220	140
Commercial and industrial Halls—light work, standing activity	1.2	70	Machine work, sawing	B.1, Standing, light activity	1.8	1.6	1.7	0.5	178	108
Spectator area (theater and event venues)	1.2	70	Resting, Seated, quiet	A.5, Auditorium	1	1.2	1.1	0.1	115	45
Foyer (theater and event venues)	1.2	70	Resting, Standing, relaxed	B.1, Standing, light activity	1.2	1.6	1.4	0.2	146	76
Stage (theater and event venues)	2	90	Dancing, social	B.1, Walking, 3 km/h	3.2	2.4	2.8	0.8	293	203
Exhibition, congress	1.2	70	Office, Filing, Standing	A.5, Department store	1.4	1.6	1.5	0.3	157	87
Exhibition room and museum conservational demands	1.2	70	Resting, Standing, relaxed	B.1, Standing, light activity	1.2	1.6	1.4	0.2	146	76
Library—reading room	1.2	70	Office, Reading, seated	B.1, Sedentary activity	1	1.2	1.1	0.1	115	45
Library—open stacks	1.2	70	Office, Filing, standing	B.1, Standing, light activity	1.4	1.6	1.5	0.3	157	87
Library—magazine and depot	1.2	70	Office, Filing, standing	B.1, Standing, light activity	1.4	1.6	1.5	0.3	157	87
Gym (without spectator area)	3	120	Calisthenics/exercise	B.1, Walking, 5 km/h	3.5	3.4	3.5	0.5	366	246
Parking garages (office and private usage)	0	35	Office Walking	B.1, Walking, 2 km/h	1.7	1.9	1.8	1.8	188	153
Parking garages (public usage)	0	35	Office Walking	B.1, Walking, 2 km/h	1.7	1.9	1.8	1.8	188	153
Sauna area	1.2	70	Resting, Seated, quiet	B.1, Seated, relaxed	1	1	1	0.2	105	35
Exercise room	3	120	Office, Writing	A.5, Classroom	1	1	1	2.0	105	15
Laboratory	1.2	70	Office, Filing, Seated	B.1, Sedentary activity	1.2	1.2	1.2	0.0	125	55
Examination- or treatment room	1.2	70	Office, Filing, standing	B.1, Standing, light activity	1.4	1.6	1.5	0.3	157	87
Special care area	1.2	70	Office, Lifting/packing	B.1, Standing, medium activity	2.1	2	2.1	0.9	220	150
Corridors in the general care area	1.2	70	Office Walking	B.1, Walking, 2 km/h	1.7	1.9	1.8	0.6	188	118
Medical and therapeutic practices	1.2	70	Office, Filing, Seated	B.1, Standing, light activity	1.2	1.6	1.4	0.2	146	76
Storehouse, logistics building	2	90	Office, Lifting/packing	B.1, Standing, medium activity	2.1	2	2.1	0.1	220	130
Living	1.2	70	Resting, Seated, quiet	B.1, Sedentary activity	1	1.2	1.1	0.1	115	45
Classroom	1	70	Office, Writing	A.5, Classroom	1	1.2	1.1	0.1	115	45

Appendix A.2. Clothing Parameters

Table A2. Deriving clothing parameters for archetypal enrichment of thermal zones based on ASHRAE Fundamentals and ISO 7730. Columns: (1) Room type according to TEASER templates, (2) Chosen clothing type according to ASHRAE Fundamentals, Chapter 9 Table 7, (3) Chosen clothing type according to ISO 7730, (4) ASHRAE clothing (clo), (5) ISO 7730 clothing (clo), (6) Resulting combined clothing parameter (clo), (7) Chosen surrounding insulation type, (8) Surrounding insulation (clo).

(1) Room Type	(2) ASHRAE Fundamentals	(3) ISO 7730	(4)	(5)	(6) Clothing Insulation (clo)	(7) Surrounding Insulation Description	(8) Surrounding Insulation (clo)
Single office	Trousers, long-sleeved shirt	Underwear, shirt, trousers, socks, shoes	0.61	0.7	0.66	ISO 7730, C.3 Executive chair	0.15
Group Office (between 2 and 6 employees)	Trousers, long-sleeved shirt	Underwear, shirt, trousers, socks, shoes	0.61	0.7	0.66	ISO 7730, C.3 Standard office chair	0.1
Open-plan Office (7 or more employees)	Trousers, long-sleeved shirt	Underwear, shirt, trousers, socks, shoes	0.61	0.7	0.66	ISO 7730, C.3 Standard office chair	0.1
Meeting, Conference, seminar	Trousers, long-sleeved shirt, suit jacket	Underwear, shirt, trousers, socks, shoes	0.96	0.7	0.83	ISO 7730, C.3 Wooden stool	0.01
Main Hall, Reception	Trousers, long-sleeved shirt, suit jacket	Underwear with short sleeves and legs, shirt, trousers, jacket, socks, shoes	0.96	1	0.98	ISO 7730, C.3 Standard office chair	0.1
Retail, department store	Trousers, long-sleeved shirt, long-sleeved sweater, T-shirt	Panties, shirt, trousers, jacket, socks, shoes	1.01	1	1.01	None	
Retail with cooling	Trousers, long-sleeved shirt, long-sleeved sweater, T-shirt	Panties, shirt, trousers, jacket, socks, shoes	1.01	1	1.01	None	
Class room (school), group room (kindergarden)	Trousers, long-sleeved shirt	Underwear, shirt, trousers, socks, shoes	0.61	0.7	0.66	ISO 7730, C.3 Wooden stool	0.01
Lecture hall, auditorium	Trousers, long-sleeved shirt	Underwear, shirt, trousers, socks, shoes	0.61	0.7	0.66	ISO 7730, C.3 Wooden stool	0.01
Bed room	Walking shorts, short-sleeved shirt	Panties, T-shirt, shorts, light socks, sandals	0.36	0.3	0.33	Average based on Zhang et al. [65]	2
Hotel room	Walking shorts, short-sleeved shirt	Panties, T-shirt, shorts, light socks, sandals	0.36	0.3	0.33	Average based on Zhang et al. [65]	2
Canteen	Trousers, long-sleeved shirt	Underwear, shirt, trousers, socks, shoes	0.61	0.7	0.66	ISO 7730, C.3 Wooden stool	0.01

Table A2. *Cont.*

(1) Room Type	(2) ASHRAE Fundamentals	(3) ISO 7730	(4)	(5)	(6) Clothing Insulation (clo)	(7) Surrounding Insulation Description	(8) Surrounding Insulation (clo)
Restaurant	Trousers, long-sleeved shirt	Underwear, shirt, trousers, socks, shoes	0.61	0.7	0.66	ISO 7730, C.3 Wooden stool	0.01
Kitchen in non-residential buildings	Long-sleeved overalls, T-Shirt	Underpants, shirt, trousers, smock, socks, shoes	0.72	0.9	0.81	None	
Kitchen—preparations, storage	Long-sleeved overalls, T-Shirt	Underpants, shirt, trousers, smock, socks, shoes	0.72	0.9	0.81	None	
WC and sanitary rooms in non-residential buildings	Trousers, long-sleeved shirt	Underwear, shirt, trousers, socks, shoes	0.61	0.7	0.66	None	
Further common rooms	Trousers, long-sleeved shirt	Underwear, shirt, trousers, socks, shoes	0.61	0.7	0.66	ISO 7730, C.3 Wooden stool	0.01
Auxiliary areas (without common rooms)	Trousers, long-sleeved shirt	Underwear, shirt, trousers, socks, shoes	0.61	0.7	0.66	None	
Traffic area	Trousers, long-sleeved shirt	Underwear, shirt, trousers, socks, shoes	0.61	0.7	0.66	None	
Stock, technical equipment, archives	Trousers, long-sleeved shirt	Underwear, shirt, trousers, socks, shoes	0.61	0.7	0.66	None	
Data center	Trousers, long-sleeved shirt	Underwear, shirt, trousers, socks, shoes	0.61	0.7	0.66	None	
Commercial and industrial Halls—heavy work, standing activity	Trousers, long-sleeved shirt	Underwear, shirt, trousers, socks, shoes	0.57	0.7	0.64	None	
Commercial and industrial Halls—medium work, standing activity	Trousers, long-sleeved shirt	Underwear, shirt, trousers, socks, shoes	0.57	0.7	0.64	None	
Commercial and industrial Halls—light work, standing activity	Trousers, long-sleeved shirt	Underwear, shirt, trousers, socks, shoes	0.57	0.7	0.64	None	
Spectator area (theater and event venues)	Trousers, long-sleeved shirt, suit jacket	Underwear, shirt, trousers, socks, shoes	0.96	0.7	0.83	ISO 7730, C.3 Wooden stool	0.01
Foyer (theater and event venues)	Trousers, long-sleeved shirt, suit jacket	Underwear, shirt, trousers, socks, shoes	0.96	0.7	0.83	None	
Stage (theater and event venues)	Trousers, long-sleeved shirt	Underwear, shirt, trousers, socks, shoes	0.61	0.7	0.66	None	
Exhibition, congress	Trousers, long-sleeved shirt, suit jacket	Underwear, shirt, trousers, socks, shoes	0.96	0.7	0.83	None	
Exhibition room and museum conservational demands	Trousers, long-sleeved shirt, suit jacket	Underwear, shirt, trousers, socks, shoes	0.96	0.7	0.83	None	

Table A2. *Cont.*

(1) Room Type	(2) ASHRAE Fundamentals	(3) ISO 7730	(4)	(5)	(6) Clothing Insulation (clo)	(7) Surrounding Insulation Description	(8) Surrounding Insulation (clo)
Library—reading room	Trousers, long-sleeved shirt	Underwear, shirt, trousers, socks, shoes	0.61	0.7	0.66	ISO 7730, C.3 Wooden stool	0.01
Library—open stacks	Trousers, long-sleeved shirt	Underwear, shirt, trousers, socks, shoes	0.61	0.7	0.66	None	
Library—magazine and depot	Trousers, long-sleeved shirt	Underwear, shirt, trousers, socks, shoes	0.61	0.7	0.66	None	
Gym (without spectator area)	Walking shorts, short-sleeved shirt	Underpants, shirt with short sleeves, light trousers, light socks, shoes	0.36	0.5	0.43	None	
Parking garages (office and private usage)	Trousers, long-sleeved shirt, long-sleeved sweater, T-shirt	Panties, shirt, trousers, jacket, socks, shoes	1.01	1	1.01	None	
Parking garages (public usage)	Trousers, long-sleeved shirt, long-sleeved sweater, T-shirt	Panties, shirt, trousers, jacket, socks, shoes	1.01	1	1.01	None	
Sauna area	Not applicable	Not applicable	0	0	0	ISO 7730, C.3 Wooden stool	0.01
Exercise room	Trousers, long-sleeved shirt	Underwear, shirt, trousers, socks, shoes	0.61	0.7	0.66	ISO 7730, C.3 Wooden stool	0.01
Laboratory	Long-sleeved overalls, T-Shirt	Underpants, shirt, trousers, smock, socks, shoes	0.72	0.9	0.81	ISO 7730, C.3 Wooden stool	0.01
Examination- or treatment room	Long-sleeved overalls, T-Shirt	Underpants, shirt, trousers, smock, socks, shoes	0.72	0.9	0.81	ISO 7730, C.3 Wooden stool	0.01
Special care area	Long-sleeved overalls, T-Shirt	Underpants, shirt, trousers, smock, socks, shoes	0.72	0.9	0.81	Average based on Zhang et al. [65]	2
Corridors in the general care area	Long-sleeved overalls, T-Shirt	Underpants, shirt, trousers, smock, socks, shoes	0.72	0.9	0.81	None	
Medical and therapeutic practices	Long-sleeved overalls, T-Shirt	Underpants, shirt, trousers, smock, socks, shoes	0.72	0.9	0.81	ISO 7730, C.3 Wooden stool	0.01
Storehouse, logistics building	Long-sleeved overalls, T-Shirt	Underwear, shirt, trousers, socks, shoes	0.72	0.7	0.71	None	
Living	Trousers, long-sleeved shirt	Underwear, shirt, trousers, socks, shoes	0.61	0.7	0.66	ISO 7730, C.3 Executive chair	0.15
Classroom	Trousers, long-sleeved shirt	Underwear, shirt, trousers, socks, shoes	0.61	0.7	0.66	ISO 7730, C.3 Wooden stool	0.01

A.3. Simulation Setup for Use Case 1: FZK-Haus

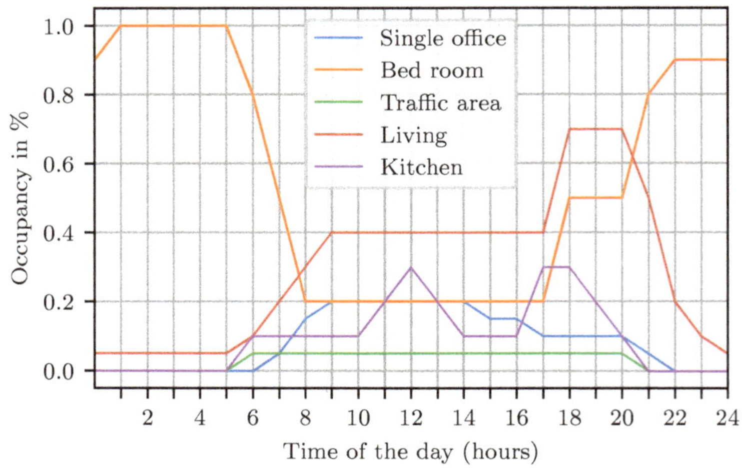

Figure A1. Occupancy profiles derived from Mitra et al. [67].

Appendix A.4. Additional Simulation Results for Use Case 1: FZK-Haus

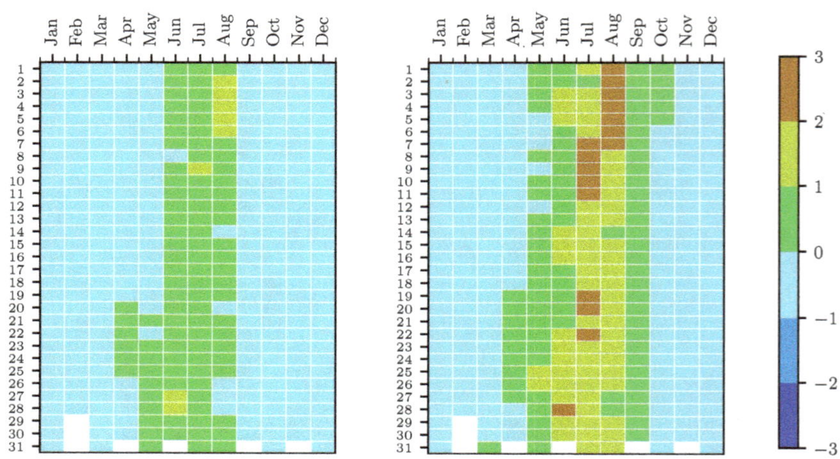

(**a**) TMYx (2007–2021): Single office (**b**) SSP5-8.5 (2080): Single office

Figure A2. 24 h mean PMV values for thermal zone "Single office" for TMYx (2007–2021) and SSP5-8.5 (2080).

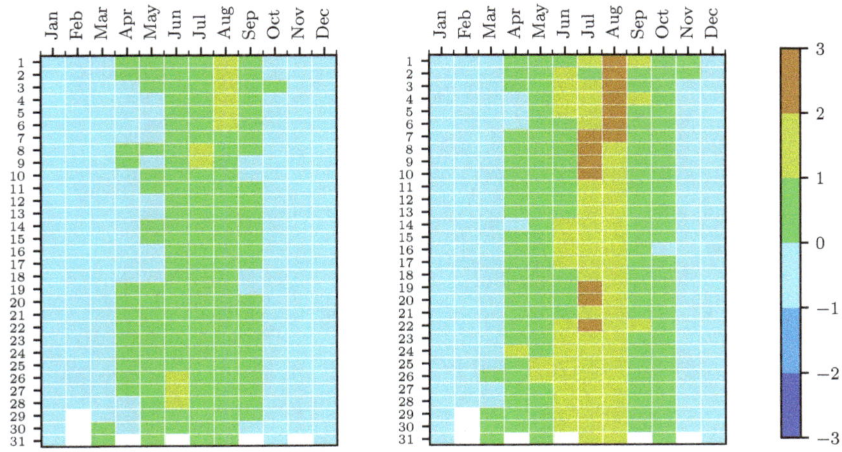

(a) TMYx (2007–2021): Traffic area (b) SSP5-8.5 (2080): Traffic area

Figure A3. 24 h mean PMV values for thermal zone "Traffic area" for TMYx (2007–2021) and SSP5-8.5 (2080).

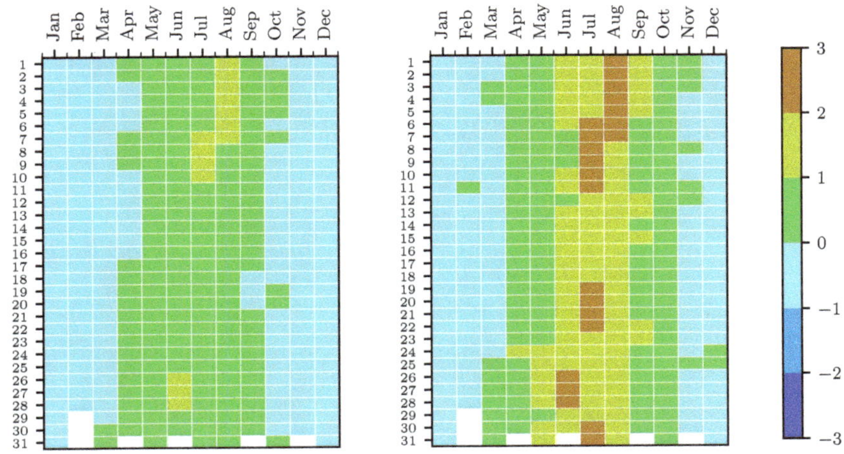

(a) TMYx (2007–2021): Bedroom (b) SSP5-8.5 (2080): Bedroom

Figure A4. 24 h mean PMV values for thermal zone "Bedroom" for TMYx (2007–2021) and SSP5-8.5 (2080).

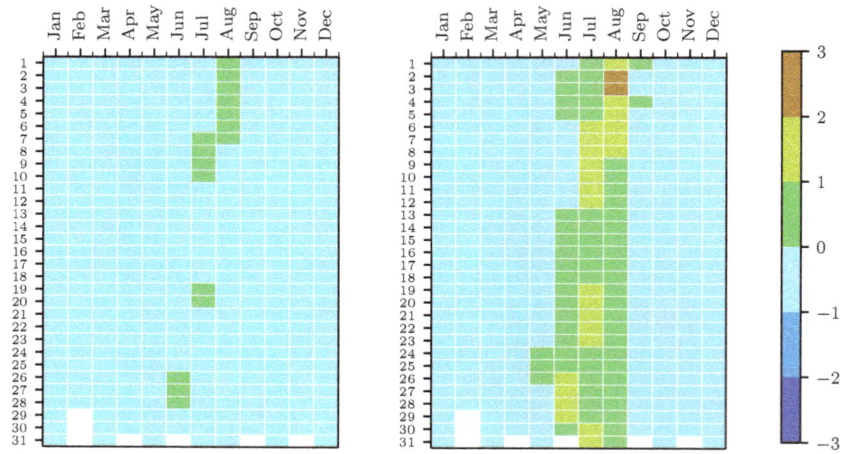

(a) TMYx (2007–2021): Bathroom

(b) SSP5-8.5 (2080): Bathroom

Figure A5. 24 h mean PMV values for thermal zone "Bathroom" for TMYx (2007–2021) and SSP5-8.5 (2080).

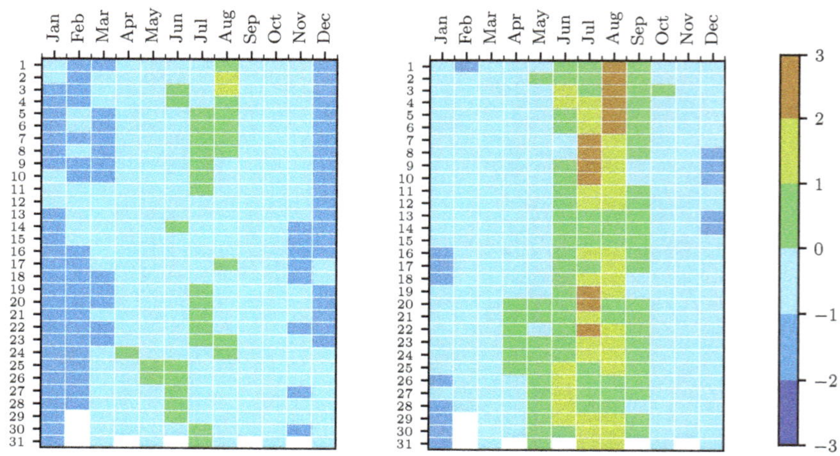

(a) TMYx (2007–2021): Living 2

(b) SSP5-8.5 (2080): Living 2

Figure A6. 24 h mean PMV values for thermal zone "Living 2" for TMYx (2007–2021) and SSP5-8.5 (2080).

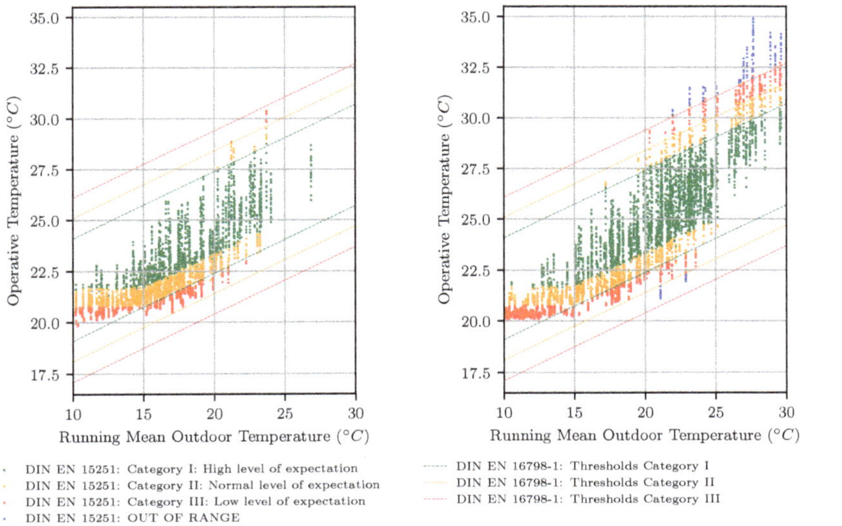

(**a**) TMYx (2007–2021): Living2 (**b**) SSP5-8.5 (2080): Living2

Figure A7. Adaptive comfort according to DIN EN 16798-1/DIN EN 15251 for zone "Living2"for TMYx (2007–2021) and SSP5-8.5 (2080).

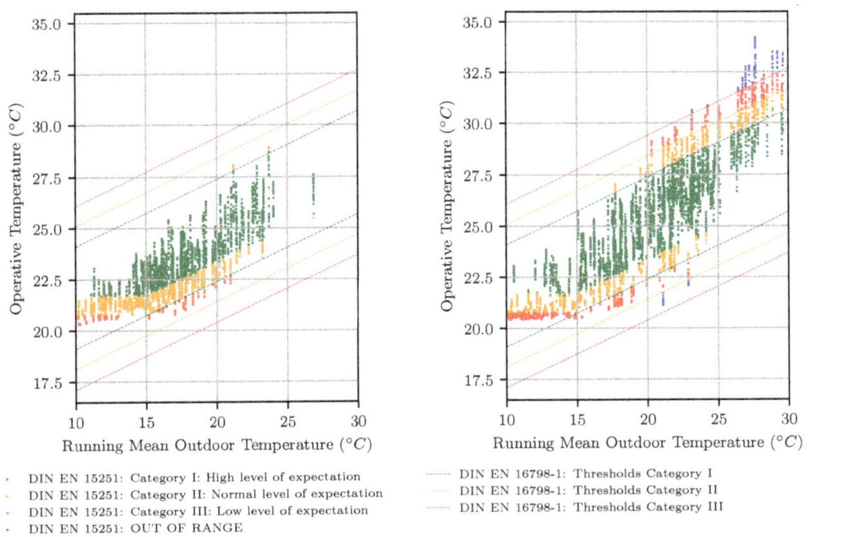

(**a**) TMYx (2007–2021): Bedroom (**b**) SSP5-8.5 (2080): Bedroom

Figure A8. Adaptive comfort according to DIN EN 16798-1/DIN EN 15251 for zone "Bedroom" for TMYx (2007–2021) and SSP5-8.5 (2080).

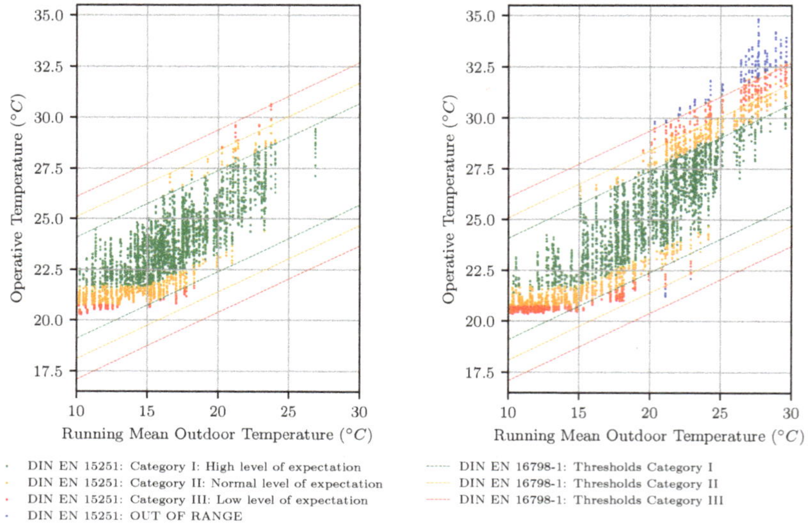

(**a**) TMYx (2007–2021): Single office (**b**) SSP5-8.5 (2080): Single office

Figure A9. Adaptive comfort according to DIN EN 16798-1/DIN EN 15251 for zone "Single office" for TMYx (2007–2021) and SSP5-8.5 (2080).

References

1. Holmes, M.J.; Hacker, J.N. Climate change, thermal comfort and energy: Meeting the design challenges of the 21st century. *Energy Build.* **2007**, *39*, 802–814. [CrossRef]
2. Bell, N.; Bilbao, J.; Kay, M.; Sproul, A. Future climate scenarios and their impact on heating, ventilation and air-conditioning system design and performance for commercial buildings for 2050. *Renew. Sustain. Energy Rev.* **2022**, *162*, 112363. [CrossRef]
3. Klepeis, N.E.; Nelson, W.C.; Ott, W.R.; Robinson, J.P.; Tsang, A.M.; Switzer, P.; Behar, J.V.; Hern, S.C.; Engelmann, W.H. The National Human Activity Pattern Survey (NHAPS): A resource for assessing exposure to environmental pollutants. *J. Expo. Sci. Environ. Epidemiol.* **2001**, *11*, 231–252. [CrossRef] [PubMed]
4. Krause, C.; Schulz, C. Aufenthaltszeiten der deutschen Bevölkerung im Innenraum, im Freien, im Straßenverkehr [German population's time spent indoors, outdoors, in traffic]. *Umweltmed. Forsch. Und Prax.* **1998**, *3*, 249.
5. Zomorodian, Z.S.; Tahsildoost, M.; Hafezi, M. Thermal comfort in educational buildings: A review article. *Renew. Sustain. Energy Rev.* **2016**, *59*, 895–906. [CrossRef]
6. Antoniadou, P.; Papadopoulos, A.M. Occupants' thermal comfort: State of the art and the prospects of personalized assessment in office buildings. *Energy Build.* **2017**, *153*, 136–149. [CrossRef]
7. Andargie, M.S.; Touchie, M.; O'Brien, W. A review of factors affecting occupant comfort in multi-unit residential buildings. *Build. Environ.* **2019**, *160*, 106182. [CrossRef]
8. *ISO 16739*; Industry Foundation Classes (IFC) for Data Sharing in the Construction and Facility Management Industries—Part 1: Data Schema. Technical Report 16739; International Organization for Standardization: Geneva, Switzerland, 2018.
9. Jansen, D.; Mehrfeld, P.; Müller, D.; Fichter, E.; Richter, V.; Barz, A.; Brunkhorst, J.; Dahncke, M.; Jahangiri, P.; Warnecke, C.; et al. BIM2SIM—Development of semi-automated methods for the generation of simulation models using Building Information Modeling. In Proceedings of the Building Simulation 2021: 17th Conference of International Building Performance Simulation Association, Bruges, Belgium, 1–3 September 2021. [CrossRef]
10. Müller, D.; Jansen, D.; van Treeck, C.; Fichter, E.; Richter, V.; Lüdemann, B.; Jahangiri, P.; Brunkhorst, J.; Dahnke, M.; Warnecke, C. *BIM2SIM—Methodenentwicklung zur Erstellung von Simulationsmodellen aus Daten des Building Information Modeling: Gemeinsamer Endbericht: Berichtszeitraum: 1 May 2018–31 October 2021*; RWTH Aachen University: Aachen, Germany, 2021. [CrossRef]
11. Fanger, P.O. *Thermal Comfort. Analysis and Applications in Environmental Engineering*; Danish Technical Press: Copenhagen, Denmark, 1970.
12. Barone, G.; Buonomano, A.; Forzano, C.; Palombo, A. Enhancing trains envelope–heating, ventilation, and air conditioning systems: A new dynamic simulation approach for energy, economic, environmental impact and thermal comfort analyses. *Energy* **2020**, *204*, 117833. [CrossRef]
13. Limane, A.; Fellouah, H.; Galanis, N. Three-dimensional OpenFOAM simulation to evaluate the thermal comfort of occupants, indoor air quality and heat losses inside an indoor swimming pool. *Energy Build.* **2018**, *167*, 49–68. [CrossRef]
14. Luo, M.; Guo, J.; Feng, X.; Chen, W. Studying occupant's heat exposure and thermal comfort in the kitchen through full-scale experiments and CFD simulations. *Indoor Built Environ.* **2023**, *32*, 928–943. [CrossRef]

15. Li, G.; Chi, L.; Guo, J.; Liu, C.; Luo, Y.; She, C. Numerical simulation on indoor thermal comfort of a new integrated rural heating system. *Procedia Eng.* **2015**, *121*, 1111–1117. [CrossRef]
16. Mahecha Zambrano, J.; Filippi Oberegger, U.; Salvalai, G. Towards integrating occupant behaviour modelling in simulation-aided building design: Reasons, challenges and solutions. *Energy Build.* **2021**, *253*, 111498. [CrossRef]
17. Gritzki, R.; Rösler, M. Komfort für Passivhaus-Büros- Planungsunterstützung mit Hilfe gekoppelter Gebäude-, Anlagen-und Strömungssimulation. *Bauphysik* **2013**, *35*, 8–15. [CrossRef]
18. Barbosa, R.; Vicente, R.; Santos, R. Climate change and thermal comfort in Southern Europe housing: A case study from Lisbon. *Build. Environ.* **2015**, *92*, 440–451. [CrossRef]
19. Escandón, R.; Suárez, R.; Sendra, J.J.; Ascione, F.; Bianco, N.; Mauro, G.M. Predicting the impact of climate change on thermal comfort in a building category: The Case of Linear-type Social Housing Stock in Southern Spain. *Energies* **2019**, *12*, 2238. [CrossRef]
20. Yan, H.; Ji, G.; Yan, K. Data-driven prediction and optimization of residential building performance in Singapore considering the impact of climate change. *Build. Environ.* **2022**, *226*, 109735. [CrossRef]
21. Siu, C.Y.; O'Brien, W.; Touchie, M.; Armstrong, M.; Laouadi, A.; Gaur, A.; Jandaghian, Z.; Macdonald, I. Evaluating thermal resilience of building designs using building performance simulation—A review of existing practices. *Build. Environ.* **2023**, *234*, 110124. [CrossRef]
22. Katavoutas, G.; Founda, D.; Kitsara, G.; Giannakopoulos, C. Climate change and thermal comfort in top tourist destinations—The case of Santorini (Greece). *Sustainability* **2021**, *13*, 9107. [CrossRef]
23. Mahadevia, D.; Pathak, M.; Bhatia, N.; Patel, S. Climate Change, Heat Waves and Thermal Comfort—Reflections on Housing Policy in India. *Environ. Urban. ASIA* **2020**, *11*, 29–50. [CrossRef]
24. Thapa, S.; Rijal, H.B.; Pasut, W.; Singh, R.; Indraganti, M.; Bansal, A.K.; Panda, G.K. Simulation of thermal comfort and energy demand in buildings of sub-Himalayan eastern India-Impact of climate change at mid (2050) and distant (2080) future. *J. Build. Eng.* **2023**, *68*, 106068. [CrossRef]
25. ASHRAE Standard 55. *ANSI/ASHRAE Standard 55: Thermal Environmental Conditions for Human Occupancy*; Technical report; ASHRAE Special Publications: Peachtree Corners, GA, USA, 2017.
26. *ISO 7730*; Ergonomics of the Thermal Environment—Analytical Determination and Interpretation of Thermal Comfort Using Calculation of the PMV and PPD Indices and Local Thermal Comfort Criteria. International Organization for Standardization: Geneva, Switzerland, 2005.
27. Van Hoof, J. Forty years of Fanger's model of thermal comfort: Comfort for all? *Indoor Air* **2008**, *18*, 182–201. [CrossRef] [PubMed]
28. Charef, R.; Emmitt, S.; Alaka, H.; Fouchal, F. Building Information Modelling adoption in the European Union: An overview. *J. Build. Eng.* **2019**, *25*, 100777. [CrossRef]
29. Ying, H.; Lee, S. A rule-based system to automatically validate IFC second-level space boundaries for building energy analysis. *Autom. Constr.* **2021**, *127*, 103724. [CrossRef]
30. Richter, V.; Malhotra, A.; Fichter, E.; Hochberger, A.; Frisch, J.; van Treeck, C. Validation of IFC-based Geometric Input for Building Energy Performance Simulation. In Proceedings of the 2022 Building Performance Modeling Conference and SimBuild Co-Organized by ASHRAE and IBPSA-USA, Chicago, IL, USA, 14–16 September 2022. [CrossRef]
31. Di Biccari, C.; Calcerano, F.; D'Uffizi, F.; Esposito, A.; Campari, M.; Gigliarelli, E. Building information modeling and building performance simulation interoperability: State-of-the-art and trends in current literature. *Adv. Eng. Inform.* **2022**, *54*, 101753. [CrossRef]
32. Bazjanac, V. Space Boundary Requirements for Modeling of Building Geometry for Energy and Other Performance Simulation. In Proceedings of the CIB W78 2010: 27th International Conference, Cairo, Egypt, 16–18 November 2010; International Council for Research and Innovation in Building and Construction: Kanata, ON, Canada, 2010.
33. van Treeck, C.; Rank, E. Dimensional reduction of 3D building models using graph theory and its application in building energy simulation. *Eng. Comput.* **2007**, *23*, 109–122. [CrossRef]
34. Rose, C.M.; Bazjanac, V. An algorithm to generate space boundaries for building energy simulation. *Eng. Comput.* **2015**, *31*, 271–280. [CrossRef]
35. Jones, N.L.; McCrone, C.J.; Walter, B.J.; Pratt, K.B.; Greenberg, D.P. Automated Translation and Thermal Zoning of Digital Building Models for Energy Analysis. In Proceedings of the BS2013: 13th Conference of International Building Performance Simulation Association, Chambéry, France, 26–28 August 2013.
36. Lilis, G.N.; Giannakis, G.I.; Rovas, D.V. Automatic generation of second-level space boundary topology from IFC geometry inputs. *Autom. Constr.* **2017**, *76*, 108–124. [CrossRef]
37. Ying, H.; Lee, S. Generating second-level space boundaries from large-scale IFC-compliant building information models using multiple geometry representations. *Autom. Constr.* **2021**, *126*, 103659. [CrossRef]
38. Fichter, E.; Richter, V.; Frisch, J.; van Treeck, C. Automatic generation of second level space boundary geometry from IFC models. In Proceedings of the Building Simulation 2021: 17th Conference of International Building Performance Simulation Association, Bruges, Belgium, 1–3 September 2021. [CrossRef]
39. Mediavilla, A.; Elguezabal, P.; Lasarte, N. Graph-Based methodology for Multi-Scale generation of energy analysis models from IFC. *Energy Build.* **2023**, *282*, 112795. [CrossRef]

40. Giannakis, G.I.; Katsigarakis, K.I.; Lilis, G.N.; Rovas, D.V. A Workflow for Automated Building Energy Performance Model Generation Using BIM. In *Proceedings of the Building Simulation 2019*; Corrado, V., Ed.; International Building Performance Simulation Association: Rome, Italy, 2019.
41. Nytsch-Geusen, C.; Rädler, J.; Thorade, M.; Ribas Tugores, C. BIM2Modelica—An open source toolchain for generating and simulating thermal multi-zone building models by using structured data from BIM models. In Proceedings of the 13th International Modelica Conference, Regensburg, Germany, 4–6 March 2019. Linköing University Electronic Press: Linköing, Sweden, 2019; pp. 33–38. [CrossRef]
42. O'Donnell, J.; See, R.; Rose, C.; Maile, T.; Bazjanac, V. SimModel: A domain data model for whole building energy simulation. In Proceedings of the Building Simulation 2011: 12th Conference of International Building Performance Simulation Association, Sydney, Australia, 14–16 November 2011.
43. Andriamamonjy, A.; Saelens, D.; Klein, R. An automated IFC-based workflow for building energy performance simulation with Modelica. *Autom. Constr.* **2018**, *91*, 166–181. [CrossRef]
44. Cao, J. SimModel Transformation Middleware for Modelica-based Building Energy Modeling and Simulation. Ph.D. Thesis, RWTH Aachen University, Aachen, Germany, 2018.
45. Chen, Z.; Deng, Z.; Chong, A.; Chen, Y. AutoBPS-BIM: A toolkit to transfer BIM to BEM for load calculation and chiller design optimization. *Build. Simul.* **2023**. [CrossRef]
46. Remmen, P.; Lauster, M.; Mans, M.; Fuchs, M.; Osterhage, T.; Müller, D. TEASER: An open tool for urban energy modelling of building stocks. *J. Build. Perform. Simul.* **2018**, *11*, 84–98. [CrossRef]
47. *DIN V 18599-10*; Energy Efficiency of Buildings—Calculation of the Net, Final and Primary Energy Demand for Heating, Cooling, Ventilation, Domestic Hot Water and Lighting—Part 10: Boundary Conditions of Use, Climatic Data. Deutsches Institut für Normung: Berlin, Germany, September 2018.
48. *VDI 2078*; Calculation of Thermal Loads and Room Temperatures (Design Cooling Load and Annual Simulation). Verein Deutscher Ingenieure: Düsseldorf, Germany, June 2015.
49. *SIA 2024*; Standard-Nutzungsbedingungen für die Energie- und Gebäudetechnik. Schweizerischer Ingenieur- und Architekten-verein (SIA) (engl. Swiss Society of Engineers and Architects): Zürich, Switzerland, 2006.
50. Jentsch, M.F.; James, P.A.; Bourikas, L.; Bahaj, A.S. Transforming existing weather data for worldwide locations to enable energy and building performance simulation under future climates. *Renew. Energy* **2013**, *55*, 514–524. [CrossRef]
51. Tapiador, F.J.; Navarro, A.; Moreno, R.; Sánchez, J.L.; García-Ortega, E. Regional climate models: 30 years of dynamical downscaling. *Atmos. Res.* **2020**, *235*, 104785. [CrossRef]
52. Wilby, R.L.; Wigley, T. Downscaling general circulation model output: A review of methods and limitations. *Prog. Phys. Geogr. Earth Environ.* **1997**, *21*, 530–548. [CrossRef]
53. Belcher, S.E.; Hacker, J.N.; Powell, D.S. Constructing design weather data for future climates. *Build. Serv. Eng. Res. Technol.* **2005**, *26*, 49–61. [CrossRef]
54. Zeng, Z.; Kim, J.H.; Tan, H.; Hu, Y.; Rastogi, P.; Wang, J.; Muehleisen, R. A critical analysis of future weather data for building and energy modeling. In Proceedings of the Building Simulation 2023: 18th Conference of IBPSA, Shanghai, China, 4–6 September 2023.
55. Nielsen, C.N.; Kolarik, J. Utilization of Climate Files Predicting Future Weather in Dynamic Building Performance Simulation—A review. *J. Phys. Conf. Ser.* **2021**, *2069*, 012070. [CrossRef]
56. Rodrigues, E.; Fernandes, M.S.; Carvalho, D. Future weather generator for building performance research: An open-source morphing tool and an application. *Build. Environ.* **2023**, *233*, 110104. [CrossRef]
57. Jiang, A.; Liu, X.; Czarnecki, E.; Zhang, C. Hourly weather data projection due to climate change for impact assessment on building and infrastructure. *Sustain. Cities Soc.* **2019**, *50*, 101688. [CrossRef]
58. Hong, T.; Malik, J.; Krelling, A.; O'Brien, W.; Sun, K.; Lamberts, R.; Wei, M. Ten questions concerning thermal resilience of buildings and occupants for climate adaptation. *Build. Environ.* **2023**, *244*, 110806. [CrossRef]
59. *DIN EN 15251*; Indoor Environmental Input Parameters for Design and Assessment of Energy Performance of Buildings Addressing Indoor Air Quality, Thermal Environment, Lighting and Acoustics, German Version. Deutsches Institut für Normung: Berlin, Germany, December 2012.
60. *DIN EN 16798-1*; Energy Performance of Buildings—Ventilation for Buildings—Part 1: Indoor Environmental Input Parameters for Design and Assessment of Energy Performance of Buildings Addressing Indoor Air Quality, Thermal Environment, Lighting and Acoustics—Module M1-6, German Version. Deutsches Institut für Normung: Berlin, Germany, March 2022.
61. Richter, V.; Lorenz, C.L.; Syndicus, M.; Frisch, J.; van Treeck, C. Framework for automated IFC-based thermal comfort analysis based on IFC model maturity. In Proceedings of the Building Simulation 2023: 18th Conference of IBPSA, Shanghai, China, 4–6 September 2023 (*to be published*).
62. *2017 ASHRAE Handbook: Fundamentals*, si ed.; ASHRAE: Atlanta, GA, USA 2017.
63. *DIN EN ISO 8996*; Ergonomics of the Thermal Environment—Determination of Metabolic Rate (ISO 8996:2021); German Version. Deutsches Institut für Normung: Berlin, Germany, October 2022.
64. *DIN EN ISO 9920*: Ergonomics of the Thermal Environment—Estimation of Thermal Insulation and Water Vapour Resistance of a Clothing Ensemble (ISO 9920:2007, Corrected version 2008-11-01); German Version. Deutsches Institut für Normung: Berlin, Germany, October 2009.

65. Zhang, N.; Cao, B.; Zhu, Y. An effective method to determine bedding system insulation based on measured data. *Build. Simul.* **2023**, *16*, 121–132. [CrossRef]
66. EC-Earth Consortium. EC-Earth-Consortium EC-Earth3 model output prepared for CMIP6 CMIP. [CrossRef]
67. Mitra, D.; Steinmetz, N.; Chu, Y.; Cetin, K.S. Typical occupancy profiles and behaviors in residential buildings in the United States. *Energy Build.* **2020**, *210*, 109713. [CrossRef]
68. Elnagar, E.; Gendebien, S.; Georges, E.; Berardi, U.; Doutreloup, S.; Lemort, V. Framework to assess climate change impact on heating and cooling energy demands in building stock: A case study of Belgium in 2050 and 2100. *Energy Build.* **2023**, *298*, 113547. [CrossRef]

applied sciences

MDPI

Article

Improving Thermo-Energetic Consumption of Medical Center in Mexican Hot–Humid Climate Region: Case Study of San Francisco de Campeche, Mexico

Oscar May Tzuc [1,*], Gerardo Peña López [1,2], Mauricio Huchin Miss [1], Juan Edgar Andrade Durán [1], Jorge J. Chan González [1], Francisco Lezama Zárraga [1] and Mario Jiménez Torres [1,3,*]

1 Facultad de Ingeniería, Universidad Autónoma de Campeche, Campus V, Av. Humberto Lanz Cárdenas Col. Ex Hacienda Kalá, San Francisco de Campeche ZIP 24085, Campeche, Mexico; al027290@uacam.mx (G.P.L.); mihuchim@uacam.mx (M.H.M.); jeandrad@uacam.mx (J.E.A.D.); jorjchan@uacam.mx (J.J.C.G.); frlezama@uacam.mx (F.L.Z.)
2 Departamento de Robótica Computacional, Universidad Politécnica de Yucatán, Km. 4.5 Carretera Mérida-Tetiz Tablaje Catastral 7193, Ucú ZIP 97357, Yucatán, Mexico
3 Facultad de Ingeniería, Universidad Autónoma de Yucatán, Av. Industrias No Contaminantes por Anillo Periférico Norte, Apdo. Postal 150, Mérida ZIP 97310, Yucatán, Mexico
* Correspondence: oscajmay@uacam.mx (O.M.T.); majimene@uacam.mx (M.J.T.)

Abstract: An assessment of the thermal refurbishment of an outpatient medical center in a tropical location, such as the City of San Francisco de Campeche, was presented with the aim to diminish its energy consumption. A year-long energy audit of the facility was conducted to formulate and validate a numerical simulation model while scrutinizing enhancement strategies. The examined improvement alternatives encompass passive adjustments to the roof (utilizing insulating materials, applying reflective coatings, and installing a green roof), modifications to active systems incorporating inverter technology, and alterations to the walls via reflective paint. The outcomes of the simulated enhancement scenarios were assessed utilizing energy, environmental, and economic metrics: key performance index (KPI), equivalent CO_2 emission index (CEI), and net savings (NS). These results were subsequently juxtaposed against TOPSIS decision-making algorithms to ascertain the alternative that optimally balances the three options. It was identified that using reflective paint on the roof provides the best energy benefits and contributes to mitigating emissions from electricity use. Furthermore, combining this passive technology with the integration of inverter air conditioning systems offers the best economic return at the end of 15 years. For its part, the TOPSIS method indicated that by prioritizing the financial aspect, the reflective coating on the roof combined with inverter air conditioning is enough. However, adding a wall with insulating paint brings environmental and energy benefits. The results of this work serve as a starting point for the analysis of other post-occupied buildings in the region and others under tropical climatic conditions.

Keywords: energy efficiency; medical building; tropical climate; thermal restauration

Citation: May Tzuc, O.; Peña López, G.; Huchin Miss, M.; Andrade Durán, J.E.; Chan González, J.J.; Lezama Zárraga, F.; Jiménez Torres, M. Improving Thermo-Energetic Consumption of Medical Center in Mexican Hot–Humid Climate Region: Case Study of San Francisco de Campeche, Mexico. *Appl. Sci.* **2023**, *13*, 12444. https://doi.org/10.3390/app132212444

Academic Editors: David Bienvenido Huertas and Daniel Sánchez-García

Received: 26 September 2023
Revised: 13 October 2023
Accepted: 13 October 2023
Published: 17 November 2023

1. Introduction

Global energy consumption has increased in the last 40 years by over 100%, meaning that society and governments have focused on energy efficiency and sustainability to counteract its negative effects [1,2]. The primary goal is to avoid the exhaustion of energy resources, environmental issues, and increase in energy prices; however, urbanization affects this target [3]. According to the United Nations, more than 60% of the population will reside in urban areas by 2050 [4], with buildings being one of the most critical sectors in energy consumption. So, well-designed buildings are crucial for controlling current and future electrical consumption and pollutant emissions.

Buildings represent more than 30% of global energy consumption and are responsible for 27% of CO_2 emissions into the environment [5]. According to multiple analyses, this behavior will increase the products of climate change [6]. This annual requirement is primarily for cooling, heating, ventilation, and lighting [7]. Energy management can reduce energy consumption in facilities by up to 20% and impact total energy demand [8]. However, in predicting energy consumption, there are various challenges, such as ambient temperature, area and type of building, and age of the construction, to mention a few [9,10]. In this context, health centers are important, since their operation requires an efficient and high-quality service that must be met in indoor conditions that guarantee thermal comfort. In addition, health standards are increasingly regulated and have a high impact on this type of building. Meeting these requirements carries a significant energy demand that has been overlooked for many years.

Medical centers have the highest energy consumption among the leading commercial buildings; their energy requirement is allocated to air conditioning, medical equipment, and maintenance [11]. Various studies have shown that 40% of consumption in medical centers comes from air conditioning systems [12]. This consumption is estimated to be worse in countries with warm and tropical climates, characterized by high temperatures increasing the electrical demand for indoor cooling [13]. In this context, in Mexico, the annual consumption in medical centers was 3698 MJ/m^2, where 80% of the annual demand comes from regions with warm and humid climates [14]. Furthermore, according to some studies, the public sector might present an underestimation issue [14,15]. This consumption is the product of many factors, but one of the most important is the inefficient physical state of the building and inefficient equipment; for this reason, it is imperative to study different strategies that help reduce operating costs [12].

Sun, Y. et al. [11] carried out a comprehensive evaluation through satisfaction questionnaires of users for a medical center in China, improving energy efficiency. Seo et al. [16] studied the impact of modifying the building envelope using green retrofits in South Korea. Their results demonstrated that this technique improves heating systems' energy efficiency by 30%. As and Bilir [17] focused on energy efficiency and CO_2 mitigation in a hospital building, evaluating insulation materials, building orientation, lighting efficiency and window-to-wall ratios, obtaining energy consumption reductions by 50%. Similar analyses have been reported for hospitals in Spain [18], Italy [19], and Australia [20], to name a few.

The bibliographic review in this paper establishes the importance of analyzing medical centers' energy consumption and prediction. The works presented in the bibliographic review focus on modifying operating parameters, HVAC system technologies, and modifying the building envelope. Furthermore, few works address this topic for regions with a tropical climate and coastal cities, focusing mainly on temperate or Mediterranean climates. No studies focused on the thermo-energetic behavior of medical centers in Mexico, specifically in regions with a tropical climate. That said, this study analyzes and identifies alternatives for thermo-energy improvements in an outpatient medical center located in a coastal city with a tropical climate, such as San Francisco de Campeche. This work's main objective is to identify alternatives that reduce the energy and environmental impacts of a building that operates under the real conditions of medical activity and represents a viable investment alternative in the medium term. The results represent the first approach to thermal restoration for buildings in the region linked to economic, environmental, and energy aspects. In addition, it provides an overview of actions that can be taken to make thermal restoration more attractive.

The rest of the work is divided into four sections: The second section addresses the description of the study building, the characteristics of the construction materials that compose it, and the electrical consumption equipment and operational patterns of the medical center. The third section addresses the computational methodology design for modeling, calibration, and thermo-energetic restoration of the medical center based on passive and active technologies. The fourth section presents the results of the energy, environmental, and economic analysis for the thermal improvement of the building, as well as the selection of the best option that combines the three elements. Finally, section five contains the conclusions and discussion of the work.

2. Case Study Location and Building Data

2.1. Description of the Case Study Location

The case study corresponds to an ambulatory medical center for ophthalmological care located in the historical center of San Francisco de Campeche, Mexico (19°50′41″ N, 90°32′23″ W). The city's zone is a completely urbanized area (with practically no natural vegetation) comprised of masonry buildings and concrete streets, which has fostered the heat island phenomenon. According to the Köppen climate classification, the city is characterized by a hot–humid climate with rains in summer, and an average monthly temperature between 22 °C and 31 °C, with May being the hottest month and January the coldest [21]. The average solar radiation in the city is over 16.2 MJ/m^2-day, reaching values higher than 18 MJ/m^2-day at summer. The region is also characterized by a high percentage of relative humidity throughout the year, between 70% and 98% [21]. Figure 1 illustrates the perceived heat index in the city's historical center, where the blue dots indicate the ambient temperatures recorded for 2022. More than 60% of the year, the buildings in the region are within the margin of caution, which increases dangerous heat and extreme danger indices the closer one is to the summer. In addition, it is essential to emphasize that non-operational hours (night hours) are mainly those where it is outside of this risk zone. All of them drastically impact the thermal comfort of this zone buildings' occupants and increase energy use to counteract it. Therefore, the analysis of the building is critical because it forms part of a compendium of facilities that currently perform functions for which and are susceptible to thermal restorations.

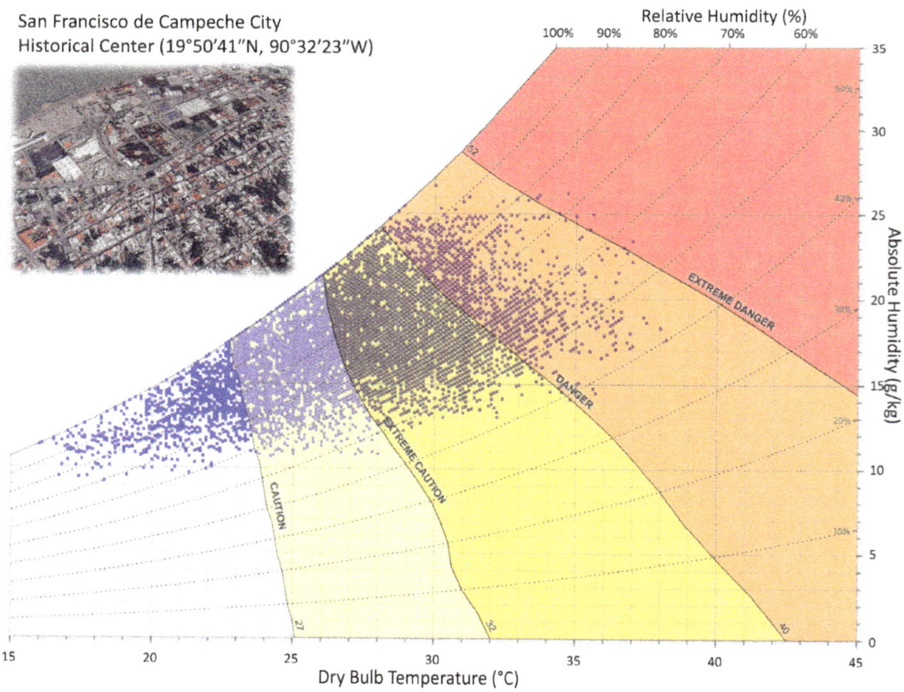

Figure 1. Psychrometric chart with the heat indices reported in the historical center of San Francisco de Campeche, Mexico for 2022.

2.2. Building Information: Outpatient Medical Center

The medical center is a single-story building facing south, divided into 11 zones with 127.63 m^2 surface construction and a volume of 521.18 m^3. A detailed description of the building's dimensions is found in Figure 2. The medical center is mainly built from masonry stone with plaster on the exterior and interior sides, a cast concrete ceiling, door and windows frames made from wood, and windows made of 3 mm clear glass (Table 1). The building is reconditioned to carry out medical consultations and specialized ophthalmological surgeries. It operates Monday to Friday from 9:00 a.m. to 1:00 p.m. and 5:00 p.m. to 9:00 p.m.; and Saturdays from 9:00 a.m. to 4:00 p.m. The facilities operate with five employees in the consulting and optical zones and six employees in the surgery zone. The medical center has an average flow of 60 to 80 monthly patients for the consultation zone and an additional 80 people who come to the operation zones. It is important to emphasize that the age of the building implies the absence of regulatory standards regarding energy saving during its construction.

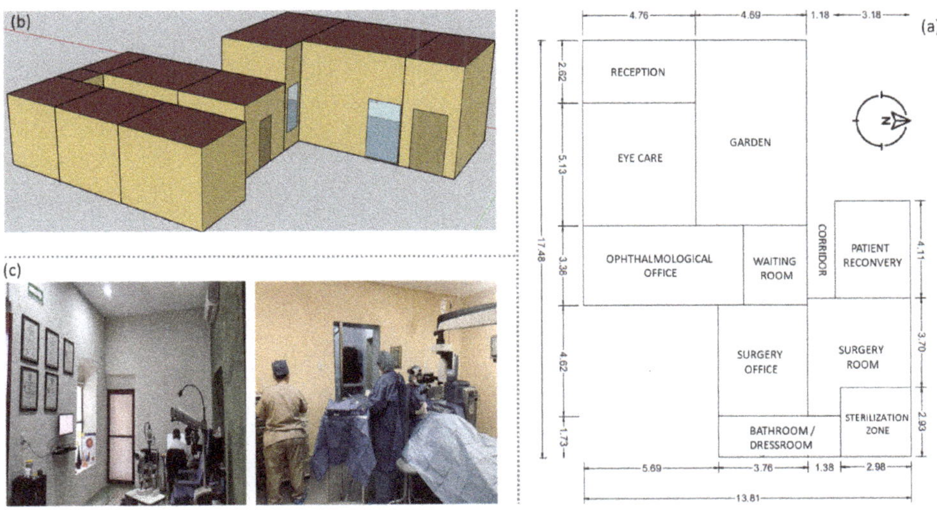

Figure 2. Dimensions and structure of the medical center: (**a**) floor plant; (**b**) computational 3D building perspective; (**c**) indoor activities of daily occupation.

Table 1. Medical center envelope elements' physical properties.

Envelope	U-Value (W/(m² K))	Total Solar Transmission (%)	Light Transmission (%)	Materials	Width (m)	Thermal Conductivity (W/(m-K))	Specific Heat (J/(kg-K))	Density (kg/m³)
Outdoor walls	1.773	Na	Na	Outdoor gypsum plaster	0.01	0.4	780	1860
				Cast concrete	0.015	0.9	980	1860
				Concrete block	0.25	0.8	1000	700
				Indoor gypsum plaster	0.02	0.5	1000	1300
Indoor walls	1.628	Na	Na	Gypsum plaster	0.01	0.4	780	1860
				Cast concrete	0.015	0.9	980	1860
				Concrete block	0.25	0.8	1000	700
Roof	1.432	Na	Na	Outdoor gypsum plaster	0.01	0.4	780	1860
				Ready-mixed concrete	0.12	0.9	900	2100
				Concrete block	0.18	0.5	1000	1400
				Indoor gypsum plaster	0.02	0.5	1000	1300
Doors	2.111	Na	Na	Wood	0.045	0.2	1760	357

Table 1. *Cont.*

Envelope	U-Value (W/(m² K))	Total Solar Transmission (%)	Light Transmission (%)	Materials	Width (m)	Thermal Conductivity (W/(m-K))	Specific Heat (J/(kg-K))	Density (kg/m³)
Windows	5.894	79.3	88.1	Clear simple glass	0.003	0.96	800	2500
Surface material		**Solar reflectance (%)**		**Thermal absorbance (%)**	**Solar absorbance (%)**		**Solar absorbance (%)**	
gypsum plaster		30		90	70		70	

An energy audit was conducted to identify the electricity devices with significant demand and establish the building's energy consumption baseline. Audit data was collected on the number of devices and operating hours of lighting equipment, HVAC systems, office equipment, and specialized medical practice systems. For the present study, HVAC systems only cover air conditioning devices, as due to regional climatic characteristics, the heating requirement for thermal comfort is null [22]. The energy audit information was collected through surveys carried out on staff, installed equipment datasheets, and measurements. Electrical data was captured by an AEMC 3945-B PowerGrid analyzer; it is a class S device designed to measure power frequency, magnitude of supplied voltage, voltage unbalance, interruptions, and electrical rapid changes with 5% precision concerning the nominal voltage. The measurement period ranged from January to December 2022. Energy audit results are presented in Table 2, which describes the electrical loads between the 11 zones of the ambulatory medical center, such as the energy final use of electricity. According to the data collected, higher electrical energy consumption is associated with using air conditioning to cool the spaces that patients and staff frequent to a greater extent. These results are consistent with the literature, where it is reported that air conditioning equipment represents about 66% of electricity consumption in regions with hot–subhumid or similar climates [23]. In facilities for the care of patients, health, or other small-sized related matters, it has been verified that the energy consumption for air conditioning is around 60% [24].

Table 2. Installed loads and electrical energy consumption of the medical center based on the energy audit.

Zone	Width (m)	Length (m)	High (m)	Electric Devices	Installed Electrical Loads (kW)			
					Lightning	Air Conditioning	Medical Equipment	Office Equipment
Ophthalmological office	3.38	6.69	4.60	11 plugs, 6 LED lamps 20 W, television, computer and printer, Minisplit	0.12	1.22	-	
Eye care	5.13	4.76	4.73	6 plugs, 7 LED lamps 20 W, computer and printer, fridge, Minisplit	0.14	1.28	-	
Reception	2.62	4.76	4.73	4 plugs, 6 LED lamps 20 W	0.12	-	-	-

Table 2. *Cont.*

Zone	Width (m)	Length (m)	High (m)	Electric Devices	Installed Electrical Loads (kW)			
					Lightning	Air Conditioning	Medical Equipment	Office Equipment
Patient recovery	4.11	3.00	3.53	10 plugs, 3 T8 lamps 32 W, Minisplit	0.10	1.24	-	-
Surgery room	3.70	4.36	3.53	14 plugs, 4 T8 lamps 32 W, Minisplit 2.52 kW, 3 medical equipment	0.13	2.52		-
Sterilization zone	2.93	2.98	3.53	8 plugs, 2 LED lamps 9 W, fridge, 2 medical equipment	0.02	-		-
Dressing room /bathroom	1.73	5.14	3.53	2 plugs, 2 LED lamps 9 W	0.02	-	-	-
Corridor and cellar	-	-	3.53	5 plugs, 10 LED lamps 9 W	0.12	-	-	-
Waiting room	3.38	2.76	3.63	2 T8 lamps 32 W, Minisplit	0.07	1.22	-	-
Surgery office	3.38	3.76	3.63	26 plugs, 2 T8 lamps 32 W, 2 medical equipment, computer, Minisplit	0.07	1.22		
Outside space	7.75	4.69	-	T8 lamp 32 W	0.03	-	-	-

Energy end use	Installed electrical loads		Average monthly electricity consumption	
	(kW)	(%)	(MJ)	(%)
Lightning	0.89	7.2	251.6	5.18
Air conditioning	8.69	70.3	3065.4	63.05
Office equipment	1.913	15.5	1510.2	31.06
Medical equipment	0.87	7.0	34.5	0.71

For its part, Figure 3 illustrates the baseline of electricity consumption and the final use of energy measured throughout 2022 for the outpatient medical center. According to Figure 3a, the highest energy consumption occurs from May to October, exceeding 1.4 MWh per month. This is associated with the summer season and the heatwave period, forcing an increase in the use of air conditioning to counteract thermal stress. However, it is essential to emphasize that in more than 60% of the year, the 1.2 MWh per month is easily exceeded, of which the eye care and ophthalmologic office represent about 76% and 71% in air conditioning and lighting consumption, respectively. Finally, in environmental matters, the baseline associated with said energy consumption is around half a ton of monthly average CO_2 equivalent emissions.

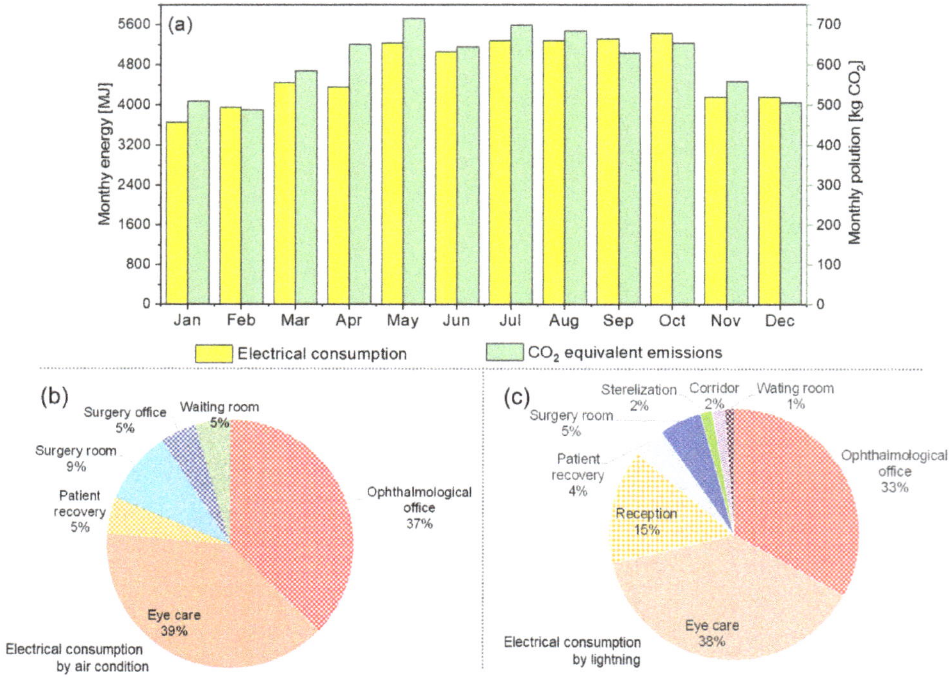

Figure 3. Summary of the results of the energy audit for the ambulatory medical center: (**a**) monthly baseline of energy consumption and equivalent CO_2 emissions, (**b**) electricity consumption associated with air conditioning equipment, (**c**) electricity consumption associated with lighting equipment.

3. Computational Methodology

Figure 4 presents the simulation process to identify the merge of passive and active technologies to improve the building's energy performance. This methodological approach is comprised of six stages. The first phase consists of developing a 3D model based on the construction properties of the medical center and the information gathered from the energy audits. Subsequently, a calibration stage is executed where the model's electric consumption simulations are contrasted with the grid analyzer measurements. Once medical center calibration was carried out, passive building technologies were evaluated to improve the building's energy performance. In the same way, active systems are integrated with high-efficiency air conditioning to maximize energy savings. At this stage, various configurations were analyzed, looking for the optimal combination for the location. Finally, in the fifth and last stage, the energy performance index (according to the ISO 50001:2018 standard) pollution mitigation, and the techno-economic feasibility of investment are contrasted through a decision-making tool to quantitatively identify the trade-off where the three indices converge. A more detailed description of the six phases is provided in the following subsections.

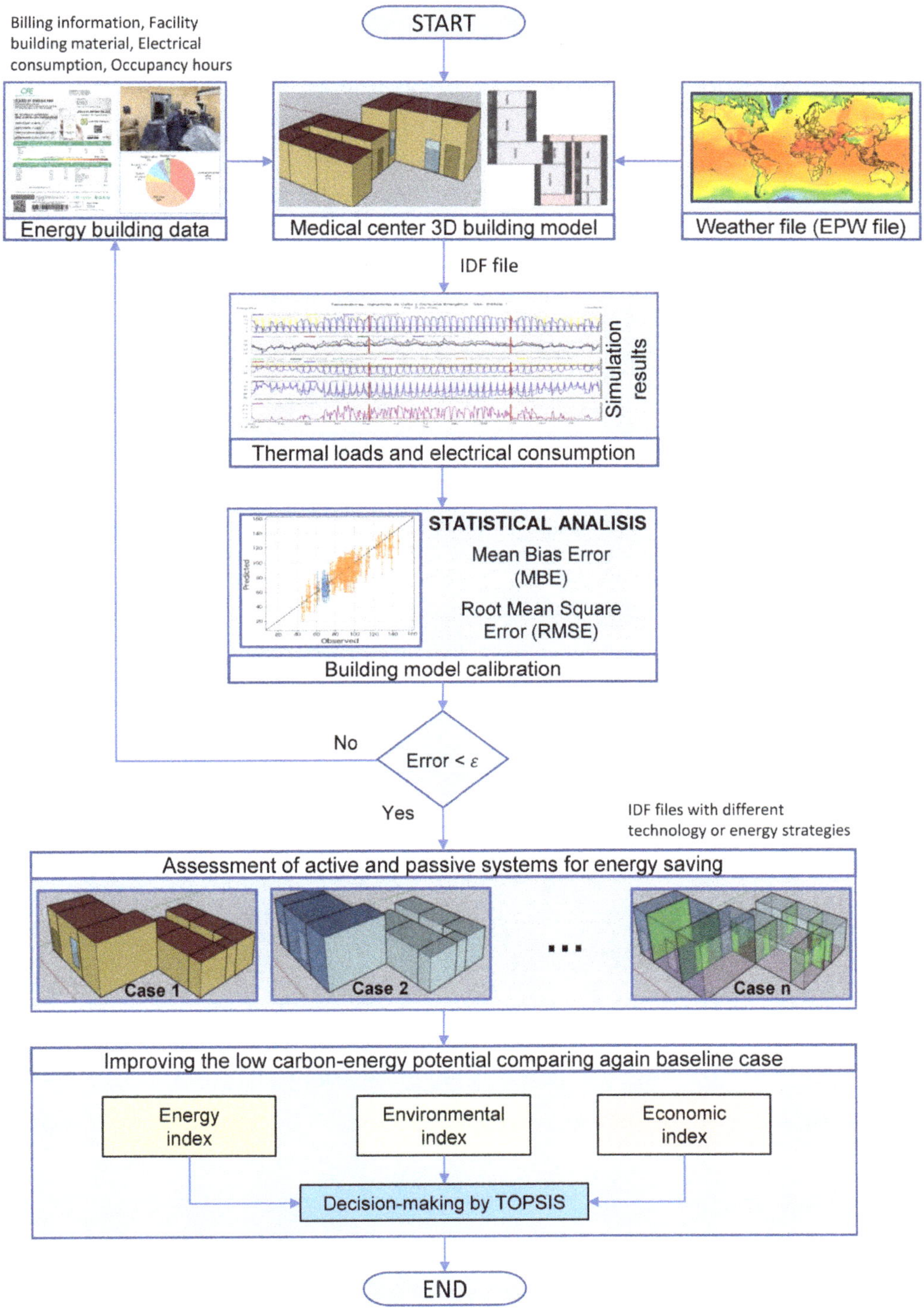

Figure 4. Computational methodology implemented to improve thermal performance in a medical center under hot-humid climate conditions.

3.1. Building Modeling

To carry out the energy analysis, a 3D model of the medical center was built aided by the simulation software Open-Studio(v. 3.4) and Energy-plus(v. 9,4). These are based on BIM methodology, specialized in predicting the indoor environment and energetic behavior of the building through heat transfer calculations and assumptions. They are designed for energy improvement, from advanced envelope strategies to developing air conditioning systems. For this purpose, the software calculates the thermal loads of the built environment and the thermal/electrical load of operation systems by using a transient heat balance [25].

For simulation, the software requires data of the building materials' thermophysical properties, schedule operation, characteristics of HVAC and lighting systems, and the weather information of the location. The model of the medical center was developed using the envelope materials properties reported in Table 1. Regarding electrical devices, it was given the information collected by energetic audits. Similarly, based on the flow of people and patients served by the medical center (described in Section 2.2), information on the occupancy percentages and hours of activity of the building linked to energy loads was programmed, as is presented in Table 3. The configuration of the HVAC systems was adjusted to a set point of 22 °C under current regulations in hospitals and health centers [26]. Finally, the climatic file used in the study comprises hourly meteorological data for 2022 obtained through the software METEONORM (v. 8.1.1) [27].

Table 3. Percentage distribution of loads in the process of energy simulation.

Loads		**Period**				
		Weekdays				
Occupancy	Schedule	00:00–7:59 h	8:00–13:59 h	14:00–16:59 h	17:00–20:59 h	21:00–23:59 h
	Percentage	0%	100%	0%	100%	0%
		Weekend (Saturday)				
	Schedule	00:00–7:59 h	8:00–15:59 h	16:00–23:59 h		
	Percentage	0%	100%	0%		
		Weekdays				
Equipment and lighting	Schedule	00:00–7:59 h	8:00–13:59 h	14:00–16:59 h	17:00–20:59 h	21:00–23:59 h
	Percentage	0%	100%	0%	100%	0%
		Weekend (Saturday)				
	Schedule	00:00–7:59 h	8:00–15:59 h	16:00–23:59 h		
	Percentage	0%	100%	0%		
		Weekdays				
Natural Ventilation	Schedule	00:00–7:59 h	8:00–13:59 h	14:00–16:59 h	17:00–20:59 h	21:00–23:59 h
	Percentage	100%	10%	10%	10%	100%
		Weekend (Saturday)				
	Schedule	00:00–7:59 h	8:00–15:59 h	16:00–17:59 h	18:00–23:59 h	
	Percentage	100%	10%	30%	0%	

3.2. Building Model Calibration

Since building energy models are complex and comprise a large number of input data, it is necessary to carry out calibration processes to corroborate and verify that they represent the actual behavior and performance of the building. The literature reports two statistical approaches commonly used to assess the quality of model calibration: the mean bias error (MBE) and the coefficient of variation of the root mean square error (Cv(RMSE)) [28].

MBE measures the overall systematic error of the building energy model and is defined as follows:

$$MBE = \frac{\sum_{i-1}^{Np}(m_i - s_i)}{\sum_{i-1}^{Np}(m_i)} \quad (1)$$

where m_i and s_i represent the measured and simulated energy data, respectively; with Np as the number of measurements analyzed. The acceptable ranges of MBE for analysis with monthly data are between 0–5%, while for hourly or daily data the tolerance is up to 10% [29]. On the other hand, the CV(RMSE) is used as a complement to prevent calibration errors due to the compensation between positive and negative values used in the MBE. The CV(RMSE) is computed following the equation:

$$CV(RMSE) = \frac{\sqrt{\sum_{i-1}^{Np}(m_i - s_i)^2/Np}}{\sum_{i-1}^{Np} m_i/Np} \quad (2)$$

According to ASHRAE and the International Performance Measurement and Verification Protocol (IPMVP), the acceptable range for CV(RMSE) using monthly data is from 0 to 15%, while hourly and daily data this extends up to 30% [29].

Figure 5 illustrates the results of the outpatient medical center model calibration concerning the cumulative monthly energy consumption over a year. In general, the overall systematic error of building model (MBE) is close to 1%, and the CV(RMSE) is 7.75%. This implies that the model balances well between overestimation and underestimation, and the CV(RMSE) confirms acceptable uncertainty range below the limit, indicating a good accuracy according to the regulations. It is essential to highlight that the largest monthly difference occurs in May, without exceeding 10%, and that, on average, the monthly differences of the model do not exceed 6%. Furthermore, the differences presented in the warmer months overestimate the monthly demand when analyzing the integration of passive technologies.

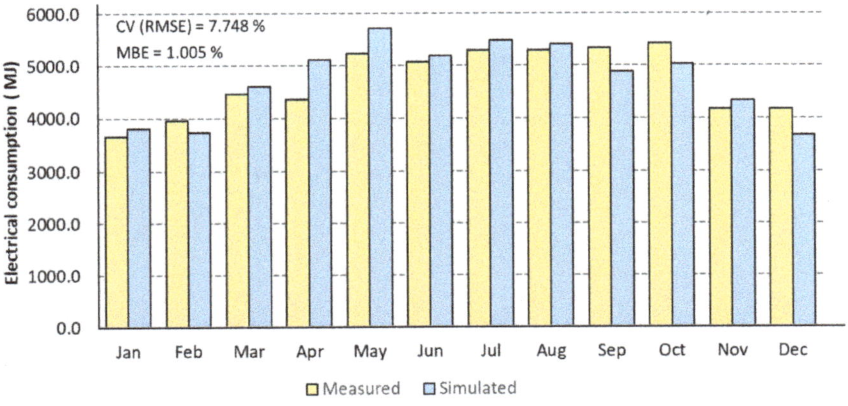

Figure 5. Comparison between measured electricity consumption data and calibrated results from the outpatient medical center.

3.3. Energy Efficiency Proposal Using Passive and Active Technologies

The integration of passive and active technologies search reduces CO_2 emissions and improve the medical center's electrical energy performance. In the case of passive systems, the literature reports various strategies to improve the energy efficiency of buildings under hot or tropical climate conditions [30]. In this work, we opted for assess thermal insulation panels, reflective coating, and green roofs because they are readily available materials in the region, require few or no structural modifications, and thermally have demonstrated advantages of use at high temperatures.

Thermal insulation using construction material was evaluated considering the expanded polystyrene (EPS) panel, a light, rigid, and formable thermoplastic. These panels are installed on the roof using plugs or adhesives and are protected with a fiberglass mesh covered with plaster. Although the labor of this material is not expensive, the price of the panels and masonry work considerably increases the initial investment cost [31]. In the case of reflective coatings, these are special paint (usually white) designed to be applied on the envelope and increase the reflectance of the roof and walls. This represents the most direct way to reduce incident solar energy and is considered the most straightforward passive measure because most of these coatings can be installed the same way as regular paint, which implies reduced implementation and labor costs [32]. In the case of green roofs, the structure consists in vegetation, soil layer, filter and drain membrane, root barrier and waterproof layer (Figure 6) The vegetation type affects the leaf area index (LAI), leaf reflectivity, leaf emissivity, and stomatal resistance [33,34]. For this study, the green roof analyzed is one of the most applied in Mexico, considering a LAI of 3.75, the reflectivity of 0.25, emissivity of 0.9, and a stomatal resistance of 175 [34].

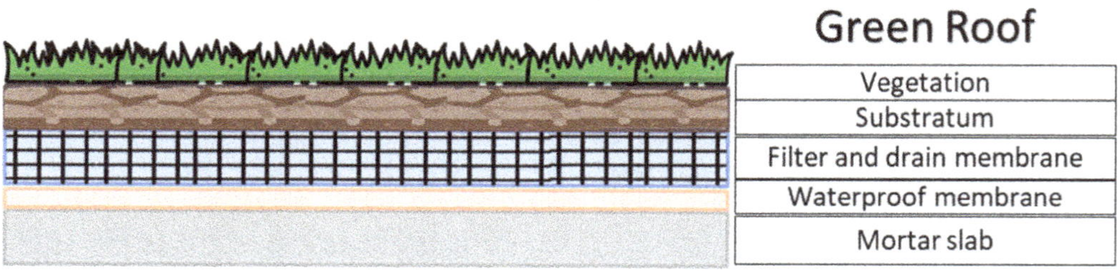

Figure 6. Typical structure of green roof layers.

Table 4 summarizes the optical and thermophysical properties of the materials considered, as well as their market and installation prices. The three materials were assessed on the calibrated model considering their application on the roof (and the exterior walls for the cases of reflective coating). Regarding active strategies, replacing conventional air conditioning equipment (on–off technology) with inverter technology was considered an energy efficiency measure. Based on this, the improvement in energy performance and economic savings were evaluated. This strategy was combined with the passive alternatives to identify the best cost–energy choice. Table 4 also contains detailed information on the current and replacement devices and their respective technological replacement costs.

Table 4. Passive and active strategies used to improve building energy efficiency and achieve thermal restauration.

Passive Thermal Restoration Technologies								
Proposed Envelope Material	**U-Value (W/(m² K))**	**Width (m)**	**Conductivity (W/(m-K))**	**Specific Heat (J/(kg-K))**	**Reflectance (%)**	**Annual Maintenance**	**Unitary Price [a] (MXN/m²)**	**Ref.**
White reflective coating	1.278	0.025	0.40	780	90	-	160	[35]
Expanded polystyrene	1.760	0.050	0.04	1400	-	-	469	[31]
Green roof	1.161	0.200	0.30	1000	25 [b]	1200 MXN	550	[36]
Active thermal restoration technologies								
Building zone	**TOR [c]**	**A/C Power (kW)**	**Average consumption (MJ/month)**	**TOR [c]**	**A/C Power (kW)**	**Average consumption (MJ/month)**	**Electrical device cost (MXN)**	**Maintenance cost (MXN/year)**
	Current conditions			Technological replacement				
Ophthalmological office	1.0	1.22	839	1.5	1.7	774	7500.00	1400.00
Eye care	1.0	1.28	882	1.5	1.71	779	7500.00	1400.00
Patient recovery	1.0	1.24	246	1.0	1.18	155	6000.00	1400.00
Surgery room	2.0	2.52	501	2.0	2.4	315	9900.00	1400.00
Surgery office	1.0	1.22	242	1.5	1.72	226	7500.00	1400.00
Waiting room	1.0	1.22	242	1.0	1.16	152	6000.00	1400.00

[a] Including installation cost; [b] Considering a LAI of 3.75, emissivity of 0.9, and a stomatal resistance of 175; [c] TOR: Tons of refrigeration.

The passive and active technologies were assessed systematically, dividing them into lots containing the thermal restoration scenarios. Lot A comprises restoration scenarios with passive modifications to the roof (insulating material, reflective coating, and green roof) or reflective paint on the walls. Lot B includes passive changes in the roof combined with efficient air conditioning systems (inverters). Finally, in Lot C, the reflective coating on the walls is added to the active and passive modifications. Table 5 enlists the 15 thermal restoration scenarios, whereas Figure 7 depicts the layer distribution of material of the 5 essential changes in the envelope from which the scenarios are derived. In the green roof, simulation is important considerer the irrigation process [37]; however, energy plus software does not have enough information to model that process. To consider the effect of irrigation, the model contemplates a scheduled irrigation once a month with an irrigation rate of 0.3 m/h, according to similar studies [36].

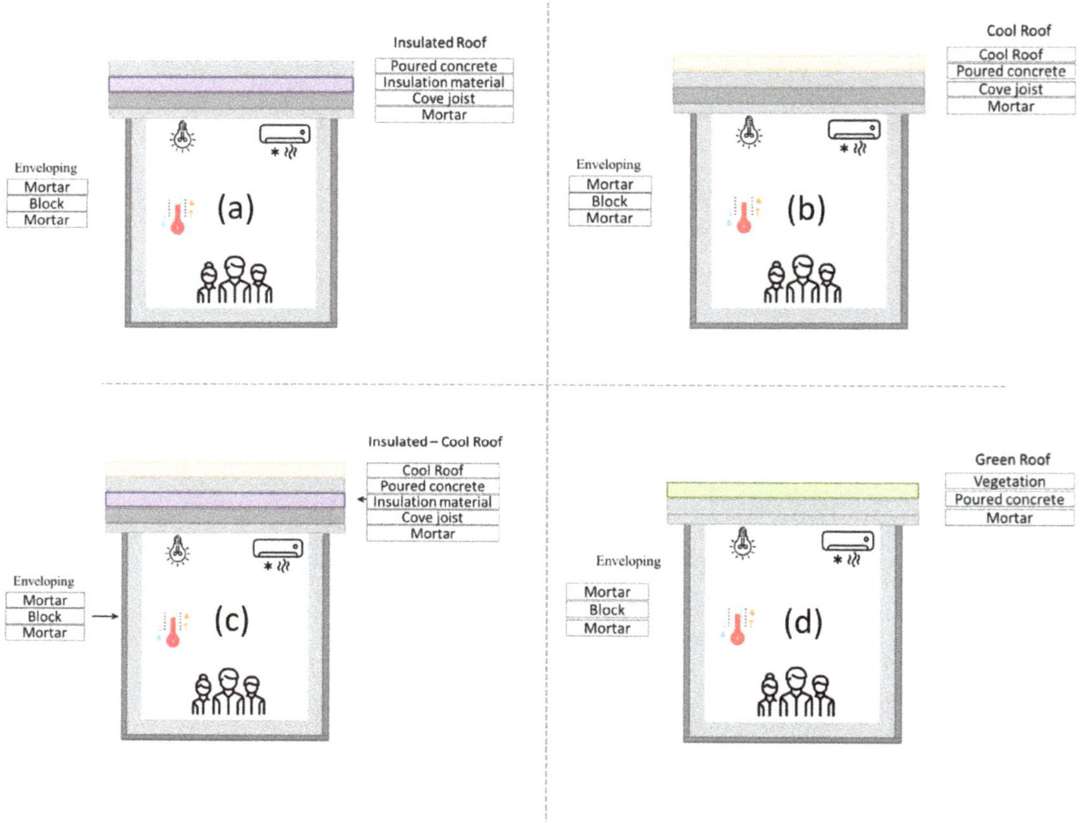

Figure 7. Description of passive modifications implemented in the envelope of the medical center: (**a**) roof with polyurethane insulation; (**b**) roof with reflective coating (This is the same material implemented in the case of the walls); (**c**) roof with polyurethane insulation and reflective coationg; (**d**) green roof.

Table 5. List of passive and active strategies to reduce energy consumption in the outpatient medical center.

Lot	Scenarios	Passive Systems		Active System
		Roof	Walls	
A	A-I	polyurethane insulation	None	Conventional air conditioner
	A-II	reflective coating	None	Conventional air conditioner
	A-III	polyurethane insulation/reflective coating	None	Conventional air conditioner
	A-IV	green roof	None	Conventional air conditioner
	A-V	None	Reflective coating	Conventional air conditioner
B	B-I	None	None	Inverter air conditioner
	B-II	polyurethane insulation	None	Inverter air conditioner
	B-III	reflective coating	None	Inverter air conditioner
	B-IV	polyurethane insulation/reflective coating	None	Inverter air conditioner
	B-V	green roof	None	Inverter air conditioner
	B-VI	None	Reflective coating	Inverter air conditioner
C	C-I	polyurethane insulation	Reflective coating	Inverter air conditioner
	C-II	reflective coating	Reflective coating	Inverter air conditioner
	C-III	polyurethane insulation/reflective coating	Reflective coating	Inverter air conditioner
	C-IV	green roof	Reflective coating	Inverter air conditioner

3.4. Energy, Environmental, and Economic Performance Indices

The improvement in building energy performance was evaluated based on energy baseline in conjunction with the key performance indicators for energy (KPI) determined by the ISO 50001:2018 standard [38], which is given as:

$$KPI = \frac{\sum_{i=1}^{a} E_{m,i}}{\sum_{i=1}^{a} E_{s,i}} \tag{3}$$

where E_m represents the actual energy consumption measured in the medical center (corresponding to the baseline) and E_s the results of electrical energy consumption estimated from the computer simulation. The comparison period is usually conducted over a month with daily or hourly "a" values. If the KPI < 1, it means that in said month, technologies implementations improve energy performance regarding the energy baseline. On the other hand, values above unity indicate that the implemented strategies are counterproductive.

Regarding the reduction in environmental impact, the carbon equivalent emissions index (CEI) was used [39]. For this work, this was defined as the monthly polluting agents emitted to the atmosphere (translated into kilograms of CO_2) derived from electrical consumption to satisfy the medical center's operational and thermal comfort needs:

$$CEI = \sum_{i=1}^{a} E_{x,i} EF \tag{4}$$

where subscript x represents the building electrical energy consumption either for the baseline (E_m) or for the energy improvement simulations (E_s). In the same way, EF is defined as the emission factor of the electric power supplied by the Mexican federal government. The emission factor (EF) value used in this study is reported in Table 6, with information obtained from the Mexican Energy Regulatory Commission (CRE).

Table 6. Costs, energy, and environmental assumptions for performance indices of the medical center.

Parameter	Value	Unit	Reference
Electricity price	1.07	MXN /MJ	[40]
Emission factor	0.012	kg CO_2/MJ	[41]
Electricity inflation rate	4.0	%	[42]
Annual discount rate	1.0–20.0	%	[43]
Maintenance annual increase rate	6.0	%	[42]
Life cycle	15	years	[31]

Finally, the cost-effectiveness of energy efficiency strategies adopted was evaluated by quantifying the economic indicators: net saving (NS). It estimates the economic benefit considering the evolution of savings and costs over time. According to Hernández-Pérez et al. [35], NS is given by:

$$NS = -I_O + \sum_{t=1}^{N} \frac{S_t(1+e_s)^t}{(1+d)^t} - \sum_{t=1}^{N} \frac{C_t(1+e_c)^t}{(1+d)^t} \quad (5)$$

where I_O is the initial investment, S_t is the saving difference in the year t, C_t is the cost difference in the year t, N is the number of years in the life cycle, d represents the discount rate, and e_s and e_c are the annual inflation rate for saving and cost, respectively. Positive values of NS indicate a most cost-effective strategy, while negative values economically favor the energy baseline. Table 6 summarizes the information on electrical and economic parameters required for the implementation of both indicators, while investment cost is described in the passive and active technologies presented in Table 4.

3.5. Selection of Best Alternative Based on Decision-Making Algorithm

A numerical tool for decision making was implemented to identify the trade-off that ensures the best overall performance of the environmental, energy, and economic aspects. For this purpose, the Technique for Order of Preference by Similarity to Ideal Solution algorithm, known as TOPSIS, was used. It works by choosing the best alternative based on the shortest and furthest Euclidean distances from the positive (v_j^+) and negative (v_j^-) ideal solutions, respectively. The ideal solution is represented by the maximum value of the proximity coefficient (CC) [44].

$$\max\left[CC_j\right] = \frac{\sqrt{\sum_{j=1}^{k}\left(v_{ij}-v_j^+\right)^2}}{\sqrt{\sum_{j=1}^{k}\left(v_{ij}-v_j^+\right)^2} - \sqrt{\sum_{j=1}^{k}\left(v_{ij}-v_j^-\right)^2}} \quad (6)$$

For this purpose, the TOPSIS algorithm requires identifying whether the variables will be improved by increasing (+) or reducing (−); also, the degree of priority of the variable needs to be indicated, where the sum of these must equal 100%. In this work, the KPI, CEI, and NS indicators were considered as the variables to be analyzed, forming a 15 × 3 decision matrix. It was instructed that the best results aim to reduce the KPI and CEI and increase the NS. The evaluation was carried out by varying the priority percentage of each variable from 80 to 50% in steps of 10%. The calculations were implemented in the Python programming environment through the PyTOPS tool [45].

4. Results

4.1. Improving the Energy Performance of the Outpatient Medical Center

Once it was verified that the model satisfactorily estimates the building's electrical demand and computed the desirable indices, the passive and active substitution scenarios were analyzed in three lots: Lot A, corresponding to cases where there are exclusively passive modifications to the roof or wall; Lot B, where passive modifications to the roof are combined with efficient air conditioning systems (inverters); and Lot C, where, in addition to the active and passive modifications, a reflective coating is added onto walls. Figure 8 illustrates the KPI results both monthly and annually for the three analyzed lots linked to electricity consumption.

For the scenarios with modifications exclusively on the roof or wall (Figure 8a), only applied passive technologies are feasible to reduce the medical center energy consumption between 5% and 11% for the hottest months (April–September). Green roof was detected as the least effective (annual KPI of 96.3%), followed by polyurethane insulation (annual KPI of 94.3%). In the case of selective coating on the façade, during the hottest months, its contribution is almost zero; however, it demonstrates exemplary performance in energy saving for winter days. Furthermore, the annual advantages in energy savings are not negligible, decreasing by around 9% annually. For its part, the analysis of the annual KPI indicates that reflective coating on the roof (Case A-II) and its combination with expanded polyurethane (Case A-III) are the ones that have the most significant benefit for the building, reducing electrical demand by 10.09% and 10.19%, respectively. It is important to highlight that the contribution of including the expanded polyurethane layer can represent a higher cost when modifying the envelope with an almost zero energy benefit of 0.1% annually and barely 1% for the hot month of the year. It is also important to emphasize that although the lower consumption in the warmer months is obtained in Case A-III, the monthly curve of Case A-II is more constant throughout the year, obtaining even better energy benefits for the winter season (November to February).

The scenarios in Lot B demonstrate the benefits of combining passive and active technologies. According to Figure 8b, the effect of passive technologies in the envelope is better appreciated when introducing inverter air conditioning systems. In summary, it is possible to reduce the electrical consumption of the outpatient medical center for the hottest months from 22% (Case B-I) to 31% (Case B-IV). Excluding Case B-I, which proposes using an inverter system without modifications to the envelope, again, selective coating on the walls (Case B-VI) and green roofs (Case B-V) have the least favorable results. At the same time, the reflective coating on the roof and its combination with insulating material were again the best thermal options, obtaining practically the same performances (KPI of 72.2% and 72.05%, respectively). The expanded polyurethane on the roof is benefits from using inverter technology, differing only 3.5% from the best Case. Finally, the results of Lot C (Figure 7c) show that the integration of selective coating on the walls improves the energy performance of the medical center by 2.45% concerning theirs homologous of the Lot B. The above implies that a white reflective coating provides the greatest thermo-energetic advantage to the envelope.

4.2. Environmental Pollution Mitigation Analysis

For its part, Figure 9 shows how the use of passive and active technologies reduces CO_2 equivalent emissions. The monthly results of Lot A are exhibited in Figure 9a. In the case of commercial thermal insulator (Case A-I polyurethane foam) or green roof (Case A-IV), both options are characterized by being mainly thermal barriers that increase resistance to heat flow from radiation. Therefore, their mitigation capacity depends on thickness, which is reflected in the figure where they express an intermediate mitigation level concerning the other cases of thermal restoration. The combinations with reflective paint on the roof (Case A-II and a-III) present the best performance in reducing emissions, with almost identical mitigation potential. In quantitative terms, implementing technologies based on the reflection of solar radiation incidents on the roof can represent mitigations above half a ton of CO_2 equivalent annually, between 719 kg and 725 kg, compared to the baseline.

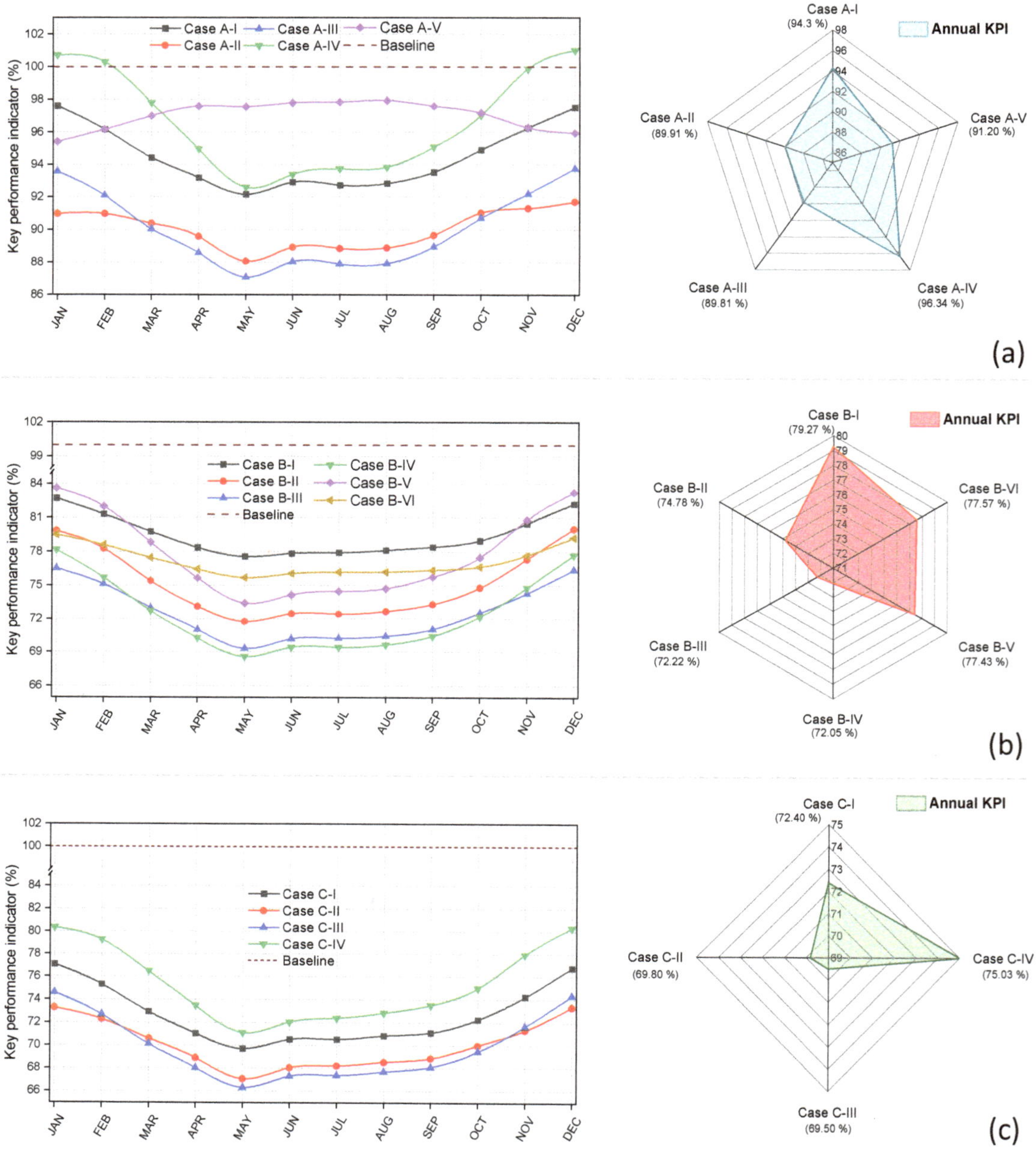

Figure 8. Monthly and annual electricity key performance index for the scenarios: (**a**) Lot A: exclusively passive modifications on the envelope; (**b**) Lot B: envelope modification combined with inverter air conditioning systems; (**c**) Lot C: active and passive modifications combined with a reflective coating on the façade.

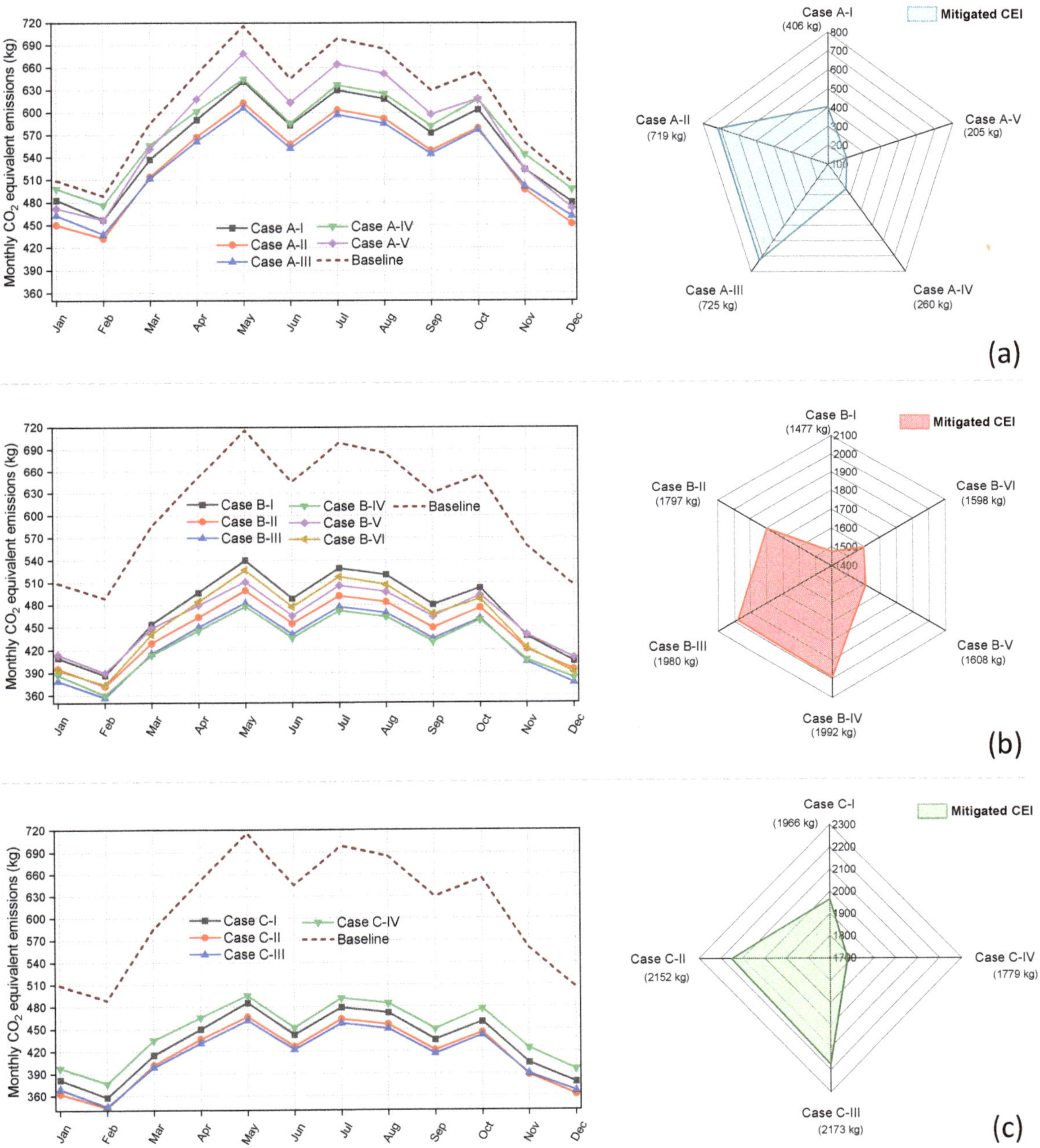

Figure 9. Monthly and annual equivalent CO_2 emission index for the scenarios: (**a**) Lot A: exclusively passive modifications on the envelope; (**b**) Lot B: envelope modification combined with inverter air conditioning systems; (**c**) Lot C: active and passive modifications combined with a reflective coating on the façade.

Regarding Lot B (Figure 9b), implementing the inverter air conditioning dramatically reduces the CEI. In the worst scenario (Case B-I), only inverter technology can mitigate nearly a ton and a half of CO_2. Under this same analysis, the integration of reflective paint on the façade or the green roof barely improves the effect achieved by the simple replacement of the active system, where although improvements in the reduction of the CEI are always desirable, it may not be economically justified to invest. For its part, by using the reflective coating on the roof with and without thermal insulation (Case B-III and B-IV, respectively), the mitigation potential can be improved by up to 26%, reaching close to two tons of reduction in CO_2 equivalent per year. This implies the impact that selective coating has on buildings in the study area. Finally, the results of Lot C (Figure 9c) show that integrating the façade with selective coating would barely reduce between 169 and 180 kg of CO_2 equivalent per year compared to its simile, presented in Figure 9b, which would not represent an attractive option from an investment perspective.

4.3. Building Net Saving Analysis

Figure 10 analyzes the economic feasibility of the scenarios. The analysis considered a discount rate range from 1% to 19% to observe the minimum rate at which it is not considered profitable to make the investment. For the case where modifications are made exclusively to the envelope of the medical center (Figure 10a), it appears that the use of reflective paint is the most attractive option. The best results are obtained with this material used on the roof (Case A-I), façade (Case A-V), and combined on the roof with thermal insulation (Case A-III), in that order. For its part, the use of the green roof proposal did not show the feasibility of implementation. The above can be explained by the reflective coating being the cheapest material of the three technologies evaluated. In addition, the Campeche area has high solar radiation, so reflecting about 90% of direct radiation contributes significantly to energy savings. It is important also to analyze the discount rate results. Given that discount values below inflation (6%) do not represent a real benefit from the investment approach, the moderate viability region was proposed, enclosed in the yellow rectangle, which is between 6 and 7%. According to the image, the use of polyurethane insulation and its simile combined with the reflective coating are not only below Cases A-II and A-V, but it is also seen that they are susceptible to falling within the zone of infeasibility upon reaching the region of moderate viability. Therefore, it may not represent an attractive investment option.

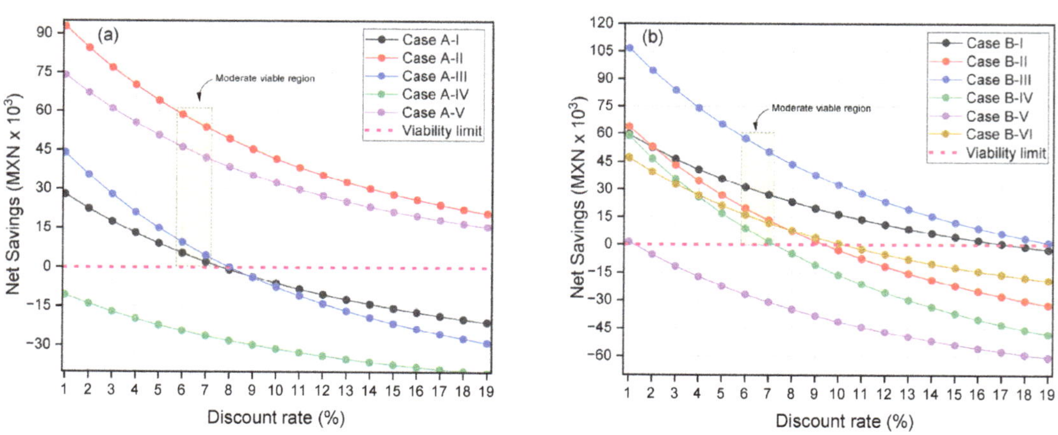

Figure 10. *Cont.*

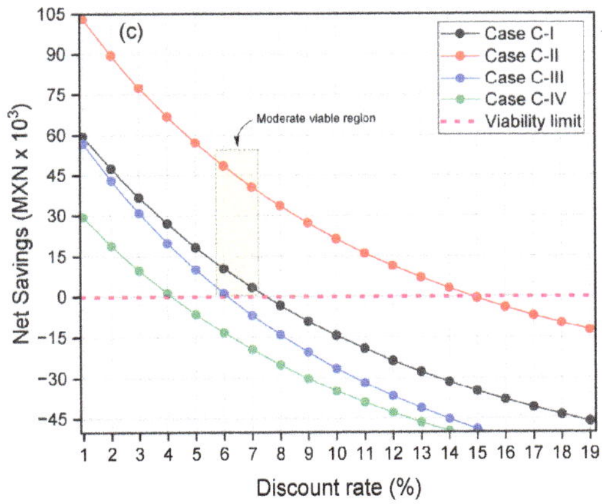

Figure 10. Net savings benefits respect discount rate for the scenarios: (**a**) Lot A: exclusively passive modifications on the envelope; (**b**) Lot B: envelope modification combined with inverter air conditioning systems; (**c**) Lot C: active and passive modifications combined with a reflective coating on the façade.

For its part, integrating technological substitution with inverter equipment and passive strategies (Figure 10b) increases the range of total net savings and the profitable alternatives for energy savings. Five of the six cases that make up Lot B may be in the moderate viability region. Of these, it is interesting that Case B-III, corresponding to the reflective coating on the roof linked to the inverter system, provides better economic performance than the simple integration of the new HVAC system alone (Case B-I). Despite case B-III, the above is a more significant initial investment by the medical center's administration. This demonstrates the economic benefits of using the correct passive system for the envelope. Finally, for Lot C, although there are certain benefits from the energy and environmental perspective, its integration entails a higher cost that is not solved in the long term, as seen in Figure 10c.

4.4. *Implementation of TOPSIS Algorithm*

Table 7 presents the results of decision making using TOPSIS. According to the results, giving up to 60% priority to the economic aspect over the environmental and energy factors means that the integration of reflective coating (especially on the roof) and inverter air conditioning represents the most attractive option. As mentioned previously, this is a product of the shrinking of the material implemented in the envelope. When the energy aspect is almost on par with economic interests, the reflective coating on the roof and wall, and inverter air conditioner (Case C-II) become more critical. For its part, all the priority percentage options converge in Case C-II for both the environmental and energy aspects. The above means in terms of compensation, the joint integration of reflective paint on the walls and ceilings together with air conditioning would represent a better environmental and energy option; it may even be a viable option as long as the medical center administrators reduce the emphasis on economic interests a little. However, in practice, the above is not very viable, as suggested by [36], promoting interest in strategies with greater impact on the thermal efficiency of the envelope that can only be developed through incentives such as carbon credits, which would help to present interest in aspects such as CEI.

Table 7. Results of integrating the TOPSIS method to identify the trade-off of the environmental, energy and economic indices.

Priority Percentage			Best Active/Passive Technological Strategy
KPI	CEI	NS	
10	10	80	Reflective coating on roof with inverter air conditioner (Case B-III)
20	10	70	Reflective coating on roof with inverter air conditioner (Case B-III)
30	10	60	Reflective coating on roof with inverter air conditioner (Case B-III)
40	10	50	Reflective coating on roof and walls with inverter air conditioner (Case C-II)
80	10	10	Reflective coating on roof and walls with inverter air conditioner (Case C-II)
70	10	20	Reflective coating on roof and walls with inverter air conditioner (Case C-II)
60	10	30	Reflective coating on roof and walls with inverter air conditioner (Case C-II)
50	10	40	Reflective coating on roof and walls with inverter air conditioner (Case C-II)
10	80	10	Reflective coating on roof and walls with inverter air conditioner (Case C-II)
10	70	20	Reflective coating on roof and walls with inverter air conditioner (Case C-II)
10	60	30	Reflective coating on roof and walls with inverter air conditioner (Case C-II)
10	50	40	Reflective coating on roof and walls with inverter air conditioner (Case C-II)

5. Discussion

As the results show, the implementation of passive systems, with a particular focus on roofing solutions, emerges as an effective strategy for enhancing the thermo-energetic performance of a building. Passive systems harness natural elements like sunlight, ventilation, and insulation to reduce the building's energy consumption. However, it is essential to acknowledge that their success is subject to a multitude of considerations. Thus, while passive systems offer great potential for energy efficiency, careful planning and the consideration of various factors are imperative for optimizing their performance in a specific context. As results show, these characteristics are predominantly evident in green roofs and polyurethane insulation performance, making them the least effective in reducing energy consumption for air conditioning. The green roof is an innovative solution with benefits; however, the high temperature and humidity that characterize Campeche affect the type of vegetation that needs to be considered, the depth of sub-stratum, and the inheriting maintenance cost. For that reason, it is essential to study aspects of inheritance vegetation that could survive extreme temperatures and humidity without requiring periodic maintenance, decreasing the cost of the system. The government has allocated multiple resources and policies in the case of polyurethane to increase their implementation as a passive measure. This is based on the hypothesis that the reduction of the global heat transfer coefficient contributes to reducing the thermo-energetic behavior of the building and is not supported by quantified data on the impact of climatic conditions on the thermophysical properties of the material. Principally, the humidity of the region is critical to consider in the application of this material because it affects the insulation effectiveness, creates a conducive environment for mold and mildew growth, and degrades the material, affecting its durability.

6. Conclusions

An analysis of the thermal restoration of an outpatient medical center in a tropical region like the City of San Francisco de Campeche was presented to reduce its energy consumption. An energy audit of the building was carried out, collecting data for one year to design and validate a numerical simulation model and verify improvement strategies. Among the improvement options analyzed are passive modifications to the roof (insulating material, reflective coating, and green roof), alterations to active systems integrating inverter technology, and changes to the walls using reflective paint. The results of the simulated improvement scenarios were analyzed using energy, environmental, and economic indicators, which were subsequently contrasted with TOPSIS decision-making algorithms to identify the alternative that best compensates the three options.

Energetically, the results showed that adding passive technologies increases energy efficiency by 11%. Under this scenario, the most significant energy savings are modifying the roof, compared to placing reflective materials on the walls. However, combining both strategies improves the building performance mainly in the winter season. Integrating inverter technology in air conditioning equipment and roof coatings improved energy performance, with inverter-reflective coating technique being the most attractive, with 28% savings. This saving results in the mitigation of around two tons of CO_2 per year from an environmental perspective. The economic results indicate that all the options that integrate reflective paint generate the best net savings. The above is derived from the affordable value of this passive technology. However, many options that have energy and environmental benefits do not achieve economic rewards. The above is an indication of the need to implement strategies that promote better use of energy with an environmental perspective through carbon credits as in other countries, which Mexico currently does not have.

Finally, the TOPSIS approach indicates that integrating reflective paint on the roof and walls with inverter air conditioner would represent the most attractive option in most cases. However, by prioritizing the economic aspect, it remains in the background.

Based on these results, the following conclusions are drawn:

- Due to the climatic conditions of high temperature and humidity, the green roof and polyurethane insulation present the lowest energy savings.
- Green roofs present multiple advantages from the environmental perspective; however, the installation cost, maintenance, irrigation, and the region directly affect their performance.
- Implementing inverter technologies in air conditioning significantly improves the energy consumption in buildings, specifically in medical centers where a stable temperature is necessary.
- Although there is a national policy regarding target temperatures in medical spaces, a range is not established according to the region's conditions, which is why it is imperative to promote a policy that results in better energy consumption practices.

The study results make it straightforward for a coastal city in a tropical climate such as San Francisco de Campeche; the reflective paint on the roof is the best option, not only from an economic perspective. However, it is better than energy-saving strategies currently being implemented at the national level, such as using expanded polystyrene for roofing buildings, where service benefits are both energetic, economic, and environmental. The study results are the first approach to analyzing structures under actual operating conditions in coastal cities with a tropical climate and allow us to provide an overview of where to promote specific technologies and regulations of regional passive strategies for energy saving in the region.

Author Contributions: Conceptualization, O.M.T. and M.H.M.; methodology, O.M.T.; software, G.P.L.; validation and formal analysis M.J.T.; investigation, G.P.L.; resources, J.E.A.D.; data curation, J.J.C.G.; writing—original draft preparation, G.P.L.; writing—review and editing, O.M.T., M.J.T.; visualization, M.H.M.; supervision, F.L.Z.; project administration, F.L.Z. and J.J.C.G.; funding acquisition, J.E.A.D. All authors have read and agreed to the published version of the manuscript.

Funding: This work is part of the internal project of the Autonomous University of Campeche 014/UAC/2023, financed by its resources. The APC was funded by Autonomous University of Campeche (UACAM).

Institutional Review Board Statement: Not applicable.

Informed Consent Statement: Not applicable.

Data Availability Statement: The data presented in this study are available on request from the corresponding author. The data are not publicly available as they correspond to an analysis of a private building; however, the authors are able to share the information upon request.

Acknowledgments: The authors would like to acknowledge the Thematic Network 723RT0151 "Red Iberoamericana de Eficiencia y Salubridad en Edificios" (IBERESE) financed by the call for Thematic Networks of the CYTED Program for 2022, for supporting this research. The second and seventh authors thank CONAHCYT for their respective postgraduate scholarships. The first, fifth and sixth authors thank CONAHCYT for the economic stimulus of the national research system (SNII), through which research by public university professors in Mexico is supported.

Conflicts of Interest: The authors declare no conflict of interest.

References

1. Li, C.; Gao, X.; Wu, J.; Wu, K. Demand prediction and regulation zoning of urban-industrial land: Evidence from Beijing-Tianjin-Hebei Urban Agglomeration, China. *Environ. Monit. Assess.* **2019**, *191*, 412. [CrossRef] [PubMed]
2. Shi, Y.; Yan, Z.; Li, C.; Li, C. Energy consumption and building layouts of public hospital buildings: A survey of 30 buildings in the cold region of China. *Sustain. Cities Soc.* **2021**, *74*, 103247. [CrossRef]
3. Sadeghian, O.; Moradzadeh, A.; Mohammadi-Ivatloo, B.; Abapour, M.; Anvari-Moghaddam, A.; Lim, J.S.; Marquez, F.P.G. A comprehensive review on energy saving options and saving potential in low voltage electricity distribution networks: Building and public lighting. *Sustain. Cities Soc.* **2021**, *72*, 103064. [CrossRef]
4. Cho, K.; Yang, J.; Kim, T.; Jang, W. Influence of building characteristics and renovation techniques on the energy-saving performances of EU smart city projects. *Energy Build.* **2021**, *252*, 111477. [CrossRef]
5. Global Alliance for Buildings and Constructuion, International Energy Agency, United Nations Environment Programme, 2019 Global Status Report for Buildings and Construction: Towards a Zero-Emissions, Efficient and Resilient Buildings and Construction Sector, Paris. 2019. Available online: https://www.iea.org/reports/global-status-report-for-buildings-and-construction-2019 (accessed on 12 October 2023).
6. Milojevic-Dupont, N.; Creutzig, F. Machine learning for geographically differentiated climate change mitigation in urban areas. *Sustain. Cities Soc.* **2021**, *64*, 102526. [CrossRef]
7. Rosas-Flores, J.A.; Rosas-Flores, D. Potential energy savings and mitigation of emissions by insulation for residential buildings in Mexico. *Energy Build.* **2019**, *209*, 109698. [CrossRef]
8. Lee, S.-J.; Song, S.-Y. Energy efficiency, visual comfort, and thermal comfort of suspended particle device smart windows in a residential building: A full-scale experimental study. *Energy Build.* **2023**, *298*, 113514. [CrossRef]
9. Zhang, Y.; Zhang, Y.; Zhang, C. Effect of physical, environmental, and social factors on prediction of building energy consumption for public buildings based on real-world big data. *Energy* **2022**, *261*, 125286. [CrossRef]
10. Guo, Y.; Uhde, H.; Wen, W. Uncertainty of energy consumption and CO_2 emissions in the building sector in China. *Sustain. Cities Soc.* **2023**, *97*, 104728. [CrossRef]
11. Sun, Y.; Kojima, S.; Nakaohkubo, K.; Zhao, J.; Ni, S. Analysis and Evaluation of Indoor Environment, Occupant Satisfaction, and Energy Consumption in General Hospital in China. *Buildings* **2023**, *13*, 1675. [CrossRef]
12. Rahman, N.M.A.; Lim, C.H.; Fazlizan, A. Optimizing the energy saving potential of public hospital through a systematic approach for green building certification in Malaysia. *J. Build. Eng.* **2021**, *43*, 103088. [CrossRef]
13. Gamero-Salinas, J.; Monge-Barrio, A.; Kishnani, N.; López-Fidalgo, J.; Sánchez-Ostiz, A. Passive cooling design strategies as adaptation measures for lowering the indoor overheating risk in tropical climates. *Energy Build.* **2021**, *252*, 111417. [CrossRef]
14. Lorentzen, D.M.C.; McNeil, M.A. Electricity demand of non-residential buildings in Mexico. *Sustain. Cities Soc.* **2020**, *59*, 102165. [CrossRef]
15. Xu, T.; Zhang, Y.; Shi, L.; Feng, Y.; Ke, X.; Zhang, C. A comprehensive evaluation framework of energy and resources consumption of public buildings: Case study, People's Bank of China. *Appl. Energy* **2023**, *351*, 121869. [CrossRef]
16. Seo, R.-S.; Jung, G.-J.; Rhee, K.-N. Impact of green retrofits on heating energy consumption of apartment buildings based on nationwide energy database in South Korea. *Energy Build.* **2023**, *292*, 113142. [CrossRef]
17. As, M.; Bilir, T. Enhancing energy efficiency and cost-effectiveness while reducing CO_2 emissions in a hospital building. *J. Build. Eng.* **2023**, *78*, 107792. [CrossRef]
18. García-Sanz-Calcedo, J.; Al-Kassir, A.; Yusaf, T. Economic and Environmental Impact of Energy Saving in Healthcare Buildings. *Appl. Sci.* **2018**, *8*, 440. [CrossRef]
19. De Masi, R.F.; Del Regno, N.; Gigante, A.; Ruggiero, S.; Russo, A.; Tariello, F.; Vanoli, G.P. The Importance of Investing in the Energy Refurbishment of Hospitals: Results of a Case Study in a Mediterranean Climate. *Sustainability* **2023**, *15*, 11450. [CrossRef]
20. Liu, A.; Ma, Y.; Miller, W.; Xia, B.; Zedan, S.; Bonney, B. Energy Analysis and Forecast of a Major Modern Hospital. *Buildings* **2022**, *12*, 1116. [CrossRef]
21. Instituto Nacional de Estadistica y Geografía, Aspectos Geográficos: Campeche 2021,Publisher by INEGI, Aguascalientes. 2022. Available online: https://www.inegi.org.mx/contenidos/app/areasgeograficas/resumen/resumen_04.pdf (accessed on 12 October 2023).
22. May-Tzuc, O.; Jiménez-Torres, M.A.; Cruz y Cruz, A.; Canul-Turriza, R.; Andrade-Durán, J.E.; Noh-Pat, F. Feasibility of the adaptive thermal comfort model under warm sub- humid climate conditions: Cooling energy savings in Campeche, Mexico. *Rev. Hábitat Sustentable* **2023**, *13*, 120–131. [CrossRef]

23. Villar-Ramos, M.M.; Hernández-Pérez, I.; Aguilar-Castro, K.M.; Zavala-Guillén, I.; Macias-Melo, E.V.; Hernández-López, I.; Serrano-Arellano, J. A Review of Thermally Activated Building Systems (TABS) as an Alternative for Improving the Indoor Environment of Buildings. *Energies* **2022**, *15*, 6179. [CrossRef]
24. Sukarno, R.; Putra, N.; Hakim, I.I.; Rachman, F.F.; Mahlia, T.M.I. Utilizing heat pipe heat exchanger to reduce the energy consumption of airborne infection isolation hospital room HVAC system. *J. Build. Eng.* **2021**, *35*, 102116. [CrossRef]
25. DesignBuilder Software LTd., DesignBuilder. 2022. Available online: https://designbuilder.co.uk/ (accessed on 15 February 2023).
26. Schurk, D. Conditioning for the environment of critical care hospital operating rooms. *ASHRAE J.* **2019**, *61*, 16. Available online: https://www.ashrae.org/technical-resources/ashrae-journal (accessed on 12 October 2023).
27. Remund, J.; Müller, S.; Schmutz, M.; Barsotti, D.; Graf, P.; Cattin, R. Meteonorm 8.1 Manual (Software). 2022, Volume 63. Available online: https://meteonorm.com/assets/downloads/mn81_software.pdf (accessed on 12 October 2023).
28. Corrado, V.; Fabrizio, E. Steady-State and Dynamic Codes, Critical Review, Advantages and Disadvantages, Accuracy, and Reliability. In *Handbook of Energy Efficiency in Buildings*; Asdrubali, F., Desideri, U., Eds.; Elsevier: Amsterdam, The Netherlands, 2019; pp. 263–294. [CrossRef]
29. Torres, M.J.; Bienvenido-Huertas, D.; Tzuc, O.M.; Bassam, A.; Castellanos, L.J.R.; Flota-Bañuelos, M. Assessment of climate change's impact on energy demand in Mexican buildings: Projection in single-family houses based on Representative Concentration Pathways. *Energy Sustain. Dev.* **2023**, *72*, 185–201. [CrossRef]
30. Hu, M.; Zhang, K.; Nguyen, Q.; Tasdizen, T. The effects of passive design on indoor thermal comfort and energy savings for residential buildings in hot climates: A systematic review. *Urban Clim.* **2023**, *49*, 101466. [CrossRef]
31. CONUEE, Catálogo de Tecnologías. Tecnologías Energeticamente Eficientes Para la Envolvente Térmica de las Edificaciones, Mexico City, Mexico. 2017. Available online: https://www.conuee.gob.mx/transparencia/nuevaestrategia/index.html (accessed on 12 October 2023).
32. Hernández-Pérez, I. Influence of Traditional and Solar Reflective Coatings on the Heat Transfer of Building Roofs in Mexico. *Appl. Sci.* **2021**, *11*, 3263. [CrossRef]
33. Mihalakakou, G.; Souliotis, M.; Papadaki, M.; Menounou, P.; Dimopoulos, P.; Kolokotsa, D.; Paravantis, J.A.; Tsangrassoulis, A.; Panaras, G.; Giannakopoulos, E.; et al. Green roofs as a nature-based solution for improving urban sustainability: Progress and perspectives. *Renew. Sustain. Energy Rev.* **2023**, *180*, 113306. [CrossRef]
34. Mousavi, S.; Gheibi, M.; Wacławek, S.; Behzadian, K. A novel smart framework for optimal design of green roofs in buildings conforming with energy conservation and thermal comfort. *Energy Build.* **2023**, *291*, 113111. [CrossRef]
35. Hernández-Pérez, I.; Xamán, J.; Macías-Melo, E.; Aguilar-Castro, K.; Zavala-Guillén, I.; Hernández-López, I.; Simá, E. Experimental thermal evaluation of building roofs with conventional and reflective coatings. *Energy Build.* **2018**, *158*, 569–579. [CrossRef]
36. De la Cruz-Uribe, A.; Jesús-Castañeda, M.; Bolívar-Fuentes, R.C.; Laines-Canepa, J.R.; Hernández-Barajas, J.R. Análisis beneficio-costo de techos verdes extensivos en condiciones del trópico húmedo en Villahermosa, México. *Ecosistemas Y Recur. Agropecu.* **2023**, *10*, 3586. [CrossRef]
37. Saraeian, Z.; Farrell, C.; Williams, N.S. Green roofs sown with an annual plant mix attain high cover and functional diversity regardless of irrigation frequency. *Urban For. Urban Green.* **2022**, *73*, 127594. [CrossRef]
38. Poveda-Orjuela, P.P.; García-Díaz, J.C.; Pulido-Rojano, A.; Cañón-Zabala, G. ISO 50001: 2018 and Its Application in a Comprehensive Management System with an Energy-Performance Focus. *Energies* **2019**, *12*, 4700. [CrossRef]
39. Torres, S.G.M.; Tzuc, O.M.; Aguilar-Castro, K.M.; Téllez, M.C.; Sierra, J.O.; Cruz, A.d.R.C.-Y.; Barrera-Lao, F.J. Analysis of Energy and Environmental Indicators for Sustainable Operation of Mexican Hotels in Tropical Climate Aided by Artificial Intelligence. *Buildings* **2022**, *12*, 1155. [CrossRef]
40. Comision Federal de Electricidad, Clasificacion Tarifaria. 2023. Available online: https://app.cfe.mx/aplicaciones/ccfe/tarifas/tarifascrenegocio/tarifas/pequenademandabt.aspx (accessed on 19 September 2023).
41. Secretaría de Medio Ambiente y Recursos Naturales, Factores de Emisión del Sistema Eléctrico Nacional 2021, Ciudad de México. 2022. Available online: https://www.gob.mx/cms/uploads/attachment/file/706809/aviso_fesen_2021.pdf (accessed on 12 October 2023).
42. OECD, Inflation (CPI), OECD Data. 2023. Available online: https://data.oecd.org/price/inflation-cpi.htm (accessed on 26 July 2023).
43. Tzuc, O.M.; Bassam, A.; Anguebes-Franseschi, F.; Ricalde, L.J.; Flota-Bañuelos, M.; Téllez, M.C. Multivariate optimization applied for the economic competitiveness analysis of photothermal systems into industrial heat production: An approach based on artificial intelligence. *J. Renew. Sustain. Energy* **2020**, *12*, 055501. [CrossRef]
44. Tariq, R.; Sohani, A.; Xamán, J.; Sayyaadi, H.; Bassam, A.; Tzuc, O.M. Multi-objective optimization for the best possible thermal, electrical and overall energy performance of a novel perforated-type regenerative evaporative humidifier. *Energy Convers. Manag.* **2019**, *198*, 111802. [CrossRef]
45. Yadav, V.; Karmakar, S.; Kalbar, P.P.; Dikshit, A. PyTOPS: A Python based tool for TOPSIS. *SoftwareX* **2019**, *9*, 217–222. [CrossRef]

applied sciences

MDPI

Article

Building Energy Savings by Developing Complex Smart Windows and Their Controllers

Seong-Ki Hong [1], Sang-Ho Choi [2] and Su-Gwang Jeong [2,*]

1 Section of Industry Cooperation and Research Planning, Jeonju University, Jeonju 55069, Republic of Korea
2 Department of Architectural Engineering, Soongsil University, Seoul 06978, Republic of Korea; sangho83@ssu.ac.kr
* Correspondence: sgjeong@ssu.ac.kr; Tel.: +82-2-820-0705

Abstract: The interest in zero-energy buildings has increased in Korea recently. Following the significant increases in cooling and lighting energy consumption in offices, various studies have been conducted to implement energy-saving measures. The purpose of this study is to reduce lighting and cooling energy consumption in the summer through the dimming control of a complex smart window system. To achieve this, the optimal dimming control algorithm has been derived and applied in simulations to analyze the energy consumption for lighting and cooling. A smart window incorporates suspended particle display glass that actively responds to changes in indoor and outdoor environments and controls light transmittance. It also includes a light-guiding glass that can actively control solar reflectance. Simulations of office buildings were conducted to develop optimal control algorithms and controllers based on solar radiation. Subsequently, we installed this complex smart window in a test room along with the developed control algorithm and controller, which responded to the amount of insolation and time. To ensure the accuracy of the experiment, we constructed separate test and reference rooms. The experimental results obtained under the same conditions showed a reduction of approximately 36.9% in cooling energy consumption in the test room compared with the reference room and a 54.5% reduction in lighting energy consumption. Furthermore, based on additional simulations and experiments, we confirmed that the application of complex smart window systems in office buildings could reduce considerably the energy consumption for cooling and lighting.

Keywords: complex smart window; smart skin; zero-energy buildings; suspended particle display; energy savings; lighting-guide glass; dimming control LED; optimal control

Citation: Hong, S.-K.; Choi, S.-H.; Jeong, S.-G. Building Energy Savings by Developing Complex Smart Windows and Their Controllers. *Appl. Sci.* **2023**, *13*, 9647. https://doi.org/10.3390/app13179647

Academic Editors: Maria da Glória Gomes, David Bienvenido Huertas and Daniel Sánchez-García

Received: 31 July 2023
Revised: 23 August 2023
Accepted: 25 August 2023
Published: 25 August 2023

1. Introduction

Korea is currently strengthening its green building roadmap. As part of this effort, we are striving to implement energy-saving measures with windows based on the "Energy-saving Design Standards" to minimize the energy demand for mandatory zero-energy buildings in the public sector by 2020 and the private sector by 2025 [1]. Notably, office buildings account for 60–70% of the total energy consumed by the building sector, making energy savings in offices an increasingly important issue [2,3]. However, the development of smart envelope and Energy Management System (EMS)-linked technologies to reduce the cooling load of office buildings, which are the most thermally vulnerable parts when implementing zero-energy buildings, remains insufficient.

Building energy consumption is increasing globally, and office energy consumption is experiencing significant growth. Consequently, many studies are underway to reduce building cooling loads. Some studies focused on active awning systems [4] and window systems using double skins [5]. However, these efforts concentrated solely on reducing the cooling load by blocking solar radiation from reaching the building. Consequently, lighting energy consumption in office buildings is increasing. To minimize heat loss through windows in conventional buildings, it is essential to produce glass that can control the transmittance of visible light radiation artificially. The need for such technological advancement has become achievable recently, thanks to the progress in the development of thin-film materials with various functionalities and research on liquid crystal materials. One of these technologies is the use of variable transmittance glass, which can artificially adjust its transmittance based on the presence or absence of light or an electric field. Such glass can be broadly categorized into polymer-dispersed liquid crystal (PDLC) technology, vacuum sputter-coated electrochromic technology, and others, depending on the type of functional material used [6,7]. These advanced materials are being pursued as next-generation, high-performance, and high-value-added products. Various foreign companies and research institutions are investing substantial budgets in their development, alongside research on control methods. Therefore, in this study, we conducted research on the reduction of lighting and cooling energy consumption in zero-energy buildings through dimming control of a developed smart window system incorporating variable transmittance glass. To achieve this, the optimal dimming control algorithm has been derived and applied in simulations to analyze the energy consumption for lighting and cooling. The developed smart skin is a composite of suspended particle device (SPD) and lighting guide glass (LGG) technologies.

To address this issue, our study aimed to develop an active technology capable of reducing both cooling and lighting energy consumption by analyzing the correlation between cooling and lighting. Therefore, in this study, we developed a complex smart window by combining SPD glass [8,9] that can control light transmittance by actively responding to changes in indoor and outdoor environments, such as indoor temperature, illuminance, outdoor temperature, and solar radiation, with a LGG [10] that is capable of actively controlling solar reflectance. In this case, SPD adjusts its transparency based on the external conditions of the building to minimize the building's energy loss through heat, while LGG refracts sunlight to enhance the incoming light, thus conserving indoor lighting energy. The refractive index of LGG has been calculated using an algorithm to achieve the optimal value by adjusting its angle based on the elevation of the sun. In addition, we incorporated a dimming LED [11] to maintain the indoor illuminance level and a controller to optimally control the complex smart window [12,13]. To achieve this, an optimal control algorithm and an EMS were developed using computer simulations [14]. Subsequently, we experimentally measured the reduction in cooling and lighting loads in an office building [15]. This study involved the development of a complex smart window system, the formulation of an SPD, LGG, and dimming LED control algorithm based on solar radiation in office building modeling using EnergyPlus and TRNSYS, and the design of a control algorithm in conjunction with a controller. Finally, we conducted a comparative experiment by installing tests and reference groups to measure energy consumption.

2. Materials and Methods

2.1. Development of Complex Smart Windows

A complex smart window is a sophisticated building envelope that responds to the external environment and adjusts the conditions of the building envelope to optimize the indoor environment. It includes sensors for detecting the external environment (outside temperature, solar radiation) and internal environment (indoor temperature, illuminance), a computer program to respond to the central EMS of the building control system, an SPD and LGG, dimming LEDs, and a driving apparatus, among other components. Essentially, it functions as a smart skin for windows. Figure 1 shows the configuration of a complex smart window.

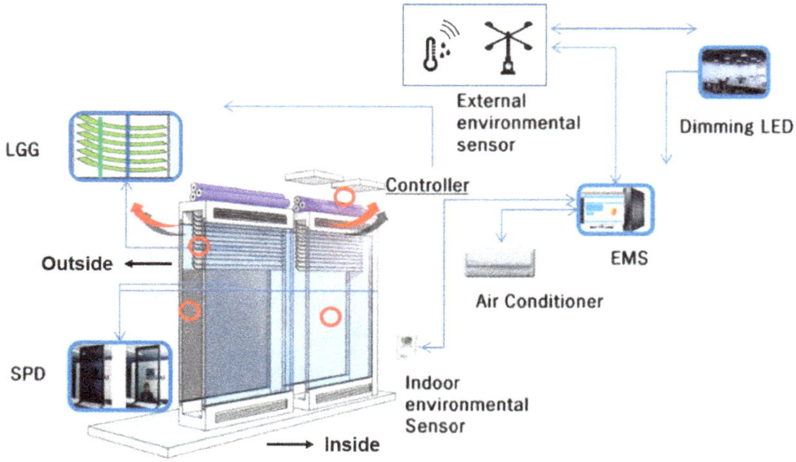

Figure 1. Configuration of the complex smart window.

Single-skin-shading and double-skin-type windows were developed [16] considering wind pressure and watertight performance [17,18] to account for problems in strength and durability against external impacts depending on the location of the configuration. A complex smart window was combined with an SPD and LGG to maintain the indoor environment at optimal conditions. SPD reduces heat loss in a building by varying the transmittance according to the external conditions of the building [19,20]. LGG refracts light from the sun to increase the amount of light entering a building, thereby reducing the lighting energy in the building. An optimal control algorithm for driving the SPD and the LGG was developed using computer simulations. Additionally, the dimming LED was designed to interlock so that it would not affect the indoor working environment and change independently based on the variation in illumination intensity by blocking and distributing sunlight through the SPD and LGG. Furthermore, a controller was developed to interface with an EMS [21] that monitors data and controls signals based on an optimal control algorithm in conjunction with the indoor and outdoor environments [22–24]. Figure 2 shows the image of complex smart window operation.

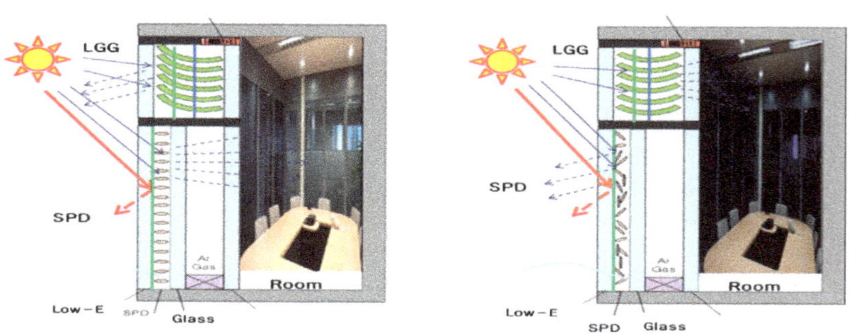

Figure 2. Complex smart window operation.

2.2. Simulations

Transmittance and dimming controls for complex smart windows are required to effectively reduce the cooling load and lighting energy during the summer. Time-dependent transmittance control is necessary to achieve optimal control of complex smart windows. To establish the correlation between transmittance control and energy consumption, basic modeling was conducted on an office building located in Seoul, and the energy consumption was calculated using DesignBuilder (version 5.5.2.003). Figure 3 shows the DesignBuilder simulation setup.

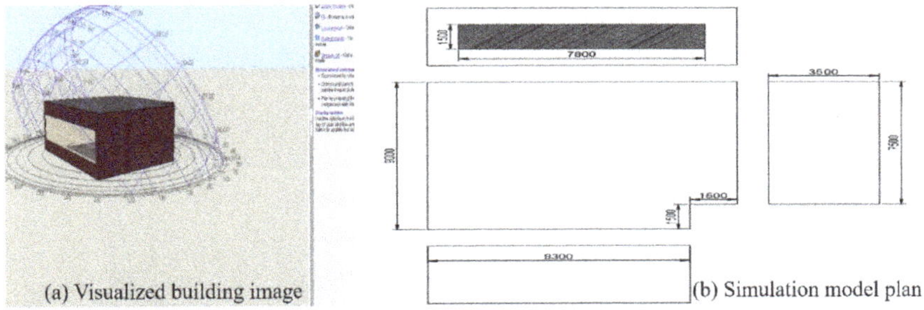

Figure 3. DesignBuilder simulation setup: (**a**) visualized building model; (**b**) simulation model plan.

The simulated office building was located in Seoul, and a window was installed at a height of 1.5 m on the south-facing side of the building, which covered 50% of the total wall area. Table 1 presents the details of the simulated buildings. For office dimming control, 700 lx was set as the standard for indoor illumination levels (range: 700–1500 lx). Regarding the SPD, a simulation was conducted by dividing the shading on the south-facing window into nine cases (with minimum and maximum shielding criteria of 0.1 and 0.9, respectively) to determine the appropriate SPD control method according to time. The weather data required for the analysis were based on the 2015 Seoul weather data provided by the Korean Meteorological Administration. A simulation was conducted for the baseline model, a model with the dimming control system was applied, and the dimming control system was applied to a general office building under natural light. All other conditions were kept constant. The results are summarized in Figure 4. In this study, the function that controls the transmittance of the smart skin over time was determined using regression analysis. This function is proposed to be a function of time to optimize the cooling load and lighting energy by controlling the smart skin.

Table 1. Simulation conditions of test building.

Simulation Condition	Contents		
Location	Jeonju City, Jeollabuk-do		
Area	85.95 m^2		
District	Residential area		
Usage	office	Illuminance	20 W/m^2
Occupancy	0.2 persons/m^2	Lighting density	500 lux
Occupancy hours	8:00–18:00	Infiltration	0.2 h^{-1} (Air exchange rate)
Ventilation	0.5 h^{-1} (Air exchange rate)	Cooling temperature	22 °C

2.3. *Pilot Plant*

To develop the smart window system and algorithm, a pilot plant was built in the laboratory to monitor changes in cooling, heating, and room lighting loads according to changes in solar radiation. The collected data were used to develop control algorithms using machine learning techniques. Figure 5 shows prototypes of the solar radiation blocking rate measuring device [25], SPD production [26], and smart window [27], which were used to construct an actual pilot plant [28–30].

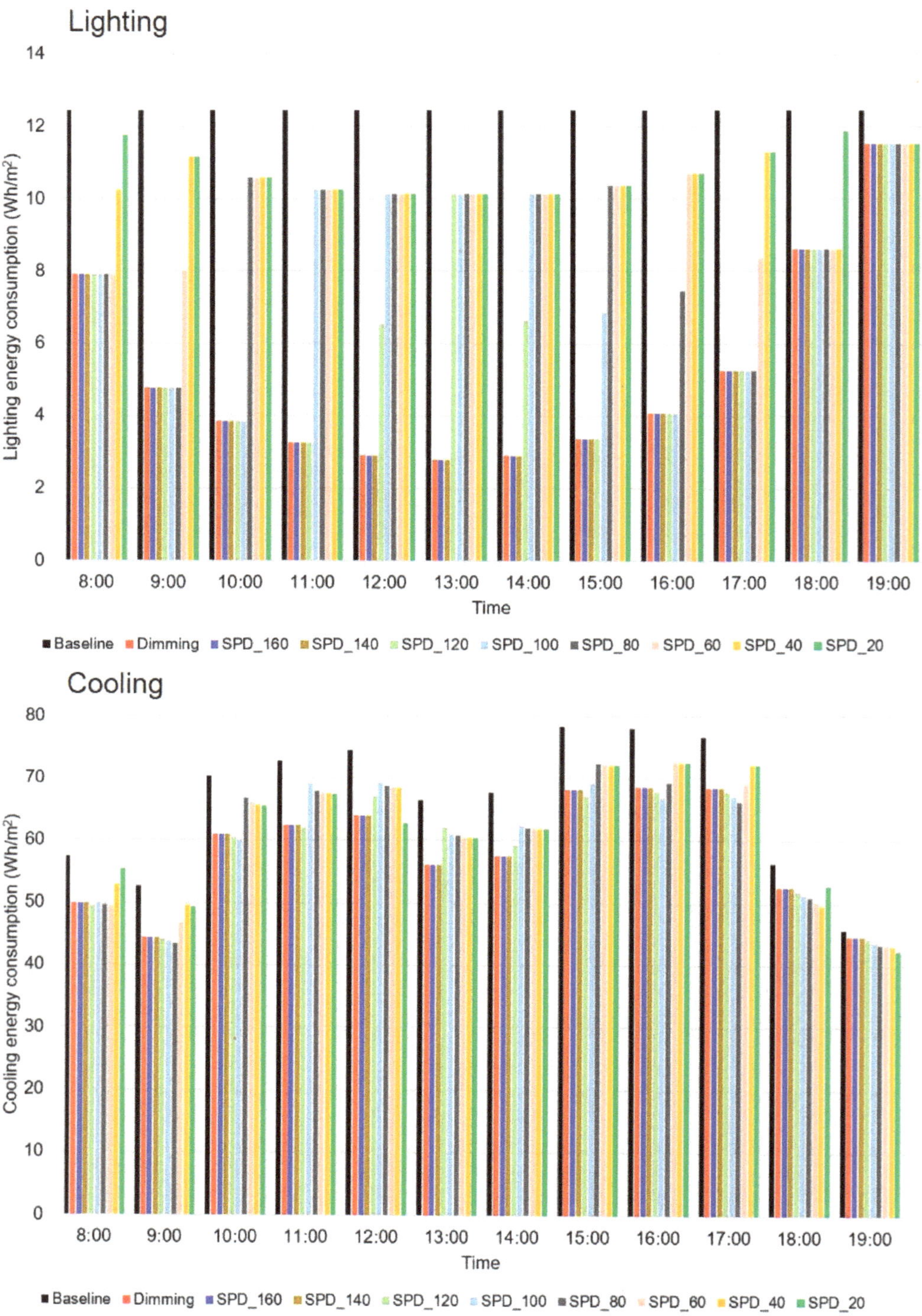

Figure 4. Simulation result according to the application of the dimming control system.

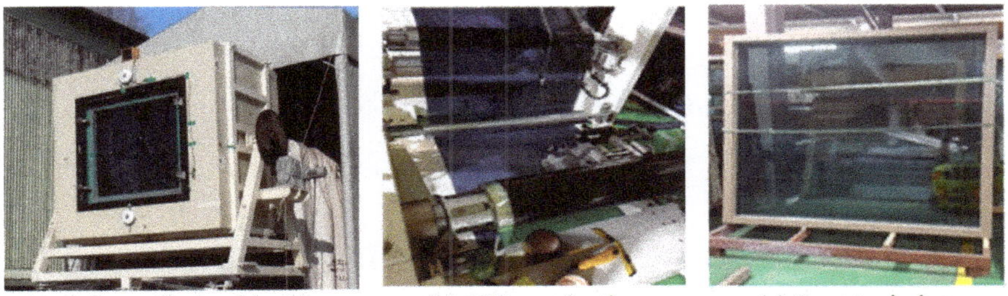

(a) Solar radiation blocking rate measuring device (b) SPD production (c) Smart window

Figure 5. Images of (**a**) a solar radiation blocking rate measuring device, (**b**) SPD production, and (**c**) a smart window.

2.4. Optimal Control Algorithm and Controller Development

In this study, the function that controls the permeability of the smart skin over time was calculated using regression analysis. Consequently, a time function that optimizes the cooling load and lighting energy by controlling the smart skin is proposed.

$$f(t) = 0.0008t^5 + 0.1121t^4 - 2.8149t^3 + 19.36t^2 - 26.65t + 71.82 \quad (1)$$

Different results were obtained when the direction of the building was changed to east, west, south, or north. Using this result, the smart skin control algorithm (implemented to achieve the maximum reduction of cooling and lighting loads in a zero-energy building) can be derived, as shown in Figure 6. In the algorithm, the output level represents the temperature and irradiation amounts obtained from the sensor attached to the SPD. The SPD output is represented as a discrete value ranging from 0 to 1, with 1 indicating complete light blocking and 0 signifying transparent control allowing light transmission.

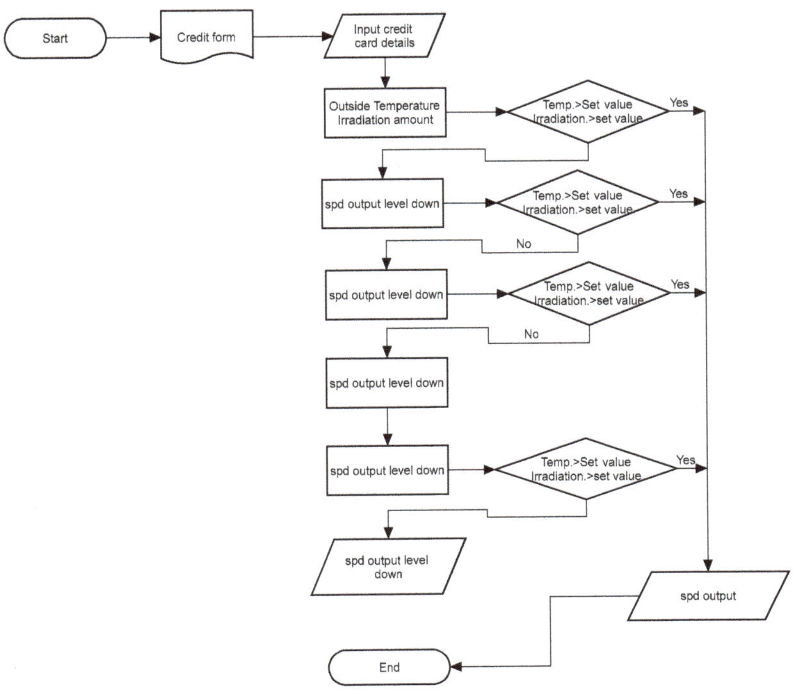

Figure 6. Control algorithm for load reduction.

One disadvantage is that the effect of the change in insolation in real time cannot be fully considered through SPD and dimming control based solely on time. To address this limitation, precise, real-time control that considers the indoor environment is necessary. Therefore, a dependent logic considering the amount of insolation is proposed as follows:

$$S(t) = -0.0185t^2 + 0.2804t - 0.0935 \tag{2}$$

where

$$t = \begin{cases} 08:00\ 1 \\ 09:00\ 2 \\ \cdots \\ 18:00\ 10 \end{cases}$$

$$|R - TR| > 100,\ X = (R - TR)/1000$$

$$R : Representative\ Insolation,\ TR : Hourly\ Insolation$$

$$S = S(t) - X(0 \le S \le 1), \tag{3}$$

In Equations (2) and (3), $S(t)$ is a control value, and t is a variable representing time. R represents the representative solar radiation, and TR represents the hourly solar radiation. Equations (2) and (3) were used to supplement the control value in cases where the hourly insolation amount had a large error compared with the representative insolation amount. The following algorithm incorporates real-time insolation from the previous algorithm, resulting in a more accurate calculation of SPD values. Therefore, this algorithm allows for the calibration of SPD values, thereby improving the accuracy of SPD values for cloudy or rainy days. Finally, through algorithmic modeling, we were able to derive the optimal SPD transmittance and dimming values, and using these, we controlled the smart window. Figure 7 shows a control method that considers real-time insolation.

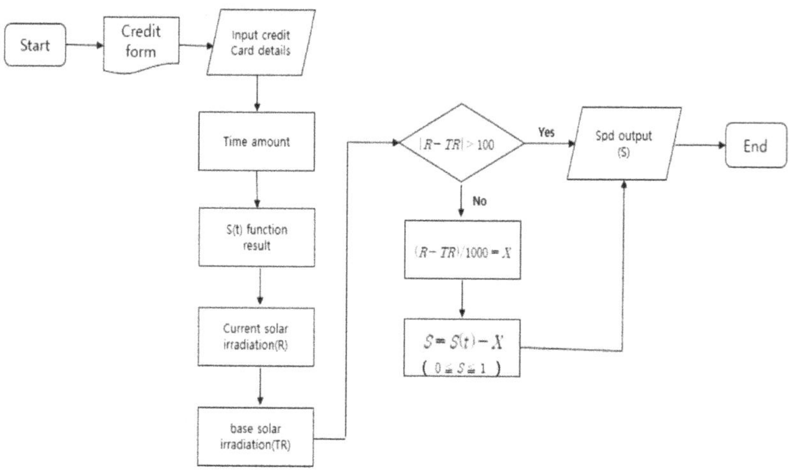

Figure 7. Control algorithm considering solar radiation for load reduction.

The development of hardware (HW) and firmware (FW) to control a complex smart window system in an office utilizes a LoRa-based wireless communication method to collect wireless sensor data. This supports LGG, SPD, and LED controls. The detailed hardware configuration is illustrated in Figure 8.

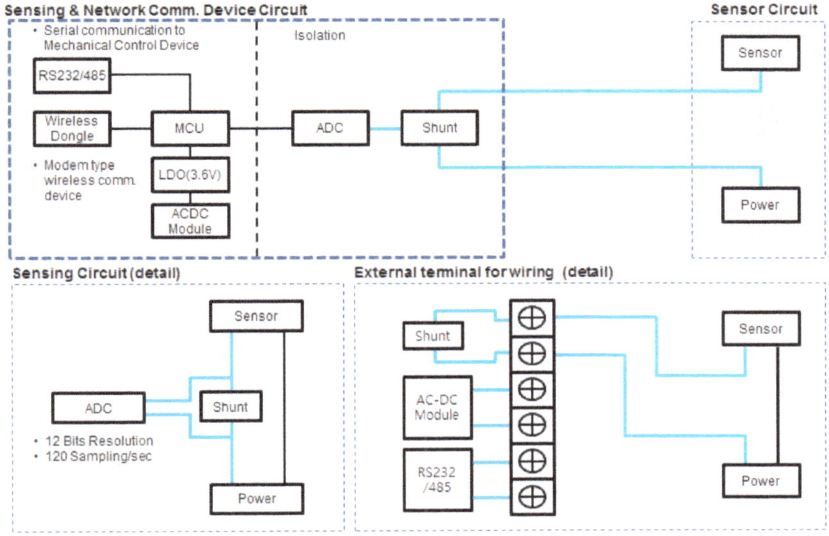

Figure 8. Hardware configuration of a complex smart window controller.

2.5. Test Building Construction and Complex Smart Window Experiment

To reduce the lighting and cooling loads in a zero-energy building, a test room was established to conduct experiments on a complex smart window, and an optimal control method was developed that applies both dimming control and variable transmittance glasses. For the experiment, two identical rooms—the test and reference—were constructed with an EPS panel structure, each with an area of 12 m^2.

The experimental setup is shown in Figure 9. Smart skins with SPD and LGG were installed in the test room, whereas ordinary glasses were installed in the reference room. Their configurations are presented in Table 2. Temperature, humidity, illuminance, and solar radiation sensors were installed in each room for monitoring purposes. The heating and cooling units were installed separately to monitor the cooling energy consumption using a meter.

Figure 9. Construction of test and reference rooms.

Table 2. Glass compositions in the conducted experiment.

Skin Type			Glass	Size (mm)	Thickness (mm)	EA (Each)
Test room	LGG		5CL + 22Ar (blind) + 5PLA ONE	1140 × 540	32	2
	SPD	inside	5CL + 16Ar (SWS) + 5PLA ONE	1140 × 800	26	4
		outside	9.75SPD + 16Ar (SWS) + 5PLA ONE		30.75	4
Reference room	top		5CL + 16Ar (SWS) + 5PLA UN	1140 × 540	26	2
	bottom		5CL + 16Ar (SWS) + 5PLA UN	1140 × 800	26	4

3. Results

3.1. Effect of a Complex Smart Window

The difference in energy consumption between the test and reference rooms is shown below. Although the current control method exhibited a high-cooling effect in the summer, it was slightly effective or ineffective in meeting the load required for heating and cooling during the shoulder season. However, when the experiment was conducted, the energy-saving effect was confirmed to be significant.

3.2. Site Experiments

The goal of this study was to reduce lighting and cooling loads in a zero-energy building by testing a smart skin and implementing an optimal control method that incorporates both dimming control and variable transmittance glass. To achieve this, the sensor and meter values were monitored using the EMS. The monitoring period for this study was limited to June–September 2000. During the test period, the energy consumption of the test and reference rooms was compared. Focus was placed on data collected from 9 am to 6 pm. This timeframe was chosen to align with standard working hours in public buildings, and changes in indoor temperature as well as the energy consumption of air conditioners and lighting were closely monitored. The illuminance level was set to 500 lx in both the test and reference rooms. In the experimental site, the indoor temperature was controlled to approximately 25 °C, and the indoor illuminance was maintained at approximately 500 lx, which is the required illuminance level for office work in a public building. The illuminance level measurement location was set at a height of approximately 0.7 m in both the test and reference rooms, based on the height of the office desk.

3.3. Comparison of Cooling and Lighting Energy Consumptions

To reduce the cooling load and lighting energy in summer effectively, an experiment was conducted using a complex smart window combined with a dimming LED to maintain indoor illumination levels. For this purpose, we simulated the reduction of cooling and lighting loads through actual monitored SPD transmittance and dimming control. This window combines SPD glass, which actively responds to changes in indoor and outdoor environments and can control light transmittance, with LGG, which can actively control solar reflectance. The results of a comparative analysis of the energy consumption of the test and reference groups in the experiment using a complex smart window are shown in the following graph. In Figures 10 and 11, the data for days 1 to 15 refer to the period from 1–15 August, which is when the cooling load is the highest during the summer season. Also, in Figures 12 and 13, the monthly data include all days of the respective month.

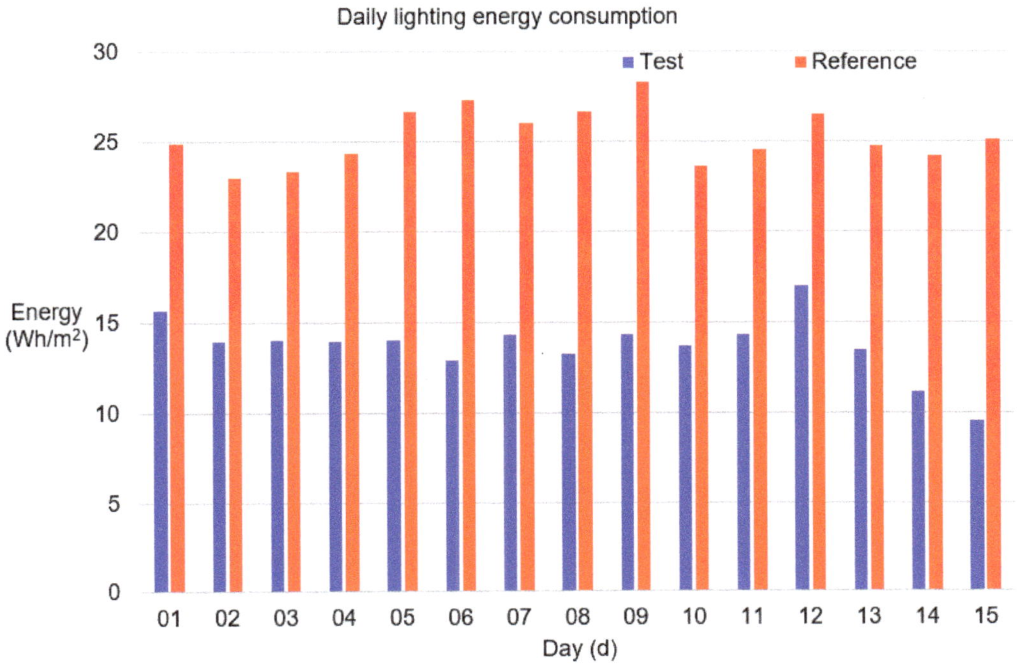

Figure 10. Experimental results of daily lighting energy consumption.

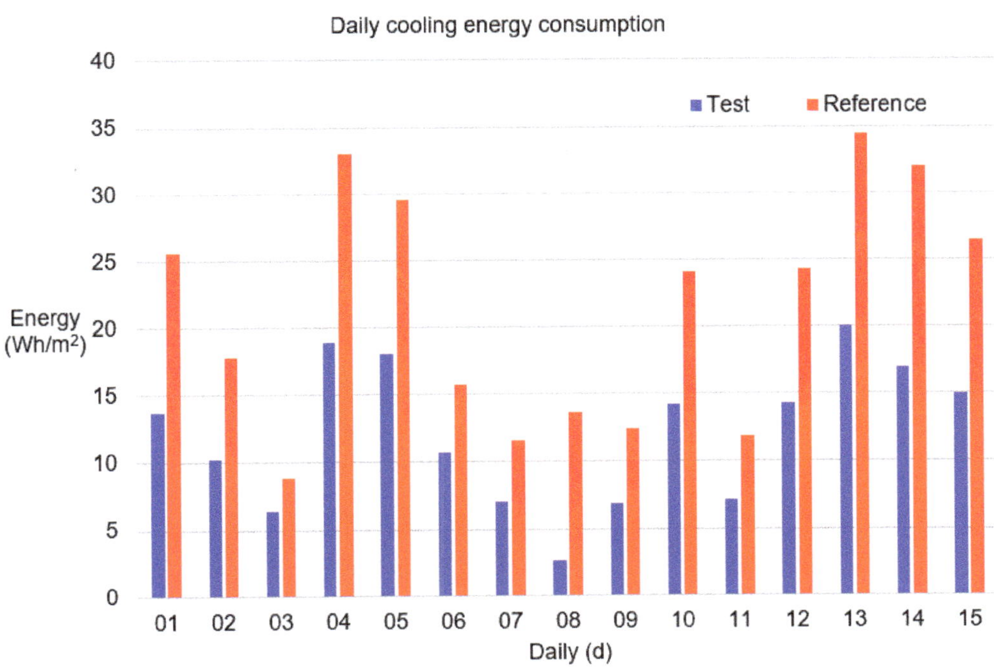

Figure 11. Experimental results of daily cooling energy consumption.

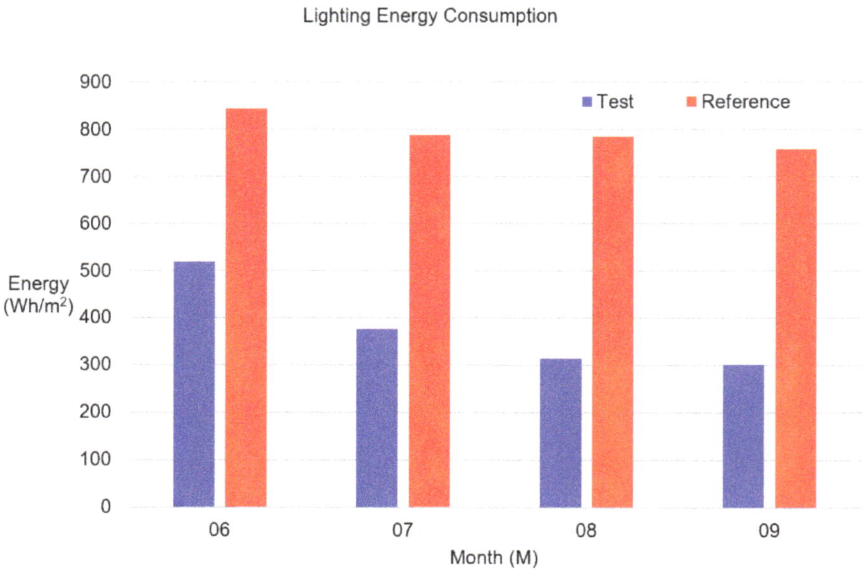

Figure 12. Experimental results of monthly lighting energy consumption.

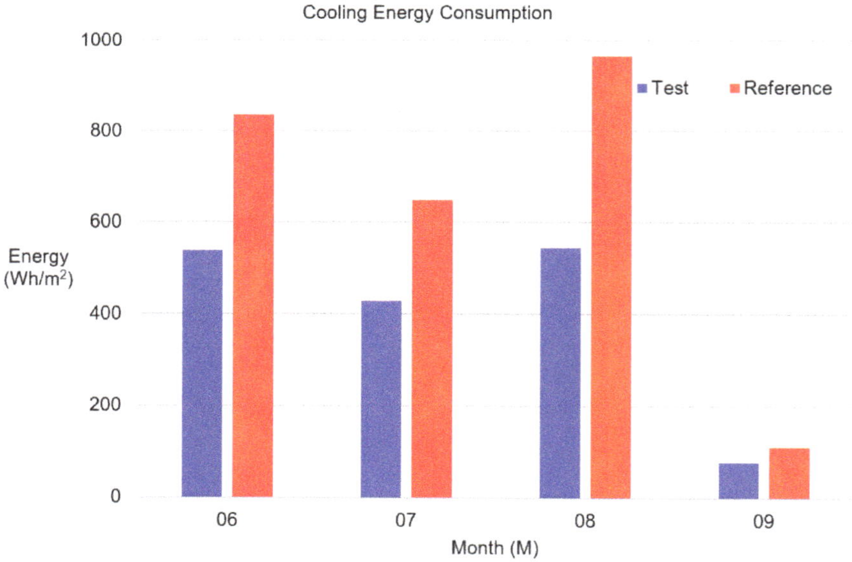

Figure 13. Experimental results of monthly cooling energy consumption.

Figure 9 displays the lighting energy consumption from 1–15 August, revealing that the Test group equipped with SPD and LGG generally exhibits lower consumption (<1.5 kWh). Figure 11 presents cooling energy consumption, indicating larger daily variations due to external temperature influences rather than lighting, yet similarly, the Test group demonstrates lower cooling energy consumption, consistent with the lighting energy results. Figures 12 and 13 depict lighting and cooling energy consumption from June to September. Overall, the Test group consistently exhibits lower energy consumption, and the highest reduction rate is observed in August, a period of heavy load.

The experiment yielded energy consumption under the same external and internal conditions, as shown in Table 3. The cooling and lighting energy consumptions of the test room were 138.6 kWh and 124.1 kWh, respectively, whereas the cooling and lighting energy consumptions of the reference room were 219.8 kWh and 272.7 kWh, respectively. By applying smart skin, energy-efficient use was achieved in parallel with dimming control. By combining LGG and the dimer, the lighting load was reduced by 54.5% and the cooling energy consumption was reduced by approximately 36.9% compared with the reference room.

Table 3. Experimental glass compositions.

Energy Consumption (kWh)		June	July	August	September	Sum	Savings
Cooling	Test	48.4	36.8	45.4	7.8	138.61	36.9%
	Reference	71.8	55.0	80.6	11.7	219.76	
Lighting	Test	43.3	31.8	27.0	21.5	124.07	54.5%
	Reference	74.5	67.6	65.6	64.7	272.66	

4. Discussion

Additional simulations were conducted (not mentioned in this study). In particular, after comparing the energy consumption of buildings with complex smart window systems based on the orientation of the office building and analyzing the differences, it was found that the highest energy savings occurred when the building faced south, whereas the lowest savings occurred when it faced north. The order of energy savings according to building direction was south, west, east, and north.

Furthermore, when comparing and simulating the amount of energy saved using different building envelope materials, such as reinforced concrete and brick, the energy-saving rate was found to be the same. However, when the window area differed from that of the wall, energy savings were proportional to the window area. Additionally, deep learning was applied to simulate the energy savings of a complex smart window system, and it was confirmed that a certain amount of energy consumption could be predicted.

In this study, to reduce cooling and lighting energy simultaneously, a complex smart window was developed by combining SPD glass, which is capable of controlling light transmittance by actively responding to changes in indoor and outdoor environments and outdoor air, with LGG, which can actively control solar reflectance. A dimming LED is also incorporated to maintain the required indoor illuminance level. Therefore, it is necessary to confirm the energy-saving effects associated with the use of this technology in various buildings. Additionally, based on this study, the developed smart window is considered effective in reducing lighting and cooling loads. However, from a long-term perspective, when applying smart skin technology, it is deemed necessary to install real-time sensors for detecting malfunctions in the SPD and LGG window systems and verifying their long-term effects.

5. Conclusions

In this study, a system was developed to reduce the lighting and cooling energy consumption in an office building during the summer period. Prior to the experiments, we developed a control algorithm to derive the optimal SPD transmittance and dimming values. The system simultaneously applied dimming control and smart skin with variable transmittance, and an optimal control method was devised to maximize energy savings. An experimental site was constructed to compare and study the actual energy consumption. The experimental period was from June to September, and the cooling period and lighting operation time were monitored from 9:00 am to 6:00 pm, aligned with the working hours of office workers. As a result of this study, it has been confirmed that the application of the developed complex smart window significantly reduces lighting and cooling loads during the summer season. The major findings are as follows:

(1) A complex smart window system was developed by combining SPD, LGG, and dimming LED to simultaneously reduce cooling and lighting loads and address the increasing energy consumption in offices during the summer period.
(2) An algorithm that can control the transmittance and dimming in parallel was developed to effectively control complex smart windows. The cooling load and lighting energy were optimized by determining a function that controlled the transmittance over time.
(3) To increase the effectiveness of the control and application of the complex smart window, an optimal control function was developed for each region using representative solar radiation data for each area. To control the complex smart window in an office, we developed HW and FW for the wireless controller using the LoRa-based wireless communication method, which supports LGG, SPD, and LED control.
(4) Based on experimentation, it was shown that energy efficiency can be achieved in parallel with dimming control when complex smart windows are used. By combining LGG and dimming, the lighting load was reduced by 54.5%.
(5) An experimental site was constructed to monitor the energy consumption in the test and reference rooms. In the test room with the smart skin system, the cooling energy consumption was reduced by approximately 36.9% compared with the reference room.

Author Contributions: Conceptualization, S.-K.H. and S.-H.C.; methodology, S.-K.H.; software, S.-K.H.; validation, S.-G.J. and S.-H.C.; investigation, S.-K.H.; data curation, S.-H.C.; writing—original draft preparation, S.-K.H.; writing—review and editing, S.-G.J.; visualization, S.-K.H.; supervision, S.-G.J.; funding acquisition, S.-K.H. All authors have read and agreed to the published version of the manuscript.

Funding: This study was supported by the Korean Institute of Energy Technology Evaluation and Planning (KETEP) and the Ministry of Trade, Industry, and Energy (MOTIE) of the Republic of Korea (No. 20192710100015C). Additionally, this research was supported by the "Regional Innovation Strategy (RIS)" through the National Research Foundation of Korea (NRF), funded by the Ministry of Education (MOE) (2021RIS-004).

Institutional Review Board Statement: Not applicable.

Informed Consent Statement: Not applicable.

Data Availability Statement: Not applicable.

Conflicts of Interest: The authors declare no conflict of interest.

References

1. Kim, S.H.; Kim, S.S.; Kim, K.W.; Cho, Y.H. A study on the proposes of energy analysis indicator by the window elements of office buildings in Korea. *Energy Build.* **2014**, *73*, 153–165. [CrossRef]
2. Cappelletti, F.; Gasparella, A.; Romagnoni, P.; Baggio, P. Analysis of the influence of installation thermal bridges on windows performance: The case of clay block walls. *Energy Build.* **2011**, *43*, 1435–1442. [CrossRef]
3. Adamus, J.; Pomada, M. Analysis of heat flow in composite structures used in window installation. *Compos. Struct.* **2018**, *202*, 127–135. [CrossRef]
4. A New Wave in the Awning Industry Following a Paradigm Shift in Building Envelopes. 2016. Available online: http://www.todayenergy.kr (accessed on 3 July 2023).
5. Seok, H.T.; Jo, J.H.; Kim, K.W. Establishing the Design Process of Double-Skin Façade Elements through Design Parameter Analysis. *J. Asian Archit. Build. Eng.* **2008**, *8*, 251–258. [CrossRef]
6. Cupelli, D.; Nicoletta, F.P.; Manfredi, S.; Vivacqua, M.; Formoso, P.; De Filpo, G.; Chidichimo, G. Self-adjusting smart windows based on polymer-dispersed liquid crystals. *Sol. Energy Mater. Sol. Cells* **2009**, *93*, 2008–2012. [CrossRef]
7. Karumuthil, S.C.; Ganesha, M.K.; Mondal, I.; Singh, A.K.; Kulkarni, G.U. Fabrication of dual-functional electrochromic smart window based on low-cost hybrid transparent electrode coated with a solution-processable polymer. *J. Mater. Chem. A* **2022**, *10*, 23265–23273. [CrossRef]
8. Tsikaloudaki, K.; Theodosiou, T.H.; Laskos, K.; Bikas, D. Assessing cooling energy performance of windows for residential buildings in the Mediterranean zone. *Energy Convers. Manag.* **2012**, *64*, 335–343. [CrossRef]

9. Ghosh, A.; Norton, B.; Duffy, A. Effect of sky conditions on light transmission through a suspended particle device switchable glazing. *Sol. Energy Mater. Sol. Cells* **2017**, *160*, 134–140. [CrossRef]
10. Kischkoweit-Lopin, M. An overview of daylighting systems. *Sol. Energy* **2002**, *73*, 77–82. [CrossRef]
11. Kim, S.H.; Kim, I.T.; Choi, A.S.; Sung, M. Evaluation of optimized PV power generation and electrical lighting energy savings from the PV blind-integrated daylight responsive dimming system using LED lighting. *Sol. Energy* **2014**, *107*, 746–757. [CrossRef]
12. Baetens, R.; Jelle, B.P.; Gustavsen, A. Properties, requirements and possibilities of smart windows for dynamic daylight and solar energy control in buildings: A state-of-the-art review. *Sol. Energy Mater. Sol. Cells* **2010**, *94*, 87–105. [CrossRef]
13. Allen, K.; Connelly, K.; Rutherford, P.; Wu, Y. Smart windows—Dynamic control of building energy performance. *Energy Build.* **2017**, *139*, 535–546. [CrossRef]
14. Beaudin, M.; Zareipour, H. Home energy management systems: A review of modelling and complexity. *Renew. Sustain. Energy Rev.* **2015**, *45*, 318–335. [CrossRef]
15. Gago, E.J.; Muneer, T.; Knez, M.; Köster, H. Natural light controls and guides in buildings. Energy saving for electrical lighting, reduction of cooling load. *Renew. Sustain. Energy Rev.* **2015**, *41*, 1–13. [CrossRef]
16. Cho, S.; Kim, S.H. Analysis of the performance of vacuum glazing in office buildings in Korea: Simulation and experimental studies. *Sustainability* **2017**, *9*, 936. [CrossRef]
17. Sun, Y.; Wu, Y.; Wilson, R. A review of thermal and optical characterisation of complex window systems and their building performance prediction. *Appl. Energy* **2018**, *222*, 729–747. [CrossRef]
18. Dussault, J.M.; Gosselin, L. Office buildings with electrochromic windows: A sensitivity analysis of design parameters on energy performance, and thermal and visual comfort. *Energy Build.* **2017**, *153*, 50–62. [CrossRef]
19. Li, C.; Tan, J.; Chow, T.T.; Qiu, Z. Experimental and theoretical study on the effect of window films on building energy consumption. *Energy Build.* **2015**, *102*, 129–138. [CrossRef]
20. Fasi, M.A.; Budaiwi, I.M. Energy performance of windows in office buildings considering daylight integration and visual comfort in hot climates. *Energy Build.* **2015**, *108*, 307–316. [CrossRef]
21. Ye, H.; Meng, X.; Xu, B. Theoretical discussions of perfect window, ideal near infrared solar spectrum regulating window and current thermochromic window. *Energy Build.* **2012**, *49*, 164–172. [CrossRef]
22. Kumar, K.; Saboor, S.; Kumar, V.; Kim, K.H. Experimental and theoretical studies of various solar control window glasses for the reduction of cooling and heating loads in buildings across different climatic regions. *Energy Build.* **2018**, *173*, 326–336. [CrossRef]
23. Ministry of Land, Transport and Maritime Affairs. *Building Energy Conservation Design Standard*; Ministry of Land, Transport and Maritime Affairs: Sejong-si, Republic of Korea, 2008.
24. Ministry of Knowledge Economy. *Energy Efficiency Equipment Operating Regulations*; Ministry of Knowledge Economy: Seoul, Republic of Korea, 2012.
25. Korea Agency for Technology and Standards, KS. Windows Set. 2015. Volume F3117. Available online: http://www.kats.go.kr/ (accessed on 3 July 2023).
26. Korea Agency for Technology and Standards, KS. Standard Test Method for Thermal Resistance for Windows and Doors. 2017, Volume F2278. Available online: http://www.kats.go.kr/ (accessed on 3 July 2023).
27. Korea Agency for Technology and Standards, KS. The Method of Air Tightness for Windows and Doors. 2018, Volume F2292. Available online: http://www.kats.go.kr/ (accessed on 3 July 2023).
28. Korea Agency for Technology and Standards. Available online: http://www.kats.go.kr/ (accessed on 3 July 2023).
29. Lawrence Berkeley National Laboratory. COMFEN. Available online: http://www.lbl.gov/ (accessed on 3 July 2023).
30. Korea Agency for Energy. Available online: http://eep.energy.or.kr/ (accessed on 3 July 2023).

applied sciences

MDPI

Review

The Green Cooling Factor: Eco-Innovative Heating, Ventilation, and Air Conditioning Solutions in Building Design

Bashar Mahmood Ali [1,*] and Mehmet Akkaş [2]

[1] Graduate School of Natural and Applied Sciences, Kastamonu University, Kastamonu 37150, Turkey
[2] Department of Mechanical Engineering, Faculty of Engineering and Architecture, Kastamonu University, Kastamonu 37150, Turkey; mehmetakkas@kastamonu.edu.tr
* Correspondence: bmali@ogr.kastamonu.edu.tr

Abstract: This research investigates the compatibility of conventional air conditioning with the principles of green building, highlighting the need for systems that enhance indoor comfort while aligning with environmental sustainability. Though proficient in regulating indoor temperatures, conventional cooling systems encounter several issues when incorporated into green buildings. These include energy waste, high running costs, and misalignment with eco-friendly practices, which may also lead to detrimental environmental effects and potentially reduce occupant comfort, particularly in retrofit situations. Given the emphasis on sustainability and energy conservation in green buildings, there is a pressing demand for heating, ventilation, and air conditioning (HVAC) solutions that support these goals. This study emphasises the critical need to reconsider traditional HVAC strategies in the face of green building advances. It advocates for the adoption of innovative HVAC technologies designed for eco-efficiency and enhanced comfort. These technologies should integrate seamlessly with sustainable construction, use greener refrigerants, and uphold environmental integrity, driving progress towards a sustainable and occupant-friendly built environment.

Keywords: HVAC; thermal comfort; green buildings; outdoor air conditioning

Citation: Ali, B.M.; Akkaş, M. The Green Cooling Factor: Eco-Innovative Heating, Ventilation, and Air Conditioning Solutions in Building Design. *Appl. Sci.* **2024**, *14*, 195. https://doi.org/10.3390/app14010195

Academic Editors: Luisa F. Cabeza, Daniel Sánchez-García and David Bienvenido Huertas

Received: 10 November 2023
Revised: 7 December 2023
Accepted: 21 December 2023
Published: 25 December 2023

1. Introduction

Recent years have seen a considerable breakthrough in outdoor air conditioning systems, altering the way we cool our living and working environments while fostering sustainability and the energy economy. This introduction provides a summary of the main advancements in outdoor air conditioning and bases its discussion on reliable sources. Outdoor air conditioning technology has advanced to solve challenges like energy usage and environmental effects. To minimise energy consumption and lessen the carbon footprint of cooling operations, modern systems use cutting-edge designs and materials [1]. These developments have produced a more environmentally friendly method of outdoor cooling. Incorporating renewable energy sources, such as solar electricity, into cooling systems is a major development in outdoor air conditioning. This method considerably lessens dependency on conventional, fossil-fuel-based power generation by using the abundant energy from the sun to power air conditioning processes [2]. Using renewable energy for outdoor cooling is an appealing breakthrough, with potential environmental advantages and lower energy costs. Additionally, the development of intelligent outdoor air conditioning systems that are IoT-connected has revolutionised how we manage and control cooling in outdoor environments. These intelligent systems adapt dynamically to shifting environmental circumstances by utilising data analytics and real-time monitoring to maximise cooling efficiency [3]. Such technology improves user comfort while also promoting energy efficiency. Finally, advancements in outdoor air conditioning systems have significantly improved customer comfort and pleasure. Outdoor cooling systems now ensure occupants breathe clean, hygienic air and maintain the proper temperature thanks to advancements in air distribution and quality control [4]. Adaptation strategies for addressing climate change can be effective. Using spectrally selective glazing, summer overheating can be diminished by 15%. Implementing high-performance glazing reduces cold discomfort by 24% and heating needs by 22%, but it can also increase warm discomfort. Shading solutions stabilise energy consumption while cutting summer discomfort by as much as 44%. Thermal insulation reduces winter discomfort and halves energy usage, although it may increase summer discomfort by up to 41%. Climate change projections for 2050–2100 indicate a potential increase in warm discomfort hours by up to 70%. However, adaptive measures can mitigate these effects: natural ventilation could slash warm discomfort by 50–60%, and adaptive temperature setpoints might lower cooling requirements by approximately 35% [5].

These developments are essential for fostering well-being in outdoor areas. In conclusion, outdoor air conditioning technology advancements have addressed energy economy, sustainability, and user comfort. These developments reshape the future of outdoor cooling by incorporating renewable energy sources, implementing intelligent systems, and emphasising air quality [3]. This makes outdoor cooling more user- and environmentally friendly. This introduction will look at the cutting-edge developments that have elevated outdoor air conditioning to a new level of adaptability, sustainability, and efficiency [2]. The mitigation of climate change and the reduction of greenhouse gas emissions are two of the most urgent problems of the twenty-first century, as shown in Figure 1.

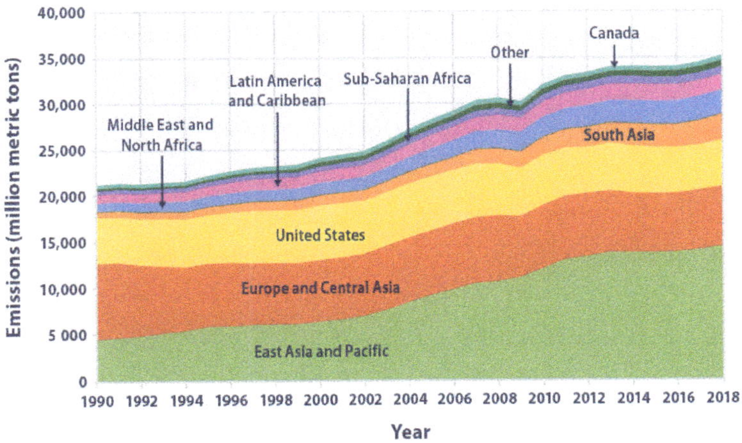

Figure 1. Greenhouse gas emissions by region [4].

Systems for cooling the outdoors are essential for this project. Modern systems use heat recovery technology and eco-friendly refrigerants to minimise their adverse environmental effects. These developments support international initiatives to reduce carbon emissions and lessen the effects of climate change. Integrating renewable energy is a critical component of contemporary outdoor air conditioning solutions. Photovoltaic panels and wind turbines are increasingly being incorporated into the design of outdoor cooling systems to use renewable energy sources to power air conditioners [6]. By switching to renewable energy, the carbon impact of outdoor cooling is drastically reduced while simultaneously lowering operational costs. A new era of outdoor air conditioning control and management has arrived with the introduction of the Internet of Things (IoT). To optimise cooling operations based on real-time weather conditions and occupancy patterns, IoT-enabled systems use sensors and data analytics [7]. With the help of this dynamic control, consumers are guaranteed a comfortable environment while consuming less energy than necessary. The mobility and scalability of outdoor air conditioning technology have also advanced. To provide on-demand cooling for various events and locations, portable outdoor cooling systems are now widely accessible and offer flexibility and cost-effectiveness [8]. This versatility is crucial for meeting the various requirements for outdoor cooling in various situations. In conclusion, outdoor air conditioning technology improvements have been made to meet user convenience, energy efficiency, and environmental issues. The outdoor cooling environment has changed into a sustainable and user-centric area with the adoption of eco-friendly refrigerants, renewable energy integration, IoT-driven intelligent systems, and the emergence of portable solutions.

Thermal comfort, a crucial aspect of the design and operation of buildings, is affected by many factors, such as air temperature, relative humidity, and air velocity [9]. Even though conventional HVAC systems are successful at maintaining a "comfort zone", they are often criticised for their high energy consumption, environmental impact, and greenhouse gas emissions [10]. To remedy these problems, green buildings strive to promote thermal comfort using energy-efficient techniques [11]. However, using conventional HVAC systems in green buildings sometimes contradicts the same sustainability ideals they intend to maintain [12]. Innovative methods such as passive design techniques, which incorporate natural ventilation and sun heating, have been investigated [13]. Moreover, modern HVAC technologies, such as Variable Refrigerant Flow (VRF) and radiant cooling systems, provide potential pathways for enhancing energy efficiency and thermal comfort [14]. The move towards human-centric methods that account for the adaptable nature of human thermal comfort is a developing trend in the literature [15]. Moreover, incorporating Internet of Things (IoT) technology allows for real-time monitoring and the adaptive management of interior conditions, providing a more dynamic approach to thermal comfort in green buildings [16]. Although heating, ventilation, and air conditioning (HVAC) systems are ubiquitous in maintaining temperature conditions, traditional systems are often criticised for their excessive energy consumption, poor indoor air quality, and large greenhouse gas emissions [10]. These limits become more troublesome in the context of green buildings, which are meant to maximise occupant comfort while reducing environmental damage [11]. The limitations of present HVAC systems are shown in Table 1.

Table 1. The limitations of HVAC systems.

Limitation	Description
Energy Consumption	HVAC systems are often significant consumers of energy, contributing to high operational costs and environmental concerns.
Installation Cost	The initial cost of purchasing and installing HVAC systems can be high, particularly for large or sophisticated systems.
Maintenance Requirements	Regular maintenance is essential for HVAC systems to operate efficiently, which can incur additional costs and downtime.
Space Requirements	Large HVAC units require considerable space, which can be a limitation in compact or densely built environments.
Noise Levels	HVAC systems, especially older or larger units, can generate noise, potentially causing disturbance in quiet environments.
Refrigerant Environmental Impact	Some HVAC systems use refrigerants that can contribute to ozone depletion and climate change if leaked.
Indoor Air Quality	If not properly maintained, HVAC systems can contribute to poor indoor air quality by circulating dust, mould, and other contaminants.
Temperature Inconsistencies	Achieving consistent temperatures in a building can be challenging, leading to hot or cold spots.
Humidity Control Limitations	Some HVAC systems may struggle to maintain optimal humidity levels, affecting comfort and indoor air quality.
Lifespan	The lifespan of HVAC systems can be limited, necessitating replacement or significant upgrades.
Adaptability	Older HVAC systems might not be easily adaptable to new technologies or changing environmental regulations.
Aesthetic Impact	Large and visible HVAC equipment can have a negative impact on building aesthetics.

To reconcile this dichotomy, scholars and practitioners have explored passive design strategies like natural ventilation, solar heating, and thermal mass, which can significantly reduce energy demand while maintaining or improving thermal comfort [17]. Advanced HVAC technologies are emerging as another solution; for example, Variable Refrigerant Flow (VRF) systems, radiant cooling systems, and chilled beams show significant promise of enhancing energy efficiency without compromising comfort [18]. The literature also increasingly focuses on adaptive and human-centric thermal comfort models, recognising that comfort is not a static, one-size-fits-all phenomenon but varies based on cultural, psychological, and individual physiological factors [19]. Internet of Things (IoT) technologies add another layer of sophistication by allowing for real-time monitoring and the adaptive control of multiple environmental parameters, thus enabling a more dynamic, responsive approach to achieving thermal comfort in green buildings, as shown in Figure 2 [20].

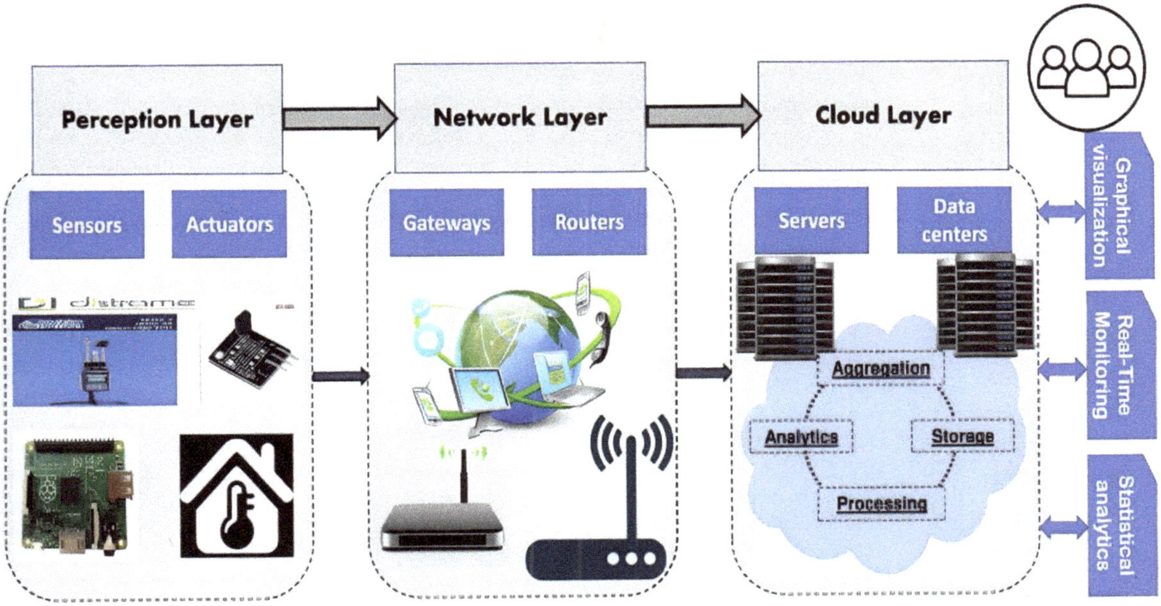

Figure 2. Internet of things (IoT) technologies [20].

Despite these developments, there are still knowledge gaps about effectively combining these varied tactics into a coherent, practical, and scalable strategy for boosting thermal comfort in green buildings. Thermal comfort, essential to the well-being of building occupants, has been widely investigated, demonstrating its dependence on parameters such as air temperature, radiant temperature, humidity, air velocity, and human characteristics, such as clothing and metabolism [21]. While traditional HVAC systems attempt to standardise these factors, they often fall short in energy efficiency and flexibility [22]. Especially in the field of green buildings, where the emphasis is placed on both comfort and environmental sustainability, traditional HVAC systems often fail to satisfy both goals [23]. The energy-intensive nature of current HVAC systems, which contributes to high operating costs and greenhouse gas emissions, is one of the greatest obstacles [21,22]. Efforts to minimise energy usage with passive design approaches such as natural ventilation and sun orientation have been reported, but they sometimes come at the expense of consistent comfort [18,19]. Emerging technologies, such as earth–air heat exchangers and phase-change materials, promise to overcome this gap by enabling temperature control without depending on energy-intensive mechanical devices [24]. Personalised thermal comfort systems, which employ wearable technology and IoT to adjust indoor settings to an individual's preferences [25], have the potential to increase comfort while lowering total energy consumption [26], since they permit more variable indoor circumstances. Moreover, machine learning methods are being investigated to forecast and adjust for occupant temperature preferences in real-time, boosting the flexibility of green building systems [27]. Despite these developments, the industry lacks comprehensive models incorporating different tactics, such as passive design, sophisticated materials, and customised systems, into a unified framework for enhancing thermal comfort in green buildings [28]. In addition, the scalability and applicability of these technologies to various climatic conditions and building types remain subjects for further study. In the dynamic realm of green building HVAC systems, the novelty of this manuscript stands out amidst a plethora of academic contributions. While several papers have traversed the technical intricacies of HVAC systems or delved into the singular facets of green buildings, this review offers an unparalleled, holistic perspective. It amalgamates discussions spanning occupant comfort, technological innovations, and market dynamics and ventures into the often-overlooked physiological and psychological dimensions governing thermal comfort. Another distinctive feature is the manuscript's exhaustive exploration of emergent HVAC technologies, such as Variable Refrigerant Flow (VRF) and phase-change materials, providing a rich, comparative analysis that might surpass many contemporaneous reviews [12]. However, this manuscript truly carves its niche in its candid exposition of the challenges plaguing the integration of traditional HVAC systems into green edifices. This, coupled with actionable insights and potential remedial measures, addresses a lacuna that remains conspicuously absent in many other works. Furthermore, introducing the adaptive comfort model, a paradigm that hinges on an occupant's experiential adaptability, infuses a fresh, human-centric perspective, balancing the often technocentric narratives of other reviews. With its judicious blend of technical depth, human considerations, and real-world implications, this manuscript distinguishes itself as a seminal contribution, poised to reshape the discourse on HVAC systems in green buildings [8]. The groundbreaking aspect of this study lies in its comprehensive approach to addressing the integration of traditional HVAC systems within the framework of green building principles. While previous research may have separately touched upon energy efficiency, occupant comfort, or the environmental impact of HVAC systems, this study is novel in its holistic examination of all these elements in tandem. Moreover, its emphasis on the economic implications of integrating traditional HVAC systems into sustainable designs provides a fresh perspective that goes beyond the environmental discourse [13]. Another pioneering feature is its exploration of eco-friendly refrigerants, a topic that, until now, has been underrepresented in mainstream research. This study also stands out for its in-depth look at retrofitting challenges, offering a unique blend of theoretical insights and practical solutions [21]. By bridging the often-separate worlds of sustainable construction and HVAC system design, this research introduces a groundbreaking narrative set to shape both industries and inspire further interdisciplinary research. In essence, the novelty of this study is its multifaceted, interdisciplinary approach, filling critical knowledge gaps and providing a roadmap for the harmonious integration of comfort, sustainability, and economic viability in the built environment. The growing interest in adaptive thermal comfort theory stems from the inadequacy of the traditional "one size fits all" comfort model in achieving widespread occupant satisfaction. This was highlighted by extensive field validation studies funded by the ASHRAE in the 1980s and 1990s. By contrast, field studies in environments employing adaptive comfort practices generally show enhanced occupant comfort and satisfaction [29]. A critical aspect of adaptive thermal comfort theory is its reliance on statistical field data analyses, making its main models somewhat opaque or "black box" in nature. This review first explores various attempts to develop more transparent, explanatory adaptive comfort models. The second focus of this review is the evolution of adaptive comfort regulatory documentation over the past 21 years. Carlucci et al. have thoroughly compared adaptive thermal comfort models frequently used in building environment standards. These models were officially incorporated into the ASHRAE standard 55 in 2004 and have been refined in subsequent updates [30].

This study aims to evaluate the efficiency and cost-effectiveness of modern HVAC systems in sustainable building design, comparing traditional and eco-innovative models. It will assess financial implications, energy efficiency, and regulatory compliance, using statistical analysis and case studies to understand their practicality in various climates and building types.

2. Methodology to Achieve the Results

This study adopted a mixed method approach, combining quantitative and qualitative research methodologies to elucidate the intricate relationship between traditional HVAC systems and green building design principles. An exhaustive literature review was conducted to understand the current state of research in the domain. Scholarly articles, conference proceedings, and industry reports were reviewed. This helped in the understanding of existing HVAC technologies, their energy consumption patterns, the evolution of green building principles, and the perceived gaps between them [22]. A quantitative analysis was performed on the energy consumption of traditional HVAC systems. Data were collected from various buildings, both residential and commercial, over one year. The data were then benchmarked against buildings incorporating green building principles and newer HVAC technologies [27]. Surveys were administered to occupants of buildings with traditional HVAC systems to gauge their comfort levels. Simultaneously, structured interviews were conducted with architects, HVAC engineers, and green building consultants. This qualitative approach provided more profound insights into the perceived challenges and opportunities in integrating HVAC systems with green building designs [18]. Several green buildings that have successfully integrated innovative HVAC systems were chosen as case studies. These provided practical insights into real-world applications and challenges. Each case study was analysed regarding energy efficiency, occupant comfort, retrofitting complexities, and the use of eco-friendly refrigerants [15]. The collected data were analysed using statistical tools and software packages. Quantitative data from energy consumption analysis and surveys were subjected to regression analysis, ANOVA, and *t*-tests to determine significant differences and patterns. Qualitative data from interviews were analysed using thematic analysis, allowing for the emergence of designs and themes related to challenges and solutions [14]. The mixed method approach provided a holistic perspective on the challenges and opportunities in integrating traditional HVAC systems with green building principles. While conventional systems posed significant challenges regarding energy inefficiency and incongruence with sustainability, innovative alternatives showed promise in bridging the gap. This study emphasises the urgency of transitioning to newer HVAC technologies that align with green building principles, ensuring energy efficiency, ecological sustainability, and occupant comfort.

The graphic in Figure 3 represents data from keywords searched on Scopus with the keywords "thermal AND comfort; AND green AND buildings". This visualisation, a 3D bubble chart, suggests a trend analysis over time, possibly evaluating the relationship between thermal comfort and green building practices from 2000 to 2030. Each bubble likely represents a dataset or a collection of studies published on the topic in a given year. As time progresses, the increasing size of the bubbles could indicate a growing number of studies or an increase in a particular metric related to thermal comfort in green buildings. The numbers within the bubbles could signify the number of research papers or the scoring of a specific index linked to the research intensity or impact in that year. For example, the jump from 82 in 2015 to 100 in 2020 could indicate a significant increase in research activities or findings on thermal comfort in green buildings.

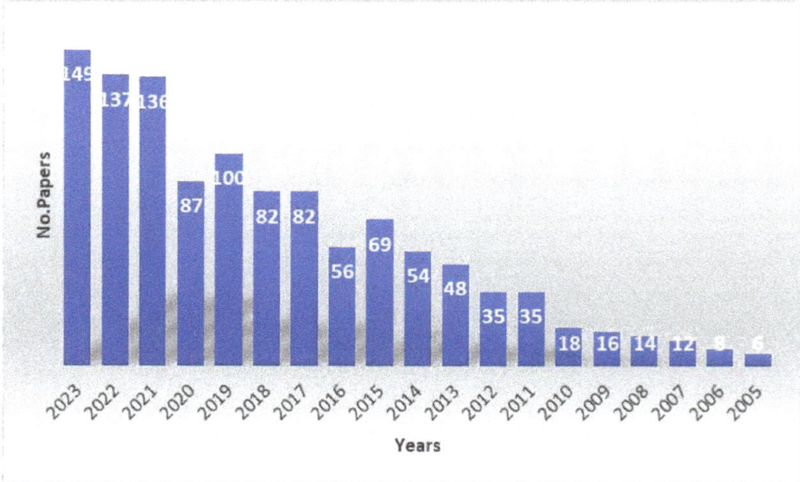

Figure 3. The graph represents publications from 2005 to 2023 containing the keywords searched.

3. Green Buildings

Green buildings, usually considered sustainable or eco-friendly structures, are a thorough and well-rounded method of building and designing, as shown in Figure 4. These buildings are skilfully designed to impact the environment as little as possible while improving energy efficiency, occupant comfort, and overall sustainability [29]. Such structures are made with a significant emphasis on reducing greenhouse gas emissions, improving indoor air quality, and conserving resources, all of which help create a more sustainable built environment. Classic HVAC (heating, ventilation, and air conditioning) systems that have historically been used for indoor climate control are often referred to as traditional air conditioning technology in the context of green buildings [30]. The effect of occupants on HVAC system operation and, consequently, on building energy consumption is a multifaceted issue that significantly impacts the efficiency and sustainability of buildings. Occupant behaviour, including how individuals use space and their personal preferences for comfort, is crucial in determining the energy performance of HVAC systems. Occupants' daily activities, such as using electronic devices, using lighting, and window operation, directly influence the indoor climate, affecting the HVAC system's workload. For instance, a higher number of occupants or increased activity levels can lead to more significant heat generation, which requires the HVAC system to work harder to maintain the desired temperature, thereby increasing energy consumption. Moreover, the individual preferences of occupants for temperature and ventilation can lead to inefficiencies in HVAC operation [18]. Different people have varying comfort levels; some prefer warmer temperatures while others prefer cooler conditions. This diversity often results in the HVAC system being adjusted frequently to meet everyone's needs, leading to inefficient operation and increased energy use. This situation is further complicated in buildings where HVAC controls are decentralised or when occupants have direct control over thermostatic settings. In such scenarios, the lack of a standardised temperature setting can result in significant energy wastage [29].

Occupant behaviour modelling has become an integral part of energy efficiency studies. More efficient HVAC systems can be designed by understanding and predicting how occupants interact with building systems and their environment. This involves using sensors and advanced analytics to monitor and analyse occupant behaviour in real-time, enabling the HVAC system to adapt dynamically to usage patterns. Such intelligent HVAC systems can optimise energy consumption by adjusting settings based on occupancy levels, time of day, and even weather conditions [30]. Furthermore, educating occupants about the impact of their behaviour on energy consumption is crucial. Implementing user-friendly interfaces for HVAC controls and providing feedback on energy usage can encourage more energy-conscious behaviour. Building designs that promote natural ventilation and daylighting can also reduce the reliance on HVAC systems, further conserving energy [18].

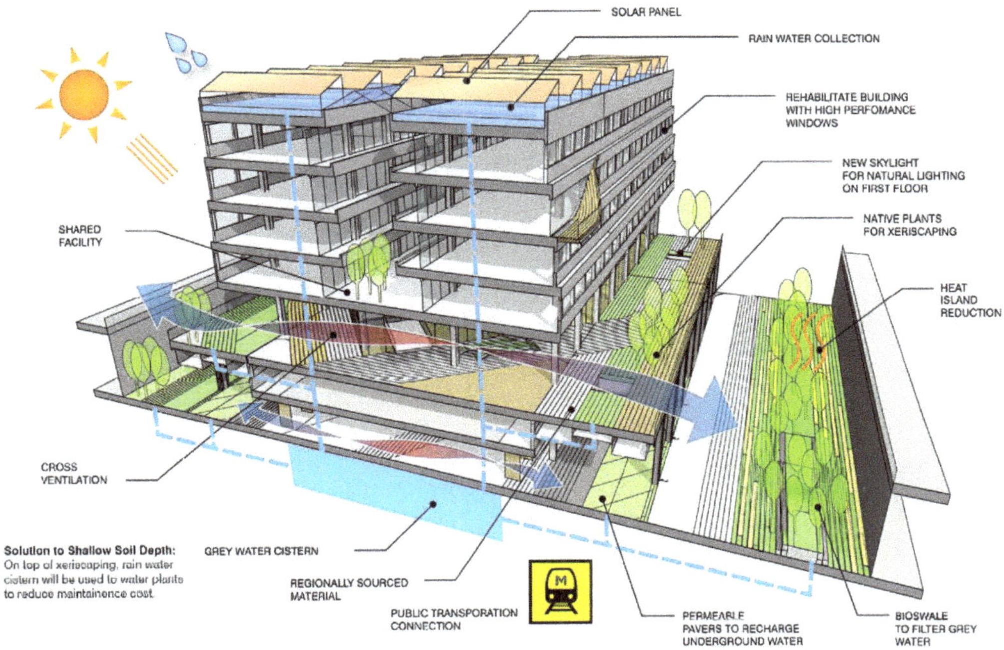

Figure 4. Green building [29].

In addition to behavioural aspects, the physical presence of occupants also affects indoor air quality (IAQ). As occupants exhale carbon dioxide and potentially introduce pollutants, HVAC systems must ensure adequate ventilation and air filtration to maintain a healthy indoor environment. This requirement often leads to a delicate balance between energy efficiency and IAQ, especially in tightly sealed, energy-efficient buildings where ventilation is primarily dependent on the HVAC system.

To cool and dehumidify indoor environments, these systems often rely on energy-intensive processes such as the mechanical compression of refrigerants. Due to their high electricity consumption and usage of refrigerants with a high global warming potential (GWP), typical HVAC systems are frequently shown to be energy inefficient and leave a significant environmental impact. Green buildings, often termed sustainable or eco-friendly structures, represent a comprehensive approach to construction and design that prioritises environmental stewardship, energy efficiency, and human well-being. These buildings are meticulously planned to minimise their ecological footprint through various strategies, including reducing greenhouse gas emissions, enhancing indoor air quality, and conserving natural resources [12,29]. The overarching goal is to create a built environment that is sustainable and conducive to occupant comfort and well-being [31]. One of the critical aspects of green buildings is their focus on energy efficiency. Traditional buildings consume significant heating, cooling, and lighting energy, contributing to approximately 40% of global energy use. On the other hand, green buildings incorporate technologies such as solar panels, energy-efficient windows, and advanced insulation materials to reduce energy consumption [31]. Table 2 shows a comparison of traditional and green buildings.

Table 2. A comparison of traditional and green buildings.

Criteria	Traditional Buildings	Green Buildings
Energy Source	Primarily rely on non-renewable energy sources like fossil fuels for heating, cooling, and power.	Utilise renewable energy sources such as solar, wind, and geothermal energy, reducing reliance on fossil fuels.
Energy Efficiency	Are generally less energy-efficient due to older technologies and materials.	Are designed with energy efficiency in mind, using advanced technologies and materials to reduce energy consumption.
Insulation	May have poor insulation, leading to higher heating and cooling needs.	Feature high-quality insulation to minimise heat loss and reduce heating and cooling needs.
Lighting	Often use inefficient lighting fixtures and bulbs, consuming more electricity.	Employ energy-efficient lighting solutions like LED and CFL and maximise natural light through design.
HVAC Systems	Use older, less efficient HVAC systems, consuming more energy.	Incorporate energy-efficient HVAC systems, often with intelligent controls to optimise performance.
Water Heating	Typically use standard water heating systems, which can be less efficient.	Use energy-efficient water heating solutions such as solar water heaters or heat pumps.
Appliances and Fixtures	Equipped with standard appliances and fixtures that consume more energy.	Are fitted with Energy Star-rated appliances and fixtures to minimise energy use.
Building Envelope	Contain conventional building materials and design, which may result in more energy loss.	Use sustainable building materials and design principles to enhance energy conservation.
Ventilation	May have less adequate ventilation, requiring more energy for air quality control.	Are designed for effective natural ventilation, reducing the need for mechanical ventilation.
Energy Management	Contain a lack of advanced energy management systems, leading to inefficient energy use.	Incorporate advanced energy management systems to monitor and optimise energy consumption.
Carbon Footprint	Have a higher carbon footprint due to higher energy consumption and reliance on fossil fuels.	Have a lower carbon footprint due to reduced energy consumption and renewable energy sources.

These features lower the building's operational costs and reduce its carbon footprint, contributing to climate change mitigation. Indoor air quality is another critical focus area for green buildings. Traditional construction materials often contain volatile organic compounds (VOCs) that can harm human health. Green buildings use low-VOC and non-toxic materials to improve indoor air quality, thereby enhancing the well-being of the occupants [32]. Water conservation is also a significant aspect of green building design. Green buildings aim to reduce water consumption and waste by using water-efficient fixtures, rainwater harvesting, and greywater recycling systems [32]. These practices conserve a vital natural resource and reduce the strain on municipal water supply systems [32]. In indoor climate control, classic HVAC (heating, ventilation, and air conditioning) systems, often referred to as traditional air conditioning technology, have been a concern in the green building discourse. These systems typically rely on energy-intensive processes such as the mechanical compression of refrigerants to cool and dehumidify indoor spaces. The high electricity consumption and use of refrigerants with a high global warming potential (GWP) make these systems both energy-inefficient and environmentally detrimental [32]. Green buildings often employ alternative climate control technologies such as natural ventilation, evaporative cooling, and ground-source heat pumps (GSHPs) to address these issues. These technologies are more energy-efficient and have a lower environmental impact than traditional HVAC systems [32]. Technological advancements, architectural practices, and ecological priorities shape the intricate relationship between HVAC systems and green building design. Central to this dynamic is the ensuring of occupant comfort without compromising sustainability principles. Traditionally, HVAC systems have significantly contributed to a building's energy consumption. However, in the realm of green buildings, which emphasise energy conservation, there is a pressing need for HVAC solutions that are both efficient and adaptive. The architectural design of a building can significantly influences its HVAC requirements. For instance, structures optimised for natural ventilation, shading, and thermal insulation can reduce the reliance on mechanical cooling or heating [19]. Integrating architectural foresight with HVAC functionalities exemplifies the symbiotic potential between the two. Furthermore, technological innovations, especially the advent of AI-driven intelligent HVAC systems, have ushered in a new era of energy efficiency [18]. These systems, equipped with sensors, can pre-emptively adjust to occupant behaviour, striking a balance between comfort and energy conservation. Yet, the environmental implications of HVAC systems, particularly concerning refrigerant use, cannot be overlooked. The shift towards eco-friendly refrigerants underscores the industry's commitment to ecological stewardship, aligning with the ethos of green buildings. While the initial investment in such advanced HVAC systems might be substantial, the tangible and intangible long-term benefits justify the costs. Reduced energy bills, enhanced indoor air quality, and the overarching advantage of a minimised environmental footprint highlight the indispensable role of HVAC systems in the future of sustainable architecture [18].

Challenges with Traditional HVAC

Sustainability, energy efficiency, and environmental stewardship are crucial for developing green buildings. Nevertheless, integrating traditional heating, ventilation, and air conditioning (HVAC) systems into these environmentally aware buildings usually creates significant challenges [32]. The low energy efficiency of the ageing HVAC systems is one of the primary challenges. They typically use a great deal of power, which might limit the energy-saving gains made achievable by green building design. The inefficiency of traditional HVAC systems is a significant issue in green buildings since the goal is to reduce energy consumption and carbon footprint [32]. Another issue is the high operational costs. For owners and occupiers of green buildings, standard HVAC systems' excessive energy consumption raises operating expenditures. These costs may discourage consumers from investing in green construction features to reduce their environmental effects and save money on energy costs [31]. Traditional HVAC systems usually need more maintenance and a shorter lifetime, increasing their long-term expenses. In addition, conventional HVAC systems are limited in their ability to suit the unique characteristics and requirements of green buildings. Green buildings often include passive design principles, such as daylighting and natural ventilation, which may interfere with the performance of traditional HVAC systems [32]. This restricted flexibility may result in waste and decreased comfort in green buildings.

Traditional HVAC systems have environmental impacts beyond energy use. These systems commonly use refrigerants with a high GWP, which may contribute to the loss of the ozone layer and worsen climate change. The selection of heating, ventilation, and air conditioning (HVAC) systems is a vital aspect of green buildings, whose primary objective is eliminating environmental damage. Moreover, although traditional HVAC systems may maintain a consistent temperature, they may not prioritise occupant comfort as much as green building designs do. In contrast to conventional HVAC systems, green buildings usually prioritise indoor air quality, natural ventilation, and thermal comfort via passive approaches [32]. This might lead to subpar interior design. Integrating green features into older buildings with traditional HVAC systems may be challenging and costly. Incorporating energy-efficient technology and renewable energy sources may entail extensive adjustments to the buildings and to the HVAC systems [32], making it more challenging for certain buildings to embrace green building standards.

Incorporating conventional heating, ventilation, and air conditioning (HVAC) systems into green buildings presents several obstacles that weaken sustainability, energy efficiency, and environmental responsibility. One of the most visible concerns is the energy inefficiency of traditional HVAC systems, which often use a disproportionate amount of power, thereby nullifying the energy-saving gains that green building designs attempt to accomplish [32]. This inefficiency is especially troublesome given that one of the critical goals of green buildings is to decrease energy usage and carbon footprint. In addition, the high energy consumption of conventional HVAC systems results in higher operating expenses for building owners and occupants, discouraging people and organisations from investing in green building features [33]. These costs may be complicated since they may balance the anticipated savings in energy expenses, diminishing the appeal of investing in a green building. Traditional HVAC systems need regular maintenance and have a shorter lifetime than contemporary, energy-efficient systems, resulting in more extraordinary long-term expenses [34]. Green buildings often incorporate novel passive design principles like daylighting and natural ventilation, which may be incompatible with the functioning of conventional HVAC systems. This lack of adaptation may lead to energy waste and poor occupant comfort, undermining the holistic approach to well-being that green buildings aim to accomplish [31]. Traditional HVAC systems often use refrigerants with a high global warming potential (GWP), which contributes to the depletion of the ozone layer and worsens climate change. This contradicts the environmental aims of green buildings, which seek to minimise such damage. Traditional HVAC systems may maintain a constant temperature. Still, they often do not prioritise other factors of occupant comfort emphasised in green buildings, such as indoor air quality and natural ventilation [32]. This may lead to inferior interior conditions, reducing inhabitants' quality of life. Lastly, it might be tough to retrofit older buildings with green features if these structures have obsolete HVAC systems. Integrating energy-efficient technology with renewable energy sources may require significant changes to the building structure and the HVAC systems, adding complexity and expense to the retrofitting process [35].

To continue the subject, the issues provided by conventional HVAC systems in green buildings extend to regulatory and legislative frameworks. Traditional HVAC systems often fail to achieve these new norms, causing compliance difficulty for building owners and developers in several jurisdictions with more strict building codes and energy efficiency regulations intended to promote sustainability [35]. This regulatory mismatch delays the permission process and exposes owners to possible legal repercussions, adding a layer of complication and expense to the construction or retrofitting process. Incompatibility between conventional HVAC systems and green building designs may result in performance discrepancies. In other words, the building may not function as effectively as first projected or anticipated, resulting in a "performance gap". This disparity may be incredibly distressing for stakeholders who invested in green building elements to attain specified energy savings and environmental objectives [35]. The performance gap may also undermine faith in green building technology and practices, thereby slowing the adoption of these vital solutions for reducing climate change and fostering sustainability [35]. Another aspect that is sometimes disregarded is the human element. Traditional HVAC systems often need specific knowledge and abilities to operate. By contrast, the controls and automation elements of contemporary, energy-efficient systems incorporated into green buildings are often more sophisticated to improve performance. Thus, moving from conventional to modern systems may require extensive retraining for facility management personnel, adding to the indirect costs and difficulties associated with implementing green building practises [36]. Furthermore, integrating renewable energy sources like solar or wind power into green buildings provides another complication when typical HVAC systems are involved. Frequently, these systems are not built for simple integration with renewable energy sources, necessitating extra equipment, control systems, and sometimes complex electrical work to make integration viable. This raises the initial construction cost and adds another layer of complexity to the building's energy management system, making the transition to a more sustainable built environment more challenging. The issues connected with integrating conventional HVAC systems into green buildings have ramifications for urban planning and the electrical grid. Inefficient HVAC systems may increase peak electrical demands, increasing problems such as energy poverty and grid instability. As cities strive to become more sustainable and resilient, the incompatibility of conventional HVAC systems with green building goals becomes a building-level problem and a systemic one that must be addressed [37]. In conclusion, integrating conventional HVAC systems into green buildings has obstacles beyond energy inefficiency and high operating costs. These obstacles include regulatory compliance, performance gaps, human issues, integration with renewable energy sources, and urban and grid-level ramifications. As green construction approaches continue to improve and gain acceptance, it becomes more critical to address these multiple difficulties. It is not only a matter of retrofitting or replacing obsolete systems but also of reconsidering how HVAC systems fit into the larger ecosystem of sustainable building practises and urban planning [35]. To comprehensively address occupant comfort within green buildings, it is imperative to employ specific metrics and methods. While this manuscript underscores the centrality of comfort, delving into its quantitative assessment can illuminate the discussion. Thermal comfort, for instance, can be measured using tools such as the predicted mean vote (PMV) and predicted percentage dissatisfied (PPD) [26]. The PMV provides an aggregate assessment of the comfort level expressed by a group of occupants, while the PPD quantifies the percentage likely to find the environment uncomfortable. By integrating these metrics, we can derive tangible insights from the performance of innovative HVAC technologies. Furthermore, the manuscript could explore how these state-of-the-art systems, through features like adaptive temperature control and humidity modulation, optimise energy consumption and enhance overall occupant comfort. Such a data-driven approach can bolster this manuscript's argument, emphasising the harmonious integration of green building principles, HVAC innovations, and the human experience [28].

4. Eco-Friendly Air-Conditioning

Recent advances in eco-friendly design and materials for outdoor air conditioning systems have been made to meet rising concerns about energy usage and environmental effects. This event marks a turning point in the industry's transformation to more eco-friendly and sustainable practices [37].

1. Using sustainable refrigerants is one of the most critical aspects of eco-friendly design. Hydrochlorofluorocarbons (HCFCs) and hydrofluorocarbons (HFCs), both potent greenhouse gases, were widely used in conventional air conditioning systems. However, more contemporary outdoor cooling systems favour using hydrofluoroolefins (HFOs) and natural refrigerants such as ammonia and carbon dioxide [38]. These alternatives are less environmentally harmful and less likely to contribute to global warming.
2. Heat recovery technology: Incorporating heat recovery technology is another breakthrough. Outdoor cooling systems may collect and reuse waste heat generated during the cooling process [39]. This reduces total energy use while enhancing energy efficiency. Systems for heat recovery are particularly beneficial in commercial and industrial settings that generate a great deal of heat, which may be utilised for activities such as water heating [37].
3. Energy-efficient components: Eco-friendly design involves using energy-efficient materials and components. High-efficiency compressors, fans, and heat exchangers are merely a few examples of the elements utilised in outdoor air conditioning systems that have reduced energy consumption [40]. These components help reduce energy use, which benefits the environment and saves users money.
4. Sustainable production methods: The production of outdoor air conditioning systems goes beyond the working stage. Manufacturers progressively embrace eco-friendly manufacturing practices [41] to decrease waste, use less energy and water during

production, and construct cooling units out of recycled or recyclable materials. These measures reduce the carbon footprint of the manufacturing process.

5. Regulatory conformity: Government regulations and industry standards have considerably impacted adopting eco-friendly design and materials in outdoor air conditioning technologies. The Montreal Protocol and its revisions have pushed the industry toward more sustainable practices by emphasising the elimination of ozone-depleting chemicals and decreasing refrigerants with a high GWP (global warming potential) [42].

The transition to environmentally friendly outdoor air conditioning systems is a multifaceted industry shift fuelled by technological innovation, sustainable manufacturing, and regulatory compliance. Green refrigerants, such as hydrofluoroolefins (HFOs) and natural chemicals, such as ammonia and carbon dioxide, have a lower global warming potential (GWP) than conventional hydrochlorofluorocarbons (HCFCs) and hydrofluorocarbons (HFCs) [43]. This development is partly the outcome of international legislation, such as the Montreal Protocol, which seeks to eradicate ozone-depleting substances. Particularly helpful in commercial and industrial settings is the implementation of heat recovery technology, which recovers waste heat produced during the cooling process to reduce overall energy consumption and boost system efficiency [35]. The emphasis on environmentally responsible design also extends to using energy-efficient components, such as high-efficiency compressors, fans, and heat exchangers, which minimise energy consumption and cut operating expenses. Beyond the operational phase, manufacturers are adopting sustainable practises in producing these systems, such as reducing waste, minimising energy and water consumption, and utilising recycled or recyclable materials, per the principles of circular economy and sustainable development [44]. Regulatory frameworks such as the Montreal Protocol have played a significant role in pushing these changes by providing the economic and legal incentives necessary for the sector to adopt more sustainable methods. In conclusion, advancements in green refrigerants, heat recovery technologies, energy-efficient components, and sustainable manufacturing practices collectively contribute to a more sustainable approach to outdoor cooling, establishing new industry standards for energy efficiency and sustainability, and providing essential solutions in the face of significant climate change-related challenges [45]. Multiple reasons, including consumer demand for sustainable goods, technology breakthroughs, and more rigorous environmental legislation, are influencing the continuous change in the air conditioning sector towards eco-friendly solutions. The transition towards green refrigerants is notable because it signals a break from the usage of HCFCs and HFCs, recognised as significant contributors to global warming and ozone depletion. Adopting alternative refrigerants such as hydrofluorooctane, ammonia, and carbon dioxide is a systemic reaction to global environmental concerns and not only a technical shift. This is bolstered by international accords such as the Montreal Protocol, which has established timelines for eliminating ozone-depleting compounds and fostered innovation in the industry. Heat recovery technology is an essential innovation in the search for environmentally friendly air conditioning [45]. These systems cut energy consumption by collecting and recycling waste heat and contribute to energy efficiency and sustainability objectives. This is particularly significant in commercial and industrial contexts, where substantial waste heat may be recycled for other energy-intensive activities, producing a more connected and efficient energy ecosystem [46]. This multifaceted approach to sustainability emphasises energy-efficient components such as high-efficiency compressors, fans, and heat exchangers. These components are intended to perform optimally, decreasing the air conditioning system's total energy consumption. This coincides with environmental goals and translates into economic gains via lower operating costs, creating a win–win outcome for both customers and the environment [47].

Sustainable manufacturing techniques are expanding the notion of environmental friendliness beyond the product to include its entire lifespan, from production to disposal. Manufacturers increasingly emphasise eliminating waste, employing recycled or recyclable materials, and decreasing energy and water usage throughout the manufacturing process. These behaviours adhere to the circular economy's ideas, encouraging a regenerative approach to production and consumption. Compliance with regulations acts as both a catalyst and a foundation for these adjustments. Laws such as the Montreal Protocol [48] have placed the sector on the road to greater sustainability by regulating the phase-out of dangerous compounds and encouraging the use of energy-efficient technology. These rules sometimes include economic incentives, such as tax breaks or subsidies, which make it financially feasible for businesses to participate in the research and development of environmentally friendly technology [44]. In conclusion, the transition to ecologically friendly outdoor air conditioning systems involves a complete endeavour involving various stakeholders, including manufacturers, regulators, and consumers. Advances in green refrigerants, heat recovery technology, energy-efficient components, and sustainable manufacturing processes lead to a more sustainable approach to outdoor cooling. These achievements are vital for facing the substantial problems faced by climate change and environmental degradation as they establish new industry standards for energy efficiency and sustainability. Figure 5 is a graphical figure to explain the eco-friendly air conditioning.

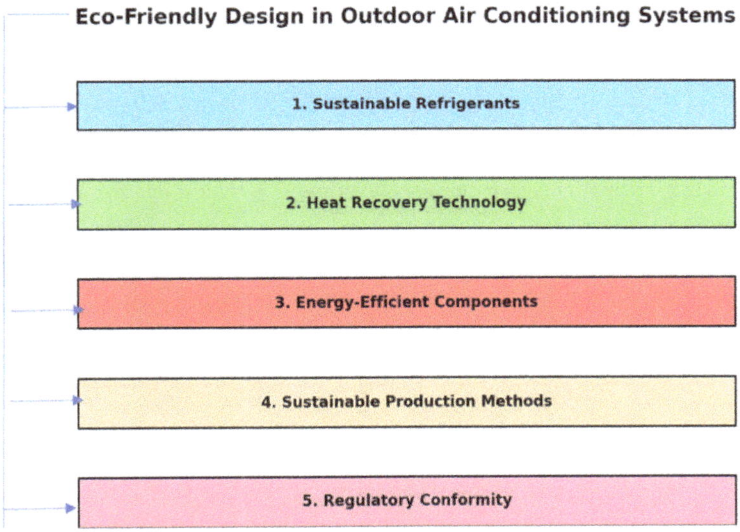

Figure 5. A representation of eco-friendly air conditioning.

5. Revolutionizing Outdoor Cooling

Renewable energy is a significant development in outdoor air conditioning systems, marking a paradigm shift toward ecologically benign and sustainable cooling techniques [49]. This technique largely depends on renewable energy sources, namely solar electricity, to power outdoor cooling equipment.

5.1. Solar Energy Integration

Solar power integration involves incorporating photovoltaic and solar thermal systems into outdoor air conditioners. Through solar panels, sunlight is transformed into energy, which is then utilised to power the cooling systems. By contrast, solar thermal systems harness the sun's energy to create heat that may be used to cool objects, such as desiccant systems or absorption chillers [49].

5.2. Reduce Fossil Fuel Use

One of the primary benefits of renewable energy integration is a significant reduction in reliance on conventional fossil fuels for power production. Frequently, the energy used by traditional cooling systems is derived from coal, natural gas, and other fossil fuels, which increases greenhouse gas emissions and harms the environment. Using solar energy and other renewable energy sources, outdoor cooling systems contribute to the battle against climate change and decrease their carbon footprint [49].

5.3. Advantages for the Environment and Sustainability

Utilising renewable energy allows outdoor cooling technologies to meet broader environmental aims. Solar energy is an endless and renewable resource that reduces the environmental impact of cooling operations. This contributes to global efforts to reduce carbon emissions and foster a cleaner future [49].

5.4. Energy-Cost Cost Savings

Renewable energy integration brings environmental advantages and possible energy cost reductions over the long run. Solar panels and related infrastructure need an initial investment, but their long-term running costs are usually less than traditional power consumption. Renewable energy is an attractive choice for outdoor cooling applications due to its low cost [49].

5.5. The Grid's Independence

Incorporating renewable energy also increases the reliability and longevity of outdoor cooling systems. By creating their power via solar panels, these systems become less dependent on the grid's dependability. This independence is essential in remote or off-grid places where a consistent electrical source is difficult to maintain [49].

5.6. Innovation in Technology

Improvements in solar panel efficiency and storage technologies have made renewable energy integration simpler. Advanced energy storage methods store additional energy during cloudy or nocturnal conditions, ensuring a continuous cooling operation. Solar panels with greater efficiency capture more solar energy [46].

A recent study confirms that using renewable energy in outdoor air conditioning systems marks a dramatic step toward sustainability and environmental stewardship. Several significant developments define this paradigm shift. First, including solar energy via photovoltaic and solar thermal systems has proven revolutionary. Photovoltaic panels turn sunlight into energy to power cooling systems, while solar thermal systems utilise the sun's heat to power cooling processes such as desiccant systems or absorption chillers [50]. This dual method optimises solar energy consumption and considerably minimises reliance on fossil fuels generally employed in conventional cooling systems [51]. The environmental advantages of this are many, helping worldwide efforts to cut carbon emissions and build a more sustainable future. Long-term energy cost reductions may compensate for the initial investment in solar panels and associated infrastructure, making renewable energy a financially feasible choice for outdoor cooling [52]. Additionally, using renewable energy sources improves the resilience and dependability of these systems by minimising their reliance on the electrical grid, which is especially advantageous in distant or off-grid areas [53]. Technological breakthroughs, such as increases in solar panel efficiency and energy storage technologies, have simplified this shift, allowing for a more efficient absorption and use of solar energy and guaranteeing ongoing cooling operations [54]. In conclusion, incorporating renewable energy into outdoor air conditioning systems is a transformative development that aligns with broader efforts to combat climate change, reduce energy costs, and enhance system resilience, setting new standards for sustainability and environmental stewardship within the industry. The dramatic trend toward incorporating renewable energy into outdoor air conditioning systems is a multidimensional phenomenon with far-reaching consequences for sustainability, environmental conservation, and energy economics [51]. This shift is backed by a growing amount of research demonstrating the practicality and advantages of employing renewable energy sources, especially solar power, in cooling technology. The integration of solar energy is accomplished via two basic methods: photovoltaic systems that convert sunlight directly into electricity and solar thermal systems that harness the sun's heat for cooling processes, such as desiccant systems and absorption chillers [55]. This dual method not only improves solar energy utilisation but also provides a flexible solution that can be customised to particular cooling demands and regional climate conditions. Traditional energy sources for cooling systems, such as coal, natural gas, and oil, will be drastically reduced due to this transition, which is one of the most important effects. By abandoning these non-renewable resources, outdoor air conditioning systems significantly reduce their carbon footprint and contribute to global efforts to combat climate change. This is a crucial breakthrough in light of the pressing need to cut greenhouse gas emissions to avert catastrophic environmental repercussions [50]. In addition, using renewable energy correlates with wider sustainability objectives, such as protecting natural resources and reducing pollution, providing a more comprehensive approach to environmental stewardship. Economically, the long-term advantages often exceed the initial expenses of installing solar panels and accompanying infrastructure. Renewable energy sources like solar electricity have lower operational costs than conventional fossil fuels, allowing for long-term cost savings [56]. The economic feasibility of renewable energy makes it an appealing alternative for home and business outdoor cooling systems. In addition, advances in energy storage technology, such as lithium-ion batteries and better control systems, have made it feasible to store extra solar energy for use during times of low sunshine, improving the efficiency and dependability of these systems. The integration of renewable energy also has the additional benefits of greater grid resilience and independence. By producing their power, outdoor cooling systems are less vulnerable to grid outages, which is especially advantageous in distant or off-grid areas where the electricity supply may be inconsistent [53]. This aspect also has significance for emergency preparation since systems independent of the grid may continue to work in the case of natural catastrophes or other interruptions to the electrical supply. Significant breakthroughs in solar panel efficiency, energy storage technologies, and intelligent control systems have facilitated this transformation via technological innovation. These advancements help efficiently collect and use solar energy and enable more complex system management, improving system performance and energy consumption [50]. In conclusion, incorporating renewable energy into outdoor air conditioning systems is a transformational and multi-dimensional innovation that tackles several crucial challenges relating to climate change, environmental conservation, and energy economics. By utilising developments in solar power technology, energy storage, and system controls, these environmentally friendly cooling solutions provide a sustainable, cost-effective, and resilient alternative to conventional systems fuelled by fossil fuels. As such, they represent a substantial advancement in establishing a more sustainable and ecologically responsible future [53].

Thermal comfort, while seemingly straightforward, is a multifaceted construct deeply rooted in physiological and psychological realms. This manuscript acknowledges its importance, but there is ample room to probe deeper into this intricate interplay of factors that govern an individual's sense of comfort within a built environment. At the forefront is temperature, which most directly aligns with our immediate perception of comfort [52]. However, this is not confined to just the ambient air temperature; the radiant temperature, determined by the warmth or coolness of surrounding surfaces, plays a critical role. Imagine a scenario where an individual is in a room with an acceptable ambient temperature but cold walls—discomfort can be palpable. Equally significant is humidity [51]. The moisture content in the air profoundly influences our thermal perception. Environments with high humidity can impede our body's natural cooling mechanism—perspiration—making us feel oppressively warm even at moderate temperatures [47]. Conversely, arid conditions can lead to skin irritation, a sense of dryness in the respiratory tract, and overall discomfort. Then, there is air velocity, a factor often overlooked but pivotal in shaping thermal comfort. The sensation of air moving across our skin can either be a boon or a bane, depending on the context. A gentle draft in a sweltering room can be the epitome of relief, but the same current in an already cold environment can make the chill almost unbearable. Beyond these physiological factors, one must consider the metabolic rate, which varies from person to person based on activities. A person engaged in rigorous physical activity generates more heat and might feel warmer than sedentary people, even if both are in the same environment [51]. Clothing insulation, or the "clo" value, further adds to this complex equation. The type and amount of clothing we wear can significantly influence our comfort, acting as a buffer between our body and the external environment. Lastly, an intriguing dimension to this discourse is the adaptive comfort model. It posits that our perception of comfort is not static but evolves based on prior experiences. Individuals accustomed to warmer climates might find a mildly warm room comfortable, while those from colder regions might perceive it as stifling. This model underscores the importance of considering the adaptive nature of human comfort, shaped by a confluence of past experiences and immediate environmental stimuli.

The role of HVAC technologies has been transformative within the ever-evolving landscape of green buildings. While this manuscript touches upon the integration challenges of traditional HVAC systems, a more profound exploration of contemporary HVAC solutions tailored for green buildings can provide invaluable depth. Several innovative technologies have emerged in the market, each promising to redefine the synergy between HVAC systems and green buildings [51]. To begin with, there is the Variable Refrigerant Flow (VRF) technology. Heralded as a game-changer, VRF systems are known for their unparalleled energy efficiency and flexibility. Unlike traditional systems that operate at constant speeds, VRF systems adjust refrigerant flow based on the exact cooling or heating needs of individual zones within a building. This dynamic adaptability conserves energy and ensures optimal comfort [48]. However, it is worth noting that the initial installation costs of VRF systems can be steep and require specialised technicians for maintenance. Next, we have earth–air heat exchangers (EAHE) [50]. These systems leverage the earth's temperatures to pre-cool or pre-heat incoming air. Especially in regions with extreme seasonal temperature variations, EAHE can be a boon, reducing the load on primary HVAC systems. However, the effectiveness of EAHE can vary based on soil properties and moisture content, making site-specific evaluations crucial. Phase-change materials (PCMs) introduce a novel approach to thermal regulation. These materials absorb or release latent heat during phase transitions, effectively acting as thermal buffers [50]. Integrated within building components, PCMs can reduce peak temperature loads, minimising the need for active cooling or heating. Their adaptability across various building types, from residential to commercial, underscores their versatility. However, selecting appropriate PCMs, based on their melting points and thermal properties, is crucial to their effective deployment [50]. Lastly, radiant cooling systems, often touted as the future of green building HVAC, provide cooling through chilled surfaces (like floors or ceilings) rather than air. The primary advantage lies in that cooling surfaces rather than air is inherently more energy efficient. Additionally, these systems enhance comfort by reducing the disparity between ambient air temperature and surface temperatures. However, a significant concern with radiant systems is the potential for condensation, especially in high-humidity environments. Proper design and integration with dehumidification systems can mitigate such risks. Outdoor air conditioning systems come in various forms to cater to different cooling needs in outdoor spaces. Misting systems use a fine water spray for cooling, ideal for restaurant patios, while evaporative coolers, effective in dry climates, use water evaporation for cooling and can be used in residential or commercial settings. Portable air conditioners offer flexibility for spot cooling in areas like patios, requiring electrical power and venting [51]. Although not air conditioners, outdoor ceiling fans enhance air circulation, creating a breezy environment. Cooling stations in public spaces provide mist or air-conditioned relief, and high-volume, low-speed (HVLS) fans are suitable for large venues like event spaces, moving air efficiently over vast areas. Additionally, ductless mini-split systems, with outdoor units linked to indoor units, offer efficient cooling for enclosed outdoor spaces like sunrooms. These diverse systems cater to specific outdoor cooling requirements, considering factors like climate, space, and energy efficiency.

6. IoT-Enabled Systems

A significant advancement in outdoor cooling is represented by smart and Internet of Things (IoT)-enabled outdoor air conditioning systems, as shown in Figure 6. These systems have various state-of-the-art features and technologies that improve their functionality, energy efficiency, and user comfort [57].

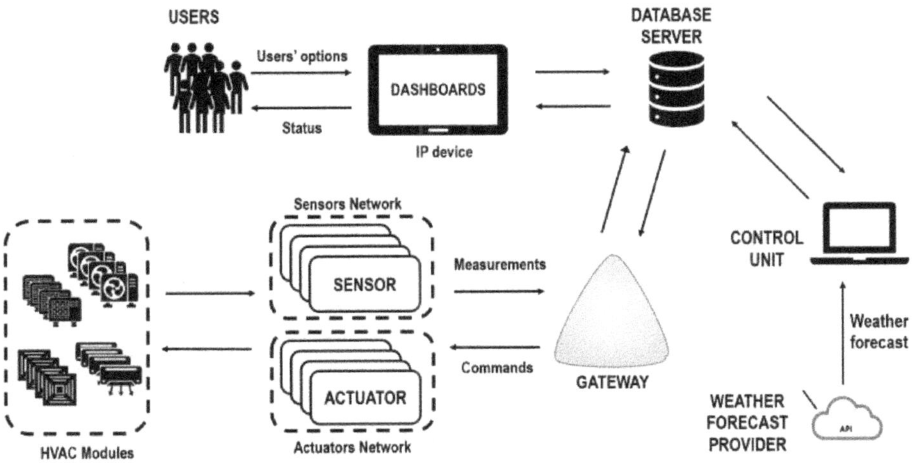

Figure 6. Internet of Things (IoT)-enabled outdoor air conditioning systems [57].

Monitoring in real time via sensors is one of the primary components of these systems, which utilise a network of sensors that collect data on various environmental properties. These sensors can measure various variables, including temperature, humidity, air quality, and occupancy. Real-time data obtained by these sensors provide the foundation for dynamic control and optimisation [58]. Data management is conducted using advanced data analytics methods, and sensor data are processed and evaluated. These strategies may identify patterns, trends, and anomalies in the data. For instance, they may detect changes in the outside temperature, which may result in adjustments to the cooling system's settings [59]. Innovative outdoor air conditioning systems can dynamically adjust their operation depending on the information gained through data analysis. For instance, the system may adjust when exterior temperatures increase significantly by boosting its cooling capability to maintain the ideal inside temperature. Alternatively, the system may operate more efficiently to save energy during periods of lower demand [60]. Additionally, these systems prioritise energy efficiency, which reduces running expenses and has a reduced effect on the environment. They may adjust the functioning of components such as pumps, fans, and compressors according to the cooling requirement, decreasing energy waste [61]. Moreover, smart systems are developed to provide the highest possible comfort for users. They can consider user preferences and adjust elements such as temperature and ventilation to create a pleasant atmosphere. They may also preserve air quality by monitoring and adjusting ventilation rates [62].

Frequently, IoT-enabled outdoor air conditioning systems can be remotely monitored and managed. Facility managers or homeowners can remotely inspect and modify system settings using a smartphone application or web interface. According to Philip et al. [63], this feature facilitates proactive system maintenance and increases usability. These systems are also capable of self-diagnosing issues and sending maintenance reminders. The method can detect component failures in advance and notify maintenance personnel, decreasing downtime and preventing costly breakdowns. Incorporating smart and Internet of Things (IoT)-enabled technology into outdoor air conditioning systems represents a significant leap in the industry, providing various cutting-edge features that improve functionality, energy efficiency, and user comfort. A network of sensors continually monitors environmental characteristics such as temperature, humidity, air quality, and occupancy, giving real-time data that serve as the foundation for dynamic management and optimisation. These sensor data are processed and analysed using sophisticated data analytics methods to find patterns, trends, and anomalies that might influence system modifications [64]. For example, when the system senses a fast increase in external temperatures, it may change its cooling capacity to maintain ideal internal temperatures. These intelligent systems emphasise energy efficiency by adapting components such as pumps, fans, and compressors to current cooling needs, minimising energy waste and saving operating costs [65]. The devices may react to the user's preferences and modify temperature and ventilation rates to produce a pleasant atmosphere [66]. Moreover, IoT-enabled systems provide remote monitoring and management through smartphone applications or web interfaces, enabling proactive system maintenance [67]. In addition, they are equipped with self-diagnostic capabilities that may detect component failures early and notify maintenance personnel, decreasing downtime and averting expensive breakdowns [68]. Overall, these intelligent and IoT-enabled outdoor air conditioning systems offer a significant advancement in outdoor cooling, closely correlating with more prominent efficiency and sustainability goals. Incorporating Internet of Things (IoT) technology into outdoor air conditioning systems is a revolutionary breakthrough with far-reaching effects on energy savings, user comfort, and system operation. This technical breakthrough is reinforced by a growing corpus of research demonstrating the many advantages of intelligent, IoT-enabled devices [69]. A network of sensors that continually monitors a variety of environmental factors, including temperature, humidity, air quality, and even occupancy, is one of the most critical aspects of these systems. This real-time data gathering is the basis for dynamic system control, allowing the air conditioning units to adapt to changing circumstances and maximise performance. This abundance of sensor data is processed using advanced data analytics methods to detect patterns, trends, and anomalies that might influence system modifications [70]. For instance, if the system senses a rapid spike in external temperatures, it may instantly change its cooling capacity to maintain a pleasant internal atmosphere. This dynamic control enables the system to predict future situations based on existing data, thus boosting efficiency. The energy efficiency of these intelligent systems is a major goal. These systems may dramatically minimise energy use by adjusting the operation of components such as pumps, fans, and compressors to the current cooling demand. This is significant not only for decreasing operating expenses but also for limiting the environmental effect of air conditioning, which is especially relevant considering the rising concerns over climate change and resource depletion [71]. Another critical emphasis of IoT-enabled air conditioning systems is user comfort. Individual user preferences may be accommodated by modifying temperature, humidity, and ventilation rates to produce a more pleasant atmosphere. In addition, they may monitor indoor air quality and change ventilation rates appropriately, promoting a better living or working environment [72].

These systems' remote monitoring and control features provide additional convenience and utility. Through smartphone applications or online interfaces, facility managers or homeowners may quickly monitor and alter system settings, allowing for more preventative system maintenance. This remote access is not only advantageous for its user-friendliness; it also provides faster reaction times in the event of system failure, hence decreasing downtime and averting expensive losses. In addition, these systems have self-diagnostic capabilities that may discover potential flaws before they become severe problems, notifying maintenance personnel and even recommending fixes. This predictive maintenance capacity is a significant improvement since it extends the life of air conditioning equipment, contributing to sustainability objectives by lowering the frequency of replacements [73]. In conclusion, incorporating Internet of Things (IoT) technology into outdoor air conditioning systems is a comprehensive innovation that solves several crucial concerns about energy efficiency, environmental sustainability, and user comfort. By integrating real-time sensor data, sophisticated analytics, dynamic control mechanisms, and remote monitoring capabilities, these intelligent systems provide a highly flexible, efficient, and user-friendly outdoor cooling solution. As a result, they are ready to establish new industry norms which correspond closely with broader goals of environmental responsibility and sustainable living. The incorporation of the Internet of Things (IoT) into HVAC systems within green architecture is symbolic of the profound technological strides the building industry is witnessing. While the authors' acknowledgement of IoT-based technologies is a step in the right direction, the depth and breadth of this integration call for a more exhaustive exploration. IoT, with its essence rooted in seamless interconnectivity, transforms HVAC systems from static entities to dynamic ecosystems that constantly communicate and adapt. This real-time data exchange is pivotal for energy conservation, a cornerstone of green buildings. Unlike traditional HVAC setups, IoT-enabled systems can discern, for instance, the occupancy of a room and modulate the cooling or heating in real-time [70]. Such nuanced adjustments, while seemingly trivial, cumulatively contribute to significant energy savings, reinforcing the sustainable ethos of green buildings. However, the advantages of IoT transcend energy efficiency. Occupant comfort, often a nuanced interplay of temperature, humidity, and individual preferences, is enhanced as interconnected sensors ensure optimal environmental conditions. Furthermore, the predictive maintenance capabilities of IoT systems herald a new era of proactive system health monitoring, pre-empting major malfunctions and ensuring uninterrupted operation. The true magic unfolds when these IoT-driven HVAC systems synergise with other building systems, from lighting to security, crafting an intelligent building ecosystem with unparalleled efficiency [71]. Moreover, when subjected to advanced analytics, the data streams from these systems provide invaluable insights into usage patterns, inefficiencies, and future energy needs. This data-centric approach, in tandem with the adaptive capabilities of IoT, positions the integration of IoT and HVAC as a linchpin in the evolution of green architecture. The discussion on IoT in HVAC systems is not just a technological narrative; it is a testament to the transformative potential of integrating digital intelligence with physical spaces, underscoring the future of sustainable and intelligent building design.

7. Portability and Scalability

Incorporating portability and scalability into outdoor air conditioning systems is a critical innovation that addresses the rising need for adaptation and flexibility in various cooling conditions. Portable units are gaining popularity because of their portability and setup simplicity, making them ideal for events, temporary workplaces, and other instances when permanent cooling systems are impracticable [25]. These devices are often lightweight and fitted with wheels or handles for easy portability, which increases their adaptability. Beneficial for temporary settings such as outdoor festivals or construction sites, the units' portability enables them to be moved as needed and to provide targeted cooling exactly where required. This versatility is a godsend for moving workstations and a cost-effective alternative to permanent systems, lowering initial and recurring costs. The fast availability and speedy installation of portable units are particularly advantageous for providing immediate respite from extreme heat, whether for emergency cooling requirements or unexpected gatherings. Scalability, another essential characteristic, permits outdoor cooling systems to adapt their cooling capacity to variable demand. This versatility is becoming more critical in contemporary system designs, particularly in locations where cooling needs may change based on time, season, or occupancy levels [15]. Scalable systems perfectly match cooling capacity to actual demand, optimising energy consumption, decreasing operating costs, and minimising environmental effects. Both transportable and scalable technologies contribute to sustainability by conforming to larger energy efficiency targets. By employing just the required cooling capacity for a particular condition, these systems reduce energy consumption and the carbon footprint, contributing to broader environmental and sustainability goals. In conclusion, the introduction of portability and scalability in outdoor air conditioning systems has ushered in a new age of adaptability, cost-efficiency, and environmental awareness. These characteristics allow for the deployment of customised cooling systems adaptable to a wide variety of applications, representing a breakthrough advance in outdoor cooling [19].

8. Economic Feasibility and Scalability

Eco-friendly HVAC technologies, pivotal in sustainable building design, markedly outperform traditional systems in energy efficiency and environmental impact [74]. Geothermal heat pumps, a standout example, are exceptionally efficient, offering an impressive energy efficiency ratio (EER) of about 10–14, compared to 9–10 for conventional systems [75]. They can deliver three to four units of energy for every one unit of electrical energy consumed. Solar-powered HVAC systems, while dependent on geographic and solar conditions, significantly cut down on electricity usage and are often capable of reducing energy bills by 20–40% [76]. Energy recovery ventilators (ERVs) enhance indoor air quality and can decrease HVAC energy consumption by up to 40%, depending on the climate and building design [76]. These systems lower operational costs and drastically reduce carbon emissions, aligning with global sustainability goals. The upfront costs for these technologies are generally higher, but the long-term savings—evidenced by their superior performance figures—are substantial, making them increasingly preferred in modern, eco-conscious construction projects.

In a detailed cost analysis of HVAC systems, contrasting traditional with eco-innovative models reveals distinct differences in financial implications and ROI [77]. Conventional HVAC systems, while cheaper to install, incur higher operational and maintenance costs over time due to less efficient energy use and frequent servicing needs. By contrast, eco-innovative HVAC systems, though more expensive initially, offer greater energy efficiency, leading to significantly lower operational costs [77]. The maintenance expenses for these advanced systems are often reduced due to their superior build quality and durability. The ROI for eco-innovative systems is favourable, as substantial long-term energy bills and maintenance savings balance the higher upfront costs. Additionally, many regions offer incentives like tax rebates for green technology adoption, enhancing the financial appeal of these systems. Another critical factor influencing the cost and implementation of HVAC systems is the impact of building codes and regulations, which vary regionally [78]. Stricter energy efficiency standards and sustainability requirements often favour eco-innovative systems, as traditional models may not comply with newer, more stringent regulations. This regulatory environment can significantly affect decision-making, pushing stakeholders towards more sustainable and compliant HVAC solutions. Therefore, while the initial investment in eco-innovative HVAC systems is higher, their long-term economic and environmental benefits and the increasing regulatory push towards sustainability present a compelling case for their adoption in modern building design [78]. When delving into the legal implications and performance of HVAC systems in different climates, a detailed understanding is crucial for sustainable building design. Legally, the implementation of HVAC systems is deeply intertwined with regional building codes and energy efficiency regulations. These laws vary widely, often dictating stringent standards that traditional HVAC systems may struggle to meet. Eco-innovative systems, designed with energy efficiency and sustainability in mind, are more likely to align with these regulations, thereby reducing the risk of legal non-compliance for building owners and developers [79]. Non-compliance can lead to penalties, legal disputes, and additional costs, making regulatory adherence a critical factor in HVAC system selection. Furthermore, these legal frameworks are dynamic and frequently updated to reflect new environmental goals or technological advancements, necessitating HVAC systems that can adapt to evolving standards. Performance in different climates is another critical aspect. Traditional HVAC systems may offer satisfactory performance in moderate temperatures but often struggle in extreme conditions, either too hot or cold, leading to inefficiencies and increased energy consumption [79]. By contrast, eco-innovative HVAC systems are typically designed for a broader range of climatic conditions. They often incorporate features like advanced insulation, smart thermostats, and renewable energy sources, enhancing their efficiency and reliability across various environmental settings. This adaptability is crucial for meeting the comfort needs of occupants and maintaining energy efficiency and reducing the overall environmental impact.

Moreover, the performance of HVAC systems in specific climates directly impacts their long-term cost-effectiveness and sustainability. Strategies that can efficiently manage cooling while minimising energy consumption are essential in hotter regions. Conversely, systems that provide effective heating with minimal energy loss are preferable in colder climates. Integrating intelligent technologies in eco-innovative systems allows for more precise control and optimisation based on external weather conditions, leading to better performance and lower energy usage. Integrating eco-innovative HVAC systems into sustainable building design poses significant challenges, primarily due to their higher initial costs compared to traditional methods, but long-term benefits like energy savings and potential incentives offset these [79]. Regulatory and compliance challenges also arise from varying building codes and energy efficiency standards, especially in retrofitting older buildings, requiring expertise in local regulations and the design of flexible systems adaptable to changing criteria. Spatial and architectural constraints in different building types necessitate innovative integration strategies and collaborative efforts between architects, engineers, and HVAC specialists. Despite these challenges, the superior performance of eco-innovative systems in terms of energy efficiency, indoor air quality, and reduced carbon footprint highlights their long-term economic and environmental viability [79]. Real-world case studies further illustrate the successes and strategies of practical implementation, reinforcing the feasibility and effectiveness of these solutions in diverse settings and presenting a comprehensive picture of the challenges and solutions in implementing sustainable HVAC technologies.

Statistically Methods from the Previous Researcher

Statistical analysis plays a pivotal role in substantiating the advancements in this field. Researchers often employ methods like regression analysis, ANOVA (analysis of variance), and time-series analysis to assess the performance of new HVAC technologies [80]. For instance, regression models are used to predict energy consumption under various scenarios, enabling a comparison between traditional and innovative systems. ANOVA helps identify the key factors influencing HVAC efficiency guiding design improvements [80]. Time-series analysis is crucial for evaluating long-term trends in energy usage and system performance, offering insights into the sustainability and cost-effectiveness of new HVAC solutions. Furthermore, using big data and advanced analytics in recent studies allows for a more comprehensive understanding of HVAC performance [81]. By analysing large datasets collected from intelligent HVAC systems, researchers can identify patterns, predict system behaviour, and suggest optimisations. This data-driven approach is instrumental in advancing HVAC technology, ensuring that new systems are innovative and empirically validated for efficiency and sustainability [82].

9. Conclusions

In conclusion, integrating conventional HVAC systems into green buildings presents a complex array of challenges that extend beyond mere energy inefficiency and high operational costs. These challenges encompass a broad spectrum of issues, including regulatory compliance, performance gaps, human factors, integration with renewable energy sources, and implications at both urban and grid levels. This paper highlights how inefficient HVAC systems can exacerbate issues like energy poverty and grid instability, adding a layer of complexity to the energy management of buildings and impacting urban planning. This problem is not just confined to individual buildings but is systemic, affecting entire cities as they endeavour to transition towards more sustainable and resilient futures. As the field of green construction evolves and gains traction, addressing these multifaceted obstacles becomes crucial. The successful integration of HVAC systems in green buildings is therefore not only a technical challenge but also involves navigating regulatory landscapes, addressing human-centric issues, and harmonising with broader urban sustainability goals.

Funding: The research is not funded by any party and is a project for a Ph.D. student.

Conflicts of Interest: The authors declare no conflict of interest.

References

1. Evans, J. 'Give Me a Laboratory, and I Will Lower Your Carbon Footprint!'–Urban Laboratories and the Pursuit of Low Carbon Futures James Evans and Andrew Karvonen. Ph.D. Thesis, University of Manchester, Manchester, UK, 2014.
2. Li, S.-F.; Liu, Z.; Wang, X.-J. A Comprehensive Review on Positive Cold Energy Storage Technologies and Applications in Air Conditioning with Phase Change Materials. *Appl. Energy* **2019**, *255*, 113667. [CrossRef]
3. Birgonul, Z. A Receptive-Responsive Tool for Customizing Occupant's Thermal Comfort and Maximizing Energy Efficiency by Blending BIM Data with Real-Time Information. *Smart Sustain. Built Environ.* **2021**, *10*, 504–535. [CrossRef]
4. Alonso, M.J.; Wolf, S.; Jørgensen, R.B.; Madsen, H.; Mathisen, H.M. A Methodology for the Selection of Pollutants for Ensuring Good Indoor Air Quality Using the De-Trended Cross-Correlation Function. *Build. Environ.* **2022**, *209*, 108668. [CrossRef]
5. Lamberti, G.; Contrada, F.; Kindinis, A. Exploring adaptive strategies to cope with climate change: The case study of Le Corbusier's Modern Architecture retrofitting. *Energy Build.* **2023**, *302*, 113756. [CrossRef]
6. Zhang, S.; Ocłoń, P.; Klemeš, J.J.; Michorczyk, P.; Pielichowska, K.; Pielichowski, K. Renewable Energy Systems for Building Heating, Cooling and Electricity Production with Thermal Energy Storage. *Renew. Sustain. Energy Rev.* **2022**, *165*, 112560. [CrossRef]
7. Yaïci, W.; Krishnamurthy, K.; Entchev, E.; Longo, M. Recent Advances in Internet of Things (IoT) Infrastructures for Building Energy Systems: A Review. *Sensors* **2021**, *21*, 2152. [CrossRef] [PubMed]
8. Goetzler, B.; Guernsey, M.; Kassuga, T.; Young, J.; Savidge, T.; Bouza, A.; Neukomm, M.; Sawyer, K. *Grid-Interactive Efficient Buildings Technical Report Series: Heating, Ventilation, and Air Conditioning (HVAC); Water Heating; Appliances; and Refrigeration*; (No. NREL/TP-5500-75473; DOE/GO-102019-5228); National Renewable Energy Laboratory (NREL): Golden, CO, USA, 2019.
9. *ANSI/ASHRAE Standard 55-2017*; Thermal Environmental Conditions for Human Occupancy. American Society of Heating, Refrigerating, and Air-Conditioning Engineers: Peachtree Corners, GA, USA, 2017; Volume 5, pp. 2–15.
10. Pérez-Lombard, L.; Ortiz, J.; Pout, C. A Review on Buildings Energy Consumption Information. *Energy Build.* **2008**, *40*, 394–398. [CrossRef]
11. Kibert, C.J. *Sustainable Construction: Green Building Design and Delivery*; John Wiley & Sons: Hoboken, NJ, USA, 2016; ISBN 1119055172.
12. Chan, A.L.S.; Chow, T.T.; Fong, K.F.; Lin, Z. Investigation on Energy Performance of Double Skin Façade in Hong Kong. *Energy Build.* **2009**, *41*, 1135–1142. [CrossRef]
13. DeKay, M.; Brown, G.Z. *Sun, Wind, and Light*; Wiley: Hoboken, NJ, USA, 2001.
14. Barber, K.A.; Krarti, M. A review of optimization based tools for design and control of building energy systems. *Renew. Sustain. Energy Rev.* **2022**, *160*, 112359. [CrossRef]
15. Nicol, J.F.; Humphreys, M.A. Adaptive Thermal Comfort and Sustainable Thermal Standards for Buildings. *Energy Build.* **2002**, *34*, 563–572. [CrossRef]
16. Tomat, V.; Ramallo-González, A.P.; Skarmeta Gómez, A.F. A comprehensive survey about thermal comfort under the IoT paradigm: Is crowdsensing the new horizon? *Sensors* **2020**, *20*, 4647. [CrossRef] [PubMed]
17. Papadakis, N.; Katsaprakakis, D.A. A Review of Energy Efficiency Interventions in Public Buildings. *Energies* **2023**, *16*, 6329. [CrossRef]
18. Goetzler, W.; Shandross, R.; Young, J.; Petritchenko, O.; Ringo, D.; McClive, S. *Energy Savings Potential and RD&D Opportunities for Commercial Building HVAC Systems*; (No. DOE/EE-1703); Navigant Consulting: Burlington, MA, USA, 2017.
19. Heschong, L.; Saxena, M. *Windows and Offices: A Study of Office Worker Performance and the Indoor Environment*; California Energy Commission: Sacramento, CA, USA, 2003.
20. Zhou, G.; Moayedi, H.; Bahiraei, M.; Lyu, Z. Employing Artificial Bee Colony and Particle Swarm Techniques for Optimizing a Neural Network in Prediction of Heating and Cooling Loads of Residential Buildings. *J. Clean. Prod.* **2020**, *254*, 120082. [CrossRef]
21. Sansaniwal, S.K.; Mathur, J.; Mathur, S. Review of practices for human thermal comfort in buildings: Present and future perspectives. *Int. J. Ambient. Energy* **2022**, *43*, 2097–2123. [CrossRef]
22. Zhou, W.; Liu, Y.; Fu, J.; Zhong, Z.; Li, Y.; Liu, Z. An Interval Scaling Algorithm and Concept Lattice Building from Extended Formal Context. In Proceedings of the 2009 International Conference on Test and Measurement, Hong Kong, China, 5–6 December 2009; IEEE: Piscataway, NJ, USA, 2009; Volume 2, pp. 409–412.
23. Ganesh, G.A.; Sinha, S.L.; Verma, T.N.; Dewangan, S.K. Investigation of indoor environment quality and factors affecting human comfort: A critical review. *Build. Environ.* **2021**, *204*, 108146. [CrossRef]
24. Santamouris, M.; Synnefa, A.; Karlessi, T. Using Advanced Cool Materials in the Urban Built Environment to Mitigate Heat Islands and Improve Thermal Comfort Conditions. *Sol. Energy* **2011**, *85*, 3085–3102. [CrossRef]

25. Kim, J.; Schiavon, S.; Brager, G. Personal Comfort Models–A New Paradigm in Thermal Comfort for Occupant-Centric Environmental Control. *Build. Environ.* **2018**, *132*, 114–124. [CrossRef]
26. Zhang, H.; Arens, E.; Zhai, Y. A Review of the Corrective Power of Personal Comfort Systems in Non-Neutral Ambient Environments. *Build. Environ.* **2015**, *91*, 15–41. [CrossRef]
27. Rupp, R.F.; Vásquez, N.G.; Lamberts, R. A Review of Human Thermal Comfort in the Built Environment. *Energy Build.* **2015**, *105*, 178–205. [CrossRef]
28. Hoes, P.; Hensen, J.L.M.; Loomans, M.G.L.C.; de Vries, B.; Bourgeois, D. User Behavior in Whole Building Simulation. *Energy Build.* **2009**, *41*, 295–302. [CrossRef]
29. De Dear, R.; Xiong, J.; Kim, J.; Cao, B. A review of adaptive thermal comfort research since 1998. *Energy Build.* **2020**, *214*, 109893. [CrossRef]
30. Carlucci, S.; Bai, L.; de Dear, R.; Yang, L. Review of adaptive thermal comfort models in built environmental regulatory documents. *Build. Environ.* **2018**, *137*, 73–89. [CrossRef]
31. Edwards, B.; Hyett, P. *Rough Guide to Sustainable Design*; RIBA: London, UK, 2001.
32. Mendell, M.J. Indoor Residential Chemical Emissions as Risk Factors for Respiratory and Allergic Effects in Children: A Review. *Indoor Air* **2007**, *17*, 259–277. [CrossRef] [PubMed]
33. Moezzi, M.; Janda, K.B.; Rotmann, S. Using Stories, Narratives, and Storytelling in Energy and Climate Change Research. *Energy Res. Soc. Sci.* **2017**, *31*, 1–10. [CrossRef]
34. Brown, M.E. *Introduction to Thermal Analysis: Techniques and Applications*; Springer: Berlin/Heidelberg, Germany, 2001.
35. Bakker, A.; Siegel, J.A.; Mendell, M.J.; Prussin, A.J.; Marr, L.C.; Peccia, J. Bacterial and Fungal Ecology on Air Conditioning Cooling Coils Is Influenced by Climate and Building Factors. *Indoor Air* **2020**, *30*, 326–334. [CrossRef] [PubMed]
36. Granderson, J.; Touzani, S.; Custodio, C.; Sohn, M.D.; Jump, D.; Fernandes, S. Accuracy of Automated Measurement and Verification (M&V) Techniques for Energy Savings in Commercial Buildings. *Appl. Energy* **2016**, *173*, 296–308.
37. Seljom, P.; Lindberg, K.B.; Tomasgard, A.; Doorman, G.; Sartori, I. The Impact of Zero Energy Buildings on the Scandinavian Energy System. *Energy* **2017**, *118*, 284–296. [CrossRef]
38. D'Ayala, D.; Wang, K.; Yan, Y.; Smith, H.; Massam, A.; Filipova, V.; Pereira, J.J. Flood Vulnerability and Risk Assessment of Urban Traditional Buildings in a Heritage District of Kuala Lumpur, Malaysia. *Nat. Hazards Earth Syst. Sci.* **2020**, *20*, 2221–2241. [CrossRef]
39. Chun, L.; Gong, G.; Peng, P.; Wan, Y.; Chua, K.J.; Fang, X.; Li, W. Research on Thermodynamic Performance of a Novel Building Cooling System Integrating Dew Point Evaporative Cooling, Air-Carrying Energy Radiant Air Conditioning and Vacuum Membrane-Based Dehumidification (DAV-Cooling System). *Energy Convers. Manag.* **2021**, *245*, 114551. [CrossRef]
40. Chua, Y.S.; Dai Pang, S.; Liew, J.Y.R.; Dai, Z. Robustness of Inter-Module Connections and Steel Modular Buildings under Column Loss Scenarios. *J. Build. Eng.* **2022**, *47*, 103888. [CrossRef]
41. Moon, H.; Yu, J.; Chua, B.-L.; Han, H. Hotel Privacy Management and Guest Trust Building: A Relational Signaling Perspective. *Int. J. Hosp. Manag.* **2022**, *102*, 103171. [CrossRef]
42. Liu, Y.; Lin, Y.; Yeoh, J.K.W.; Chua, D.K.H.; Wong, L.W.C.; Ang, M.H.; Lee, W.L.; Chew, M.Y.L. Framework for Automated UAV-Based Inspection of External Building Façades. *Autom. Cities Des. Constr. Oper. Futur. Impact* **2021**, 173–194.
43. Foster, G. Planning the Circular City: Focus on Buildings'environmental Impact. *BDC Boll. Cent. Calza Bini* **2019**, *19*, 117–123.
44. MacArthur, E. Towards the Circular Economy. *J. Ind. Ecol.* **2013**, *2*, 23–44.
45. Sorrell, S. *The Rebound Effect: An Assessment of the Evidence for Economy-Wide Energy Savings from Improved Energy Efficiency*; University of Sussex: Brighton, UK, 2007.
46. Saidur, R. Energy Consumption, Energy Savings, and Emission Analysis in Malaysian Office Buildings. *Energy Policy* **2009**, *37*, 4104–4113. [CrossRef]
47. Aynur, T.N.; Hwang, Y.; Radermacher, R. Integration of Variable Refrigerant Flow and Heat Pump Desiccant Systems for the Cooling Season. *Appl. Therm. Eng.* **2010**, *30*, 917–927. [CrossRef]
48. Allwood, J.M.; Cullen, J.M.; Carruth, M.A.; Cooper, D.R.; McBrien, M.; Milford, R.L.; Moynihan, M.C.; Patel, A.C.H. *Sustainable Materials: With Both Eyes Open*; UIT Cambridge Limited: Cambridge, UK, 2012; Volume 2012.
49. Yuan, M.; Li, Z.; Li, X.; Li, L.; Zhang, S.; Luo, X. How to Promote the Sustainable Development of Prefabricated Residential Buildings in China: A Tripartite Evolutionary Game Analysis. *J. Clean. Prod.* **2022**, *349*, 131423. [CrossRef]
50. Henning, H.-M. Solar Assisted Air Conditioning of Buildings—An Overview. *Appl. Therm. Eng.* **2007**, *27*, 1734–1749. [CrossRef]
51. Jacobson, M.Z. Review of Solutions to Global Warming, Air Pollution, and Energy Security. *Energy Environ. Sci.* **2009**, *2*, 148–173. [CrossRef]
52. Branker, K.; Pathak, M.J.M.; Pearce, J.M. A Review of Solar Photovoltaic Levelized Cost of Electricity. *Renew. Sustain. Energy Rev.* **2011**, *15*, 4470–4482. [CrossRef]
53. Parida, B.; Iniyan, S.; Goic, R. A Review of Solar Photovoltaic Technologies. *Renew. Sustain. Energy Rev.* **2011**, *15*, 1625–1636. [CrossRef]
54. Majd, E.; McCormack, M.; Davis, M.; Curriero, F.; Berman, J.; Connolly, F.; Leaf, P.; Rule, A.; Green, T.; Clemons-Erby, D. Indoor Air Quality in Inner-City Schools and Its Associations with Building Characteristics and Environmental Factors. *Environ. Res.* **2019**, *170*, 83–91. [CrossRef] [PubMed]

55. Henning, H.-M.; Pagano, T.; Mola, S.; Wiemken, E. Micro Tri-Generation System for Indoor Air Conditioning in the Mediterranean Climate. *Appl. Therm. Eng.* **2007**, *27*, 2188–2194. [CrossRef]
56. Khamidov, A.; Akhmedov, I.; Kholmirzayev, S.; Jalalov, Z.; Yusupov, S.; Umarov, I. Effectiveness of Modern Methods of Testing Building Structures. *Sci. Innov.* **2022**, *1*, 1046–1051.
57. Li, W.; Meng, W.; Au, M.H. Enhancing Collaborative Intrusion Detection via Disagreement-Based Semi-Supervised Learning in IoT Environments. *J. Netw. Comput. Appl.* **2020**, *161*, 102631. [CrossRef]
58. Shi, J.; Liu, B.; He, Z.; Liu, Y.; Jiang, J.; Xiong, T.; Shi, J. A Green Ultra-Lightweight Chemically Foamed Concrete for Building Exterior: A Feasibility Study. *J. Clean. Prod.* **2021**, *288*, 125085. [CrossRef]
59. Fan, C.; Yan, D.; Xiao, F.; Li, A.; An, J.; Kang, X. Advanced Data Analytics for Enhancing Building Performances: From Data-Driven to Big Data-Driven Approaches. In *Building Simulation*; Springer: Berlin/Heidelberg, Germany, 2021; Volume 14, pp. 3–24.
60. Liu, G.; Zhou, X.; Yan, J.; Yan, G. A Temperature and Time-Sharing Dynamic Control Approach for Space Heating of Buildings in District Heating System. *Energy* **2021**, *221*, 119835. [CrossRef]
61. Belussi, L.; Barozzi, B.; Bellazzi, A.; Danza, L.; Devitofrancesco, A.; Fanciulli, C.; Ghellere, M.; Guazzi, G.; Meroni, I.; Salamone, F. A Review of Performance of Zero Energy Buildings and Energy Efficiency Solutions. *J. Build. Eng.* **2019**, *25*, 100772. [CrossRef]
62. Smith, A.; Pitt, M. Sustainable Workplaces and Building User Comfort and Satisfaction. *J. Corp. Real Estate* **2011**, *13*, 144–156. [CrossRef]
63. Philip, A.; Islam, S.N.; Phillips, N.; Anwar, A. Optimum Energy Management for Air Conditioners in IoT-Enabled Smart Home. *Sensors* **2022**, *22*, 7102. [CrossRef]
64. Hashem, Y.; Frank, J. The Jigsaw Puzzle of MRNA Translation Initiation in Eukaryotes: A Decade of Structures Unraveling the Mechanics of the Process. *Annu. Rev. Biophys.* **2018**, *47*, 125–151. [CrossRef] [PubMed]
65. Cook, A.A.; Mısırlı, G.; Fan, Z. Anomaly Detection for IoT Time-Series Data: A Survey. *IEEE Internet Things J.* **2019**, *7*, 6481–6494. [CrossRef]
66. Al-Fuqaha, A.; Guizani, M.; Mohammadi, M.; Aledhari, M.; Ayyash, M. Internet of Things: A Survey on Enabling Technologies, Protocols, and Applications. *IEEE Commun. Surv. Tutor.* **2015**, *17*, 2347–2376. [CrossRef]
67. Whitmore, A.; Agarwal, A.; Da Xu, L. The Internet of Things—A Survey of Topics and Trends. *Inf. Syst. Front.* **2015**, *17*, 261–274. [CrossRef]
68. Atzori, L.; Iera, A.; Morabito, G. The Internet of Things: A Survey. *Comput. Netw.* **2010**, *54*, 2787–2805. [CrossRef]
69. Bashir, M.R.; Gill, A.Q. IoT Enabled Smart Buildings: A Systematic Review. In Proceedings of the 2017 Intelligent Systems Conference (IntelliSys), London, UK, 7–8 September 2017; IEEE: Piscataway, NJ, USA, 2017; pp. 151–159.
70. Marinakis, V.; Doukas, H. An Advanced IoT-Based System for Intelligent Energy Management in Buildings. *Sensors* **2018**, *18*, 610. [CrossRef]
71. Gougar, H.D.; Petti, D.A.; Demkowicz, P.A.; Windes, W.E.; Strydom, G.; Kinsey, J.C.; Ortensi, J.; Plummer, M.; Skerjanc, W.; Williamson, R.L. The US Department of Energy's High Temperature Reactor Research and Development Program–Progress as of 2019. *Nucl. Eng. Des.* **2020**, *358*, 110397. [CrossRef]
72. O'Leary, R. The Impact of Federal Court Decisions on the Policies and Administration of the US Environmental Protection Agency. In *Administrative Law*; Routledge: London, UK, 2018; pp. 259–284.
73. Manyika, J.; Chui, M.; Bisson, P.; Woetzel, J.; Dobbs, R.; Bughin, J.; Aharon, D. Unlocking the Potential of the Internet of Things. *McKinsey Glob. Inst.* **2015**, *1*, 1–5.
74. Zhong, H.; Li, Y.; Zhang, P.; Gao, S.; Liu, B.; Wang, Y.; Meng, T.; Zhou, Y.; Hou, H.; Xue, C. Hierarchically Hollow Microfibers as a Scalable and Effective Thermal Insulating Cooler for Buildings. *ACS Nano* **2021**, *15*, 10076–10083. [CrossRef]
75. Wang, Y.; Wang, X.; Yu, H.; Huang, Y.; Dong, H.; Qi, C.; Baptiste, N. Optimal design of integrated energy system considering economics, autonomy and carbon emissions. *J. Clean. Prod.* **2019**, *225*, 563–578. [CrossRef]
76. Chua, K.J.; Chou, S.K.; Yang, W.M.; Yan, J. Achieving better energy-efficient air conditioning–a review of technologies and strategies. *Appl. Energy* **2013**, *104*, 87–104. [CrossRef]
77. Cooke, P. Transition regions: Regional–national eco-innovation systems and strategies. *Prog. Plan.* **2011**, *76*, 105–146. [CrossRef]
78. Laustsen, J. *Energy Efficiency Requirements in Building Codes, Energy Efficiency Policies for New Buildings*; IEA: Paris, France, 2008.
79. Omer, A.M. Energy, environment and sustainable development. *Renew. Sustain. Energy Rev.* **2008**, *12*, 2265–2300. [CrossRef]
80. Deb, C.; Zhang, F.; Yang, J.; Lee, S.E.; Shah, K.W. A review on time series forecasting techniques for building energy consumption. *Renew. Sustain. Energy Rev.* **2017**, *74*, 902–924. [CrossRef]
81. Wei, Y.; Zhang, X.; Shi, Y.; Xia, L.; Pan, S.; Wu, J.; Han, M.; Zhao, X. A review of data-driven approaches for prediction and classification of building energy consumption. *Renew. Sustain. Energy Rev.* **2018**, *82*, 1027–1047. [CrossRef]
82. Daissaoui, A.; Boulmakoul, A.; Karim, L.; Lbath, A. IoT and big data analytics for smart buildings: A survey. *Procedia Comput. Sci.* **2020**, *170*, 161–168. [CrossRef]

MDPI AG
Grosspeteranlage 5
4052 Basel
Switzerland
Tel.: +41 61 683 77 34

Applied Sciences Editorial Office
E-mail: applsci@mdpi.com
www.mdpi.com/journal/applsci

www.ingramcontent.com/pod-product-compliance
Lightning Source LLC
Chambersburg PA
CBHW040800250926
51850CB00007BA/1039
9783725819072